Conference Proceedings of the Society for Experimental Mechanics Series

Series Editor

Tom Proulx
Society for Experimental Mechanics, Inc.
Bethel, CT, USA

More information about this series at http://www.springer.com/series/8922

Alfred Wicks
Editor

Shock & Vibration, Aircraft/Aerospace, and Energy Harvesting, Volume 9

Proceedings of the 33rd IMAC, A Conference and Exposition on Structural Dynamics, 2015

Springer

Editors
Alfred Wicks
Department of Mechanical Engineering
Virginia Polytechnic Inst. & State Uni.
Blacksburg, VA, USA

ISSN 2191-5644 ISSN 2191-5652 (electronic)
Conference Proceedings of the Society for Experimental Mechanics Series
ISBN 978-3-319-15232-5 ISBN 978-3-319-15233-2 (eBook)
DOI 10.1007/978-3-319-15233-2

Library of Congress Control Number: 2015936333

Springer Cham Heidelberg New York Dordrecht London

Printed on acid-free paper

Springer International Publishing AG Switzerland is part of Springer Science+Business Media (www.springer.com)

Preface

Shock & Vibration, Aircraft/Aerospace, Energy Harvesting represents one of ten volumes of technical papers presented at the 33rd IMAC, A Conference and Exposition on Structural Dynamics, 2015, organized by the Society for Experimental Mechanics, and held in Orlando, Florida, February 2–5, 2015. The full proceedings also include volumes on Nonlinear Dynamics; Dynamics of Civil Structures; Model Validation and Uncertainty Quantification; Dynamics of Coupled Structures; Sensors and Instrumentation; Special Topics in Structural Dynamics; Structural Health Monitoring & Damage Detection; Experimental Techniques, Rotating Machinery & Acoustics; and Topics in Modal Analysis.

Each collection presents early findings from experimental and computational investigations on an important area within Structural Dynamics. Topics represent papers on practical issues improving energy harvesting measurements, shock calibration and shock environment synthesis and applications for aircraft/aerospace structures. Topics in this volume include:

Energy Harvesting
Adaptive Support
Shock Calibration
Operating Data Applications

The organizers would like to thank the authors, presenters, session organizers, and session chairs for their participation in this track.

Blacksburg, VA, USA

Randy Allemang

Contents

1 **From Preliminary Design to Prototyping and Validation of Energy Harvester for Shoes** 1
Elvio Bonisoli, Francesco Di Monaco, Nicolò Manca, Maurizio Repetto, and Stefano Tornincasa

2 **Issues in Experimental Testing of Piezoelectric Energy Harvesters** 11
Paulo S. Varoto

3 **Innovative Piezoelectric Cantilever Beam Shape for Improved Energy Harvesting** 19
Iman Mehdipour and Francesco Braghin

4 **Energy Harvesting from Piezoelectric Stacks Using Impacting Beam** 25
Yiğit Özpak, Murat Aykan, and Mehmet Çalışkan

5 **Nonlinear 2-DOFs Vibration Energy Harvester Based on Magnetic Levitation** 39
I. Abed, N. Kacem, M.L. Bouazizi, and N. Bouhaddi

6 **Parameter Identification of Riveted Joints Using Vibration Methods** 47
Elif Altuntop, Murat Aykan, and Melin Şahin

7 **Dynamic Ground Testing: Ground Vibration Tests Through Control Surface Excitation** 55
G. Osmond, A. Azzat, S. Leroy, and O. Delverdier

8 **Adaptive Support of an Aircraft Panel** 61
Manuel Baschke and Delf Sachau

9 **Calculating the Impact Force of Supersonic Hail Stones Using SWAT-TEEM** 67
Tyler F. Schoenherr

10 **Testing and Validation of the Dynamic Inertia Measurement Method** 81
Alexander W. Chin, Claudia Y. Herrera, Natalie D. Spivey, William A. Fladung, and David Cloutier

11 **Estimation of Amplitude-Dependent Resonance and Damping in MEMS Shock Accelerometers** 105
Jason R. Foley, Thomas J. Lagoski, Jontia Brown, and Jonathan Hong

12 **Development of a Mapping Function for a Low- to High-Amplitude Input** 115
Joshua H. Campbell, Janet C. Wolfson, Jacob C. Dodson, Alain L. Beliveau, Jonathan Hong, and Greg Falbo

13 **Experimental Study of Glass Fiber Reinforced Polyester Under Repeated Impacts** 129
Ahmed M. Elmahdy, Abdelhalim M. Elhabak, Mahmoud A. Adly, and Mohamed M. Elbawab

14 **FE Modeling of Paperboard Material Using Sandwich Structure Method** 137
W. Yang, M.W. Allin, and C.J. Dehenau

15 **Evaluation of Seismic Performance of an Excavation Support System** 141
Omer F. Usluogullari, Yalcin Bulut, and Ahmet Temugan

16 **Calculating Stress and Strain from Experimental ODS Data** 149
Brian Schwarz, Shawn Richardson, and Mark Richardson

17 **Case Study: Modeling Generator Chassis Responses with ODS Data** 161
Sze Kwan Cheah

18 **Shock Calibration with Zero Shift Using a Digital Filter Technique** 169
Hideaki Nozato, Wataru Kokuyama, and Akihiro Ota

19 **Mechanical Shock Environment Synthesis for Structural Failure Elicitation** 175
Cassidy L. Fisher, Kaitlyn S. Kliewer, Gregory M. Naranjo, Stuart G. Taylor, and Kendra Van Buren

20 **Natural Frequencies of Layered Beams Using a Continuous Variation Model** 187
Arnaldo J. Mazzei and Richard A. Scott

21 **Analysis of H_∞ and H_2 Optimal Design Scheme for an Electromagnetic Damper with Shunt Resonant Circuit** 201
Wai Kei Ao and Paul Reynolds

22 **Orbit Stability Determination of Satellites Using Harmonic Force Excitation Analysis** 213
Joshua Johnson, William H. Semke, Matthew Zimmer, and Ronald Fevig

23 **Energy Harvesting Perspectives from Parametric Resonant Systems** 223
Maryam Ghandchi Tehrani, Elvio Bonisoli, and Matteo Scapolan

Chapter 1
From Preliminary Design to Prototyping and Validation of Energy Harvester for Shoes

Elvio Bonisoli, Francesco Di Monaco, Nicolò Manca, Maurizio Repetto, and Stefano Tornincasa

Abstract Powering a remote wireless sensor is a challenging task if batteries are not suitable or enough capacious and their substitution is not feasible. In this project a remote wireless sensor is placed inside training shoes with the aim to collect and transmit data to evaluate and track the performance of an athlete. The primary energy source is the impact between the shoe and the ground while walking or running. The harvester has been designed by means of a multi-physics optimization based on an integrated electromagnetic-mechanical-electric-electronic simulator. Thus an automated optimization of the device with respect to volume constraints, magnets dimensions, induction coils placement and sizes and electric/electronic coupling have been performed to increase the average power extracted from the device at different speeds. These parameters are used as starting point for the product development phase in order to obtain a consistent number of prototypes and validate the simulations on these physical demonstrators. Finally, experimental outcomes evince the expected performance and a more than satisfactory agreement with the models, confirming the feasibility of the application.

Keywords Magneto-mechanical generator • Design and optimization • Shoe mounted device • Product development Experimental application

1.1 Introduction

Powering remote wireless sensors exploiting the energy of the environment with the aim of making their life independent from some energy limited power source like an electrochemical battery, is a challenging task. In addition, batteries are often not suitable or not enough capacious, their substitution is not feasible or simply annoying and they introduce some problems due to the toxicity of their chemicals.

Due to the growing of the wearable electronics market, one of the most debated topics in the field of energy harvesting is the power supply of all those devices where the energy source is the human body motion.

The present study proposes a wearable device totally powered by an Energy Harvester (EH): an electrically autonomous bluetooth step-counter placed in the sole of training shoes. This sensor allows collecting and transmitting data to a bluetooth receiver device, as a smartphone, to evaluate and tracking the athlete performance. Figure 1.1 schematically represents the working principle: at each step energy is harvested and used to power the electric interface.

Two possible strategies are available to collect energy in shoes during a step, usually by piezoelectric or electromagnetic transducers: sole deformation caused by the contact with the ground and shocks due to the impacts of the heel on the ground.

Piezoelectric devices using sole deformation are described by [1, 2] while inertial piezoelectric generators, composed by a cantilever beam with a mass on the free end, are described in [3, 4]. References [5, 6] present two linear electromagnetic generators with one or more magnets sliding into a guide placed horizontally in the shoe. These devices do not have any kind of springs. A rotary generator activated by a harm extending under the sole is described in [7]. A generator using a flow between two pumps placed in the front and in the rear of the sole is presented in [8].

The proposed device is an electromagnetic energy harvester that exploits as primary energy source the impact of the heel on the ground during each step of walking or running activity. Differently from the devices presented in [5, 6] this one is placed in the heel with vertical sliding axis.

E. Bonisoli (✉) • F. Di Monaco • N. Manca • S. Tornincasa
Department of Management and Production Engineering, Politecnico di Torino, Corso Duca degli Abruzzi 24, 10129 Torino, Italy
e-mail: elvio.bonisoli@polito.it

M. Repetto
Department of Energy, Politecnico di Torino, Corso Duca degli Abruzzi 24, 10129 Torino, Italy

A. Wicks (ed.), *Shock & Vibration, Aircraft/Aerospace, and Energy Harvesting, Volume 9*,
Conference Proceedings of the Society for Experimental Mechanics Series, DOI 10.1007/978-3-319-15233-2_1

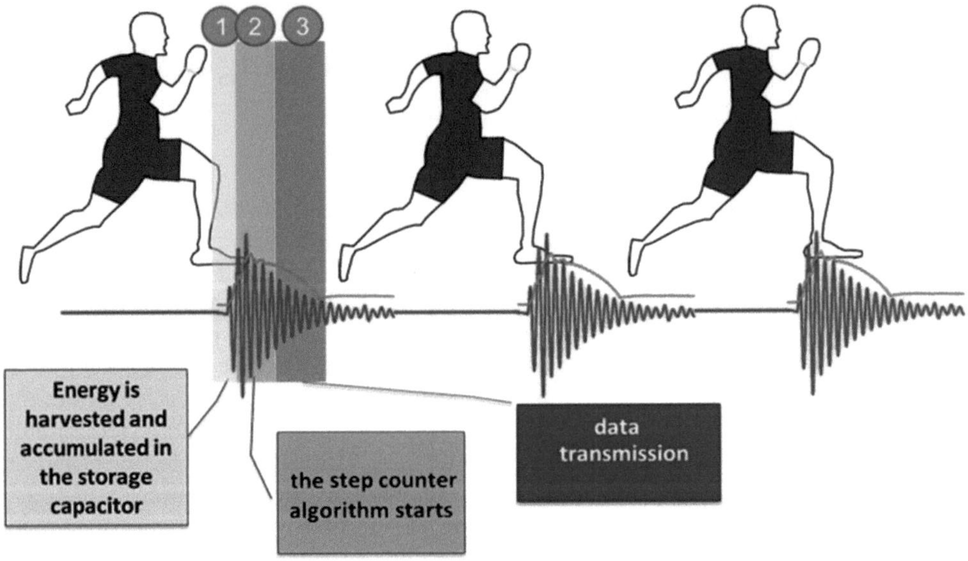

Fig. 1.1 Working principle

The harvester has been designed by means of a multi-physics optimization based on an integrated electromagnetic-mechanical-electric-electronic simulator. Thus, an automated optimization of the device with respect to volume constraints, magnet dimensions, induction coils placement and size and electric/electronic coupling have been performed to increase the average power extracted from the device at different walking speeds. These parameters represent the reference configuration for the product development phase in order to obtain a consistent number of semi-industrialized prototypes and validate the simulations on these physical demonstrators.

1.2 Device Description, Design and Optimization

The device consists of the main parts: the transducer for the vibrational energy harvesting and the conversion in electric energy, and the electric interface for step monitoring and data sending.

1.2.1 Harvester

The transducer is an electromagnetic linear generator. Its layout is shown in Fig. 1.2: a magnet can slide into a guide suspended between two springs and two coils are winded around the guide in opposite direction one each other. Inertial forces, due to the impact between the shoe housing the transducer and the ground during walking or running activity, induce motion in the magnet. The movement of the magnet causes a variation of the flux linkage in the coils and, consequently, a voltage is induced between the ends of the coils.

It is possible to represent the device as a single degree of freedom base-excited mass-spring-damper system where the transducer is considered as an inertial device, so that the forces act on its base and the mass vibrates freely. Two differential equations, one for the mechanical and one for the electrical domain, describe the system. In case of a generic electric load the two equations are:

$$\begin{cases} \frac{\mathrm{d}^2 z_\mathrm{r}}{\mathrm{d}t^2} = -\frac{\mathrm{d}^2 z_\mathrm{inp}}{\mathrm{d}t^2} - \frac{c}{m}\frac{\mathrm{d}\,z_\mathrm{r}}{\mathrm{d}t} - \frac{k}{m} z_\mathrm{r} + \frac{\lambda'}{m} i \\ \frac{\mathrm{d}\,i}{\mathrm{d}t} = -\frac{R}{L} i + \frac{V_L}{L} + \frac{\lambda'}{L}\frac{\mathrm{d}\,z_\mathrm{r}}{\mathrm{d}t} \end{cases} \tag{1.1}$$

where m is the mass of the moving magnet, k is the sum of the stiffness of the two springs, c is a generic dissipative viscous mechanical damping, z_{inp} is the base motion, z_r the relative position between magnet and base, λ' is the derivative of the magnetic flux linkage with respect to the relative displacement, R and L are respectively the resistance and inductance

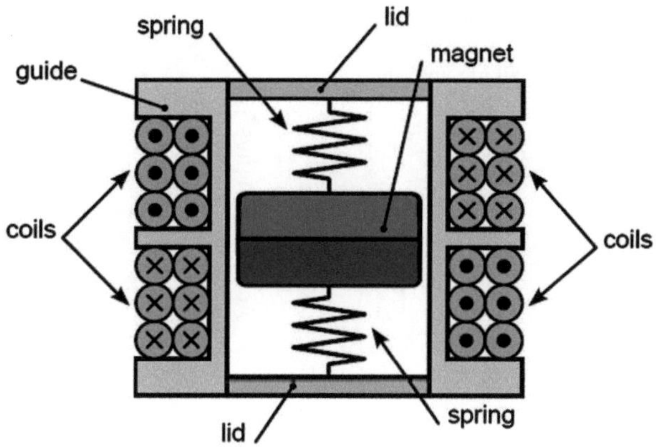

Fig. 1.2 Transducer layout

of the transducer coils, $V_L(i)$ is the voltage on the load, and i is the current [9, 10]. An accurate modelling of the system usually requires considering non-linearity of k and λ' that are function of the relative displacement [11, 12].

1.2.2 Electric Interface

The simplest electric load is a resistor directly connected to the ends of the coils of the transducer. This solution provides the maximum electrical power, but it does not allow energy storage; it follows that the EH can feed an electrical device only when the floating magnet of the transducer is moving. In order to store the recovered energy a capacitor is connected to the coils terminals and, as the provided current is alternating, a rectifier bridge is interposed between the capacitor and the transducer. Although this configuration represents the easiest way for storing, due to the voltage drops in the diodes of the rectifier, it implies very worst performance with respect to the previous case. Moreover, if the voltage between the ends of the rectifier is lower than the voltage of the capacitor, the transducer does not charge it.

To overcome the limits characterizing the bridge rectifier, an active electronic interface consisting in a step-up and a buck converter have been developed. The electronics interface is directly connected to the positive and negative terminal of the transducer. It consists of a full wave active boost converter with a transducer current control; this provides an optimum resistive load emulation independently from the signal provided by the transducer (shape and voltage level) and from the voltage stored on the output capacitor.

The interactions between the mechanical and electrical phenomena depend heavily on the power transferred by the seismic mass to the electrical load. In linear system response conditions, the optimal matching follows the well known maximum power transfer theorem. For instance, in [13] it has been demonstrated that the energy recovery in response to a sinusoidal vibration input whose frequency matches the resonant frequency of the mechanical system, is maximum in adapted load condition, namely when the resistive load R_L is:

$$R_L = R_{ADAPT} = R \tag{1.2}$$

However, due to the nonlinear effects present in the system, mainly the mechanical dissipative effects and the limited stroke of the floating magnet, the optimal resistor value is different from the one in (1.2) and it is typically larger. A theoretical analysis of this effect is present in [14], while here the best matching resistance value is calculated according to experimental evidences of the maximum power. Thus, the following equation is adopted through experimental evidence:

$$R_L = R_{OPT} = R_{ADAPT} + R_{ADD} \tag{1.3}$$

1.2.3 Application Constraints, Vibrational Input and Energy Requirements

The nature of the specific application imposes very stringent dimensional constraints to the device: a cylindrical shape volume of about Φ 27 $\times$ 16 mm that must contain the magneto-inductive energy harvester system for the power supply, the electronic interface for the data sending and the housing for the placing and the protection of the device itself in the sole of the shoe.

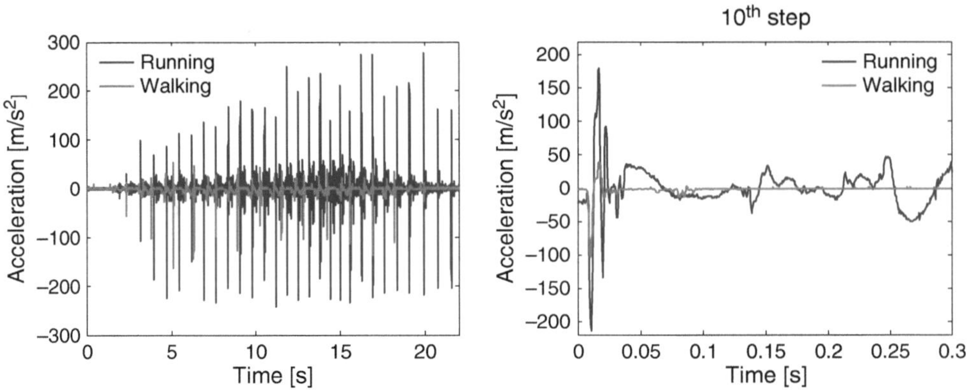

Fig. 1.3 Vertical acceleration experimentally measured in the heel during walking and running activity

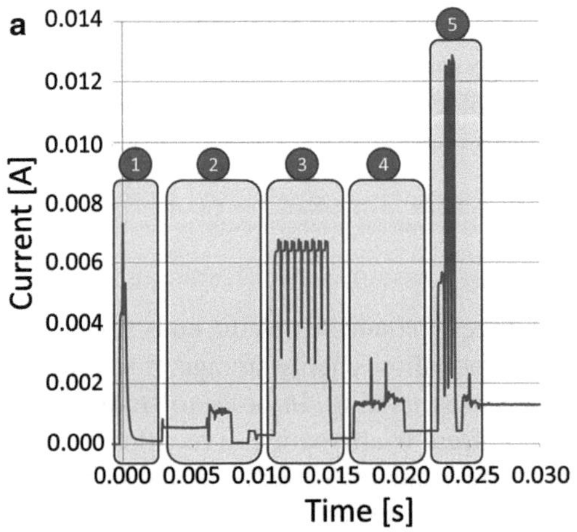

b

Power needs	
1) VDD decoupling cap charging	16µJ
2) STM32 initialization	5µJ
3) Step Counting Algorithm	24µJ
4) Bluetooth initialization	6µJ
5) RF transmission	36µJ
TOT	87µJ

Fig. 1.4 Step counter system current absorption profile (**a**) and energy demand (**b**)

As Fig. 1.3 shows, the typical input acceleration profile for walking and running consists of a very limited in duration and amplitude excitation at each step comparable to a series of impulses.

Figure 1.4 represents the shape of the current consumption of the electric interface for the via bluetooth data transmission at each step, with the energy demand corresponding to each phase.

The total energy need of 87 µJ represents the minimum energy required for the transmission of the step. Considering the harvester interface and electric components efficiency, the energy that the transducer must provide for the step detection is 104 µJ.

1.2.4 Device Optimization

The stringent constraint imposed to the dimensions, the energy requirement of the electric interface for the step detection and the low intensity energy source imply a strict coupling between the mechanical and electrical characteristics upstream and downstream of the transducer with the objective of maximize the average power extracted from the device.

The implemented optimization algorithm exploits Pattern Search algorithm [15], a well know 0th order deterministic technique extensively used in the automated optimization environment.

Figure 1.5 summarizes the optimization loop steps based on an integrated electromagnetic-mechanical-electric-electronic simulator. This is a complete Matlab/Simulink model made to study the dynamic behavior of the device integrated in the optimization algorithm in order to maximize performance. Starting from the geometrical dimensions all the operative parameters (mass, springs, motion amplitude, dimensions of coils, number of turns in coils) are calculated. Through an automatic FEM model the flux linkage as a function of the magnet position is calculated. Then the system is simulated using a proper acceleration profile to excite the device. A full description of the simulation model is available in previous articles [16, 17]. Finally, the objective function implemented in the optimization algorithm is evaluated.

Figure 1.6 shows the evolution of the parameters and the objective function. A consistent improvement of the performance is found mainly due to the tuning of the elastic characteristic. Moreover the magnetic mass is increased by the extension of

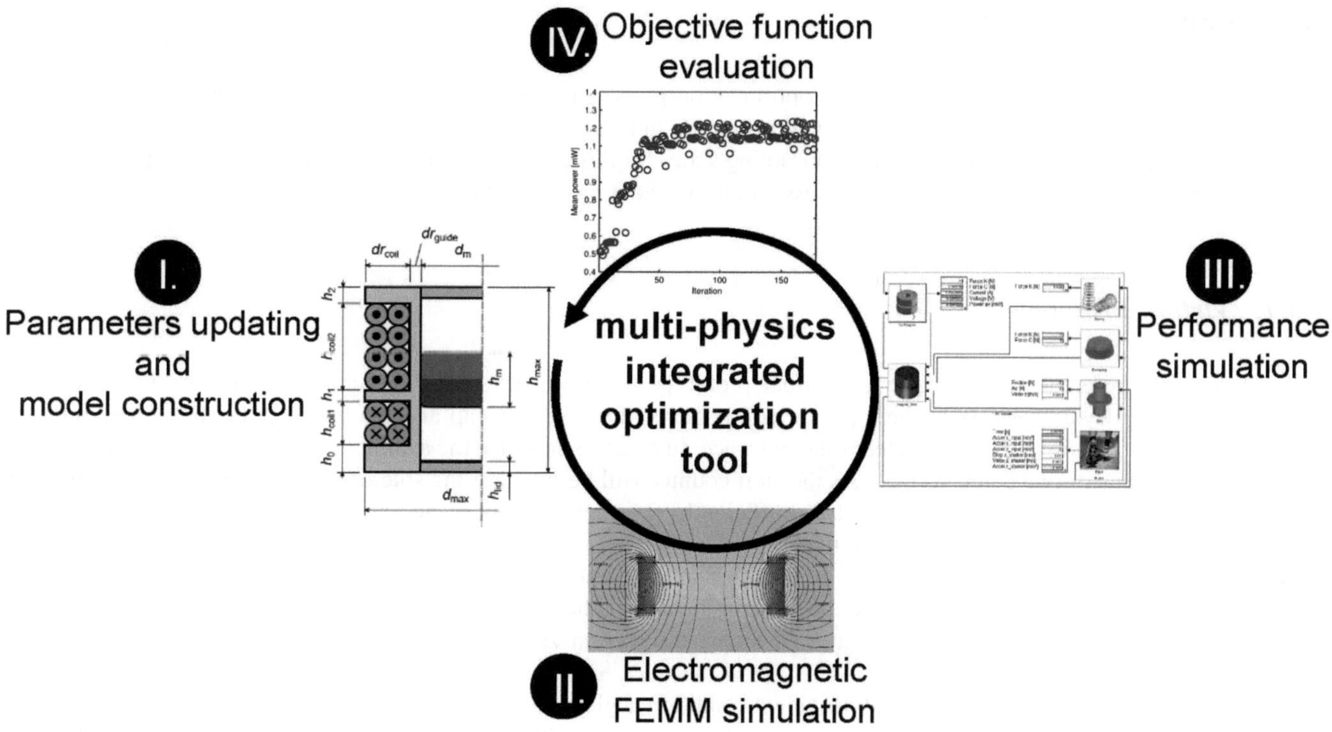

Fig. 1.5 Optimization loop

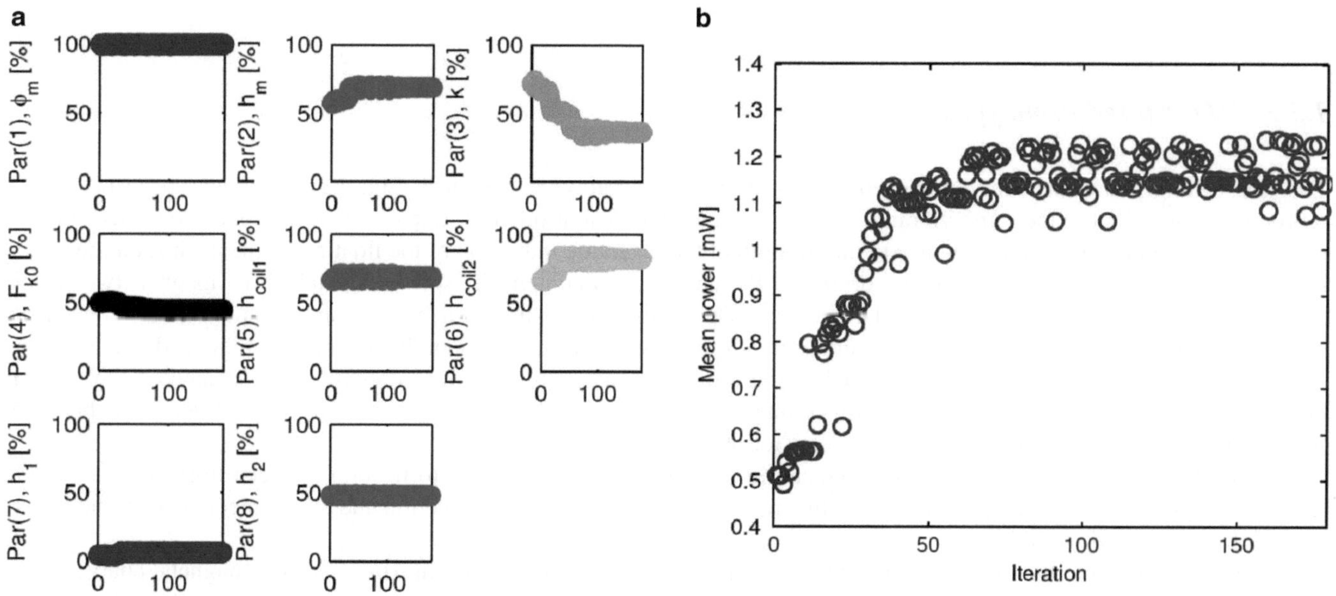

Fig. 1.6 Evolution of parameters (**a**) and objective function (**b**)

the magnet height, reducing the available stroke. The upper-coil length is also slightly increased. The simulation result emphasized the fact that initial device did not exploit all the available length of the stroke. The problem is fixed by softening its elastic characteristic and increasing the magnet height.

1.3 Product Development

The geometrical parameters resulting from the optimization process represent the reference configuration for the product development phase. The aim is to convert a virtual prototype in a real device ready for the medium-large scale production, fully functional and ready to use when placed in training shoes. In this context, a consistent number of prototypes have been made to experimentally validate the design process and prove the project feasibility.

1.3.1 Virtual Prototype

Starting from the reference configuration, a virtual prototype has been developed in according with the manufacturing requirements. The device consists of three plastic components, the case, the cap and the backcap, holding the cylindrical magnet and protecting the coils and the electronic interface. In order to optimize the available space, the elastic characteristic is obtained by means of two conic springs. As the step-counter will be insert in the sole of training shoes, a very important aspect is the design of the housing that must be sufficiently resistant to support the external load during the running and walking activity but, at the same time, not too thick in order to do not subtract space to the functional components. Therefore, 3D modelling, FEM analysis and the selection of the most suitable materials were key aspects in the product development of this device. The geometries that characterize the structural elements have been studied to minimize the internal stresses but always thinking to the ease of mounting. Figure 1.7 shows the section view of the device and an example of FEM analysis result.

Device behavior has been simulated in almost ideal condition neglecting friction. Figure 1.8 shows the simulated electrical performance in terms of average harvested power over 22 s of activity with the first step after 2.5 s, voltage trend and energy recovery with respect to the energy requirement of the tenth step. As it can be observed, the energy harvested during the step in this condition is fully sufficient to support the electric interface requirements.

1.3.2 3D Printed Prototypes

The first testing phase has been conducted on the 3D printed prototypes shown in Fig. 1.9. Tests have been performed reproducing on a shaker the same vibrational input used for the design process. Experimental evidences revealed that the springs, even if made of ductile iron, due to the strength of the magnetic induction of the floating element, shift on the lateral side of the magnet, precluded the right working of the device. To maintain the spring coaxial to the magnet, two plastic centering rings have been pasted on its flat surfaces. This allows the right linear oscillation of the floating elements but considerably reduces its available stroke resulting in a loss of performance of about 20 % when running as demonstrate by Fig. 1.10. Due to the smaller oscillation caused by the walking input, in this operative condition performance is the same as before; in fact, the stroke reduction do not affect magnet displacement when walking while strongly influence the behavior when running causing violent bumps that dissipate lot of energy.

Real performance is much lower than simulated in almost ideal condition due to friction between floating element and guide, materials efficiency, components production uncertainty, assembling imperfection and all the inconvenient that differs the real from the ideal world.

In order to overcome the loss due to the stroke reduction caused by the centering rings, ad hoc magnets, characterized by the same overall dimensions of the previous but presenting two centering holes on the flat surfaces, have been used. Exploiting larger stroke balances the smaller flux due to the smaller magnet volume. In addition, little adjustments of the guide component allowed reducing friction extending the magnet oscillation duration and increasing case resistance to the soldering high temperature.

1.3.3 Molded Prototypes

A consistent number of molded prototypes have been produced through a semi-industrialized manufacturing process, see Fig. 1.11. Plots in Fig. 1.12 demonstrate the convenience of adopting ad hoc magnets and the effectiveness of the adjustment performed on the main plastic component. No significant differences of the device performance is registered under walking

Fig. 1.7 Prototype layout (**a**) and example of FEM analysis result (**b**)

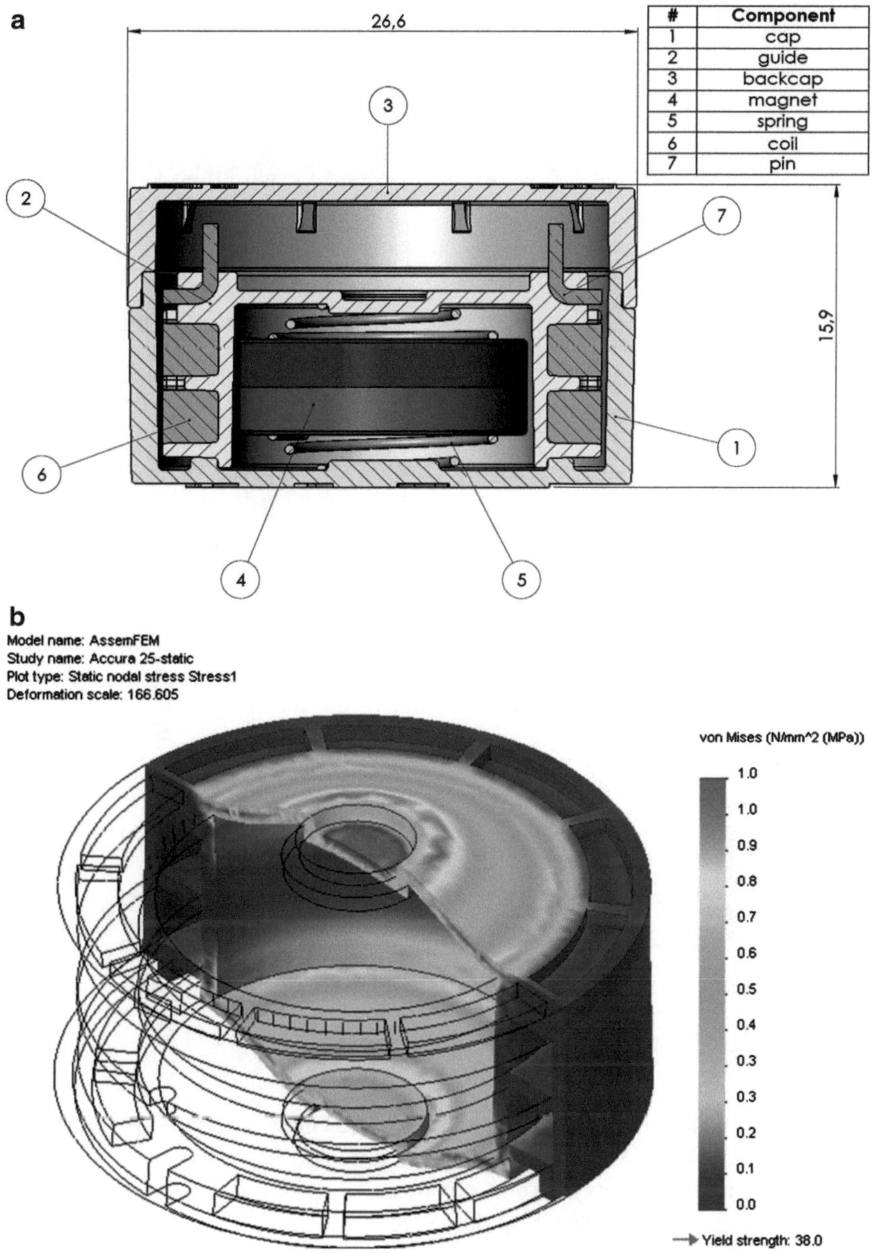

#	Component
1	cap
2	guide
3	backcap
4	magnet
5	spring
6	coil
7	pin

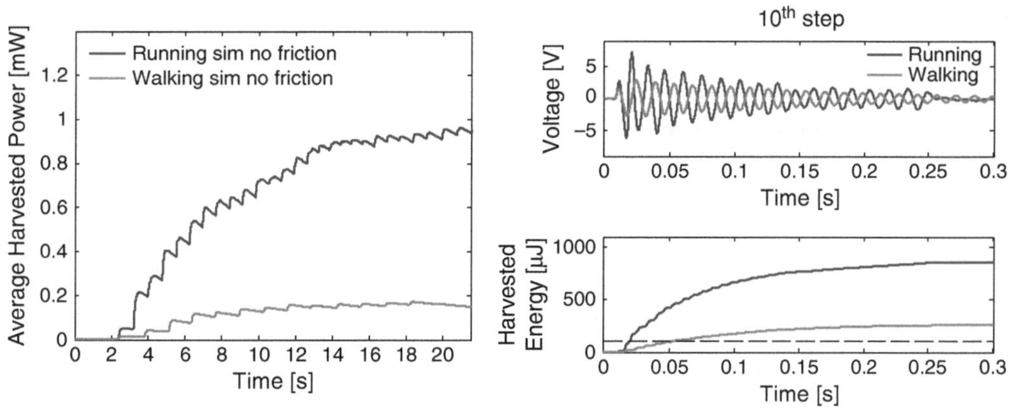

Fig. 1.8 Virtual prototype performance

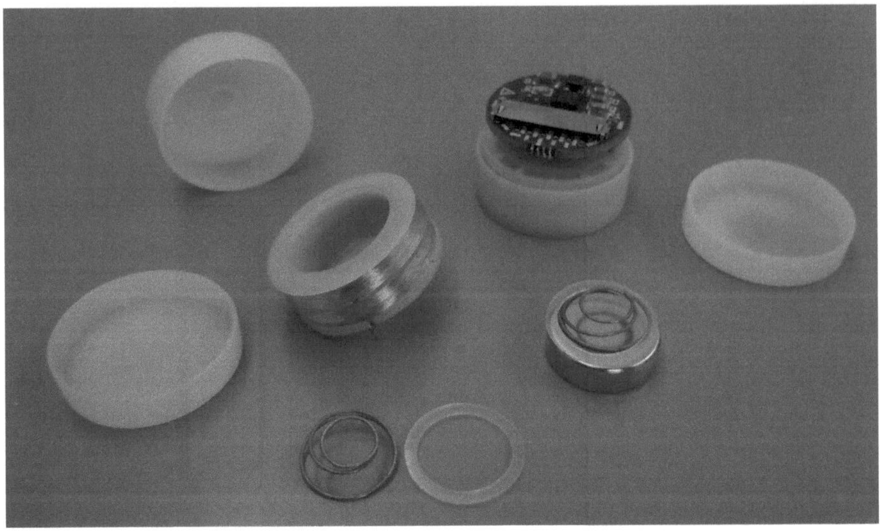

Fig. 1.9 3D printed prototypes

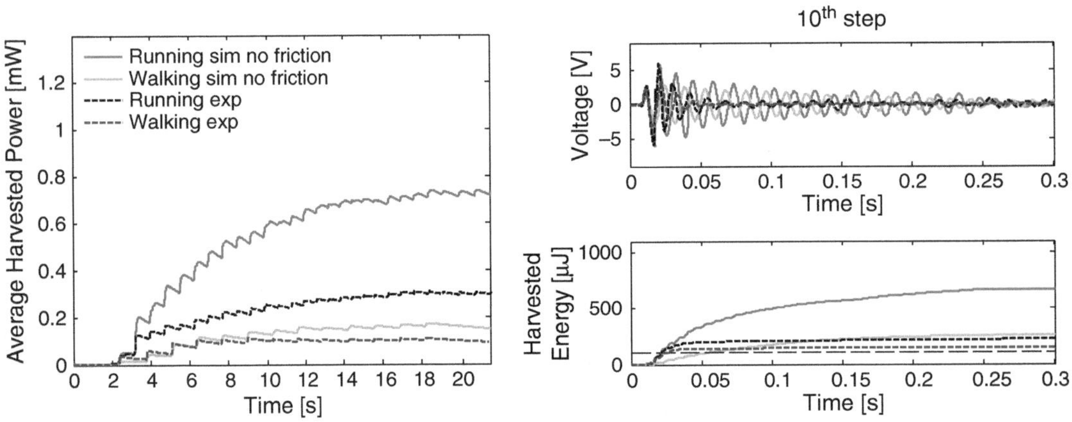

Fig. 1.10 3D printed prototypes performance

input, while, when running, a broad increase in the energy harvested is shown, both in simulation and in real test, about 40 %. The experimental recovered energy during a running step is about five times the required for the activity detection and data sending. Storing the energy surplus compensates the difference with respect to the requirements under very low intensity input allowing detecting every step during activity.

1.4 Conclusions

With the aim of feeding a bluetooth step-counter placed in the sole of a shoe for walking and running activity monitoring, a semi-industrialized set of EH prototypes with a dedicated electronic interface has been designed, produced and tested in order to demonstrate the feasibility of the project and validate the simulation model. Energy harvesting is performed with a magneto-inductive transducer designed to maximize the recovery in response to the vibrating input due to the impact between the shoe and the ground during walking or running. The electronics interface consists of a step-up and buck converter, and, by means of a full wave active boost converter with a transducer current control, provides an optimum adaptive load emulation independently from the signal provided by the transducer and from the voltage stored on the output capacitor. The device has been designed in according to the manufacturing requirements for a medium-large scale production and taking into account the severe environment in term of external load where it will be placed. Numerical and experimental results show the effectiveness of the developed EH step-counter device demonstrating that the energy recovered during the impact is

Fig. 1.11 Molded prototypes

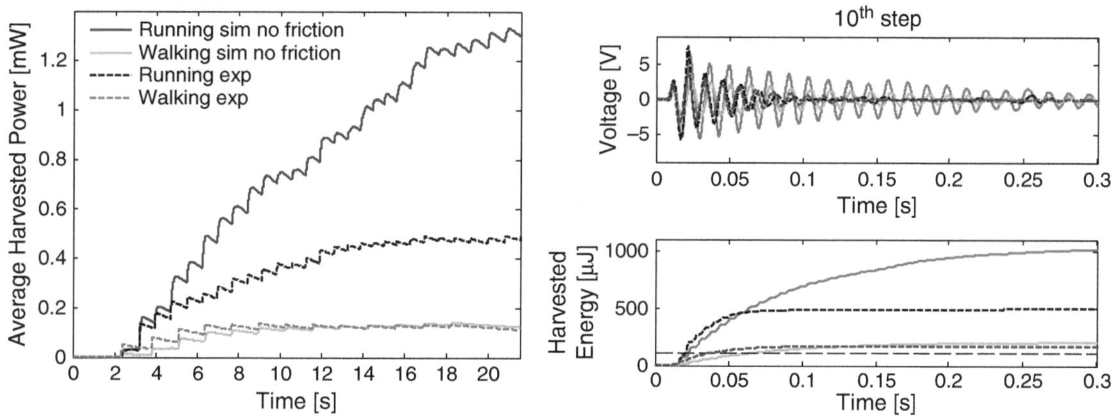

Fig. 1.12 Molded prototypes performance

maximized and it is sufficient to support the power requirement of the electronic interface allowing the step detection and the data sending also during the walking activity. Considering running activity, the harvested energy at each step is about four times the required. Storing this energy surplus compensates the difference with respect to the requirements under very low intensity input allowing detecting every step during activity.

Acknowledgment This work was performed under a research project with STMicroelectronics. The authors would like to thank Dr. Alessandro Gasparini, Dr. Stefano Ramorini and Dr. Alberto Cattani from STMicroelectronics for their enthusiasm and driving force in the project.

References

1. Shenck NS, Paradiso JA (2001) Energy scavenging with shoe-mounted piezoelectrics. IEEE Micro 21(3):30–42
2. Rocha JG, Gonçalves LM, Rocha PF, Silva MP, Lanceros-Méndez S (2010) Energy harvesting from piezoelectric materials fully integrated in footwear. IEEE Trans Ind Electron 57(3):813–819
3. Li WG, He S, Yu S (2010) Improving power density of a cantilever piezoelectric power harvester through a curved L-shaped proof mass. IEEE Trans Ind Electron 57(3):868–876
4. Moro L, Benasciutti D (2010) Harvested power and sensitivity analysis of vibrating shoe-mounted piezoelectric cantilevers. Smart Mater Struct 19(11):1–12
5. Carroll D, Duffy M (2012) Modelling, design, and testing of an electromagnetic power generator optimized for integration into shoes. J Syst Control Eng 226(2):256–270
6. Wang C, Miao D, Luk PC, Shen J, Xu C, Shi D (2010) A shoe-equipped linear generator for energy harvesting. In: 2010 IEEE international conference on sustainable energy technologies, Kandy, Sri Lanka, pp 1–6
7. Baghebani R, Ashoorirad M (2009) A power generating system for mobile electronic devices using human walking motion. Second Int Conf Comput Electr Eng 2:385–388
8. Dai D, Liu J, Zhou Y (2012) Harvesting biomechanical energy in the walking by shoe based on liquid metal magnetohydrodynamics. Front Energy 6(2):112–121
9. Nakano K, Elliott SJ, Rustighi E (2007) A unified approach to optimal conditions of power harvesting using electromagnetic and piezoelectric transducers. Smart Mater Struct 16(4):948–958

10. Beeby SP, O'Donnell T (2009) Electromagnetic energy harvesting. In: Priya S, Inman DJ (eds) Energy harvesting technologies. Springer, New York, pp 129–161
11. Tornincasa S, Repetto M, Bonisoli E, Di Monaco F (2012) Energy harvester for vehicle tires: nonlinear dynamics and experimental outcomes. J Intell Mat Syst Str 23(1):3–13
12. Tornincasa S, Repetto M, Bonisoli E, Di Monaco F (2012) Optimization of magneto-mechanical energy scavenger for automotive tire. J Intell Mater Syst Struct 23(8):2055–2064
13. Stephen NG (2006) On energy harvesting from ambient vibration. J Sound Vib 293(1–2):409–425
14. Liang J, Liao W (2012) Impedance modeling and analysis for piezoelectric energy harvesting systems. IEEE/ASME Trans Mechatron 17(6):1145–1157
15. Hooke R, Jeeves TA (1961) Direct search solution of numerical and statistic problems. J ACM 8(2):212–229
16. Bonisoli E, Canova A, Freschi F, Moos S, Repetto M, Tornincasa S (2010) Dynamic simulation of an electromechanical energy scavenging device. IEEE Trans Magn 46(8):2856–2859
17. Tornincasa S, Bonisoli E, Di Monaco F, Moos S, Repetto M, Freschi F (2011) Nonlinear dynamics of an electro-mechanical energy scavenger, vol 3, Modal analysis topics. Springer, New York, pp 339–349

Chapter 2
Issues in Experimental Testing of Piezoelectric Energy Harvesters

Paulo S. Varoto

Abstract The main goal of this article is to discuss current vibration testing procedures and their effects on the dynamics of beam type piezoelectric energy harvesters. The device is a cantilever beam partially covered by piezoelectric material with a mass at the beams free end. Governing equations of motion are derived for the harvester considering the excitation applied at its fixed boundary. Also, we consider the nonlinear constitutive piezoelectric equations in the formulation of the harvester's electromechanical model. The prototype is subjected to a series of laboratory tests in order to investigate how different testing procedures can affect the overall dynamics of the device under study. Nonlinear effects are also included in order to check their benefits on the harvester's dynamics in terms of the resulting electrical power as well as increase of usable frequency range.

Keywords Energy harvesting • Nonlinear vibrations • Piezoelectric materials • Electromechanical model • Energy scavenging

2.1 Introduction

Piezoelectric energy harvesting (PEH) has become a topic of increasing interest among several research areas as well as engineering majors, as it is the case of the aerospace and automotive industries. The possibility of converting certain amount of mechanical energy into usable electrical energy through has enabled the application of fundamental concepts from the theories of mechanical vibration and piezoelectricity in the development of new energy harvesting methodologies and devices, specially dedicated to power small electronics. Despite the difficulties just mentioned, recent contributions have shown promising results for piezoelectric energy harvesting either in employing linear [1–4] and nonlinear [5–8] modeling approaches in the mechanical to electrical conversion process or in developing new transduction and storage circuitry.

The well known cantilever beam model is probably the most commonly employed technique to model and design piezoelectric energy harvesters. In this case a metallic beam (substrate), is covered (partially or fully) by piezoelectric ceramic on both sides, forming a bimorph structure. The piezoelectric layers are connected in series or in parallel and are polled in the transverse direction in order to generate electrical signals from the bending vibrations induced to the beam by some external excitation source. The cantilever harvester is then designed to operate at its fundamental natural frequency for optimum electrical energy conversion, although in principle it can also be used to harvest energy from higher order mode shapes. The harvester's fundamental natural frequency can be tuned to a desired value by using a tip mass that is attached to the free end of the cantilever beam. The value of the tip mass is chosen such that the fundamental natural frequency of the device falls into an usable frequency range covered in most field applications (e.g. 0–100 Hz for environmental vibration signals).

From the testing viewpoint, once a prototype of a given energy harvester is available, it is subjected to a series of vibration tests, mostly in the laboratory environment in order to provide experimental data that will be used to validate the device's mathematical model. In this case, a commonly employed testing technique consists in performing transmissibility base driven tests [9]. The energy harvesting device is mounted on the armature table of an electromagnetic vibration exciter through a test fixture and the overall combined structure (harvester and fixture) is driven by an input signal that covers the frequency range of interest, that must contain at least the harvester's fundamental natural frequency. The nature of the input signal that can be used to drive the system can be selected from commonly employed excitation signals in modal and vibration testing, as it is the case of harmonic, random, pseudo-random, chirp, among others. Usually the output signals that are measured consist of the cantilever base acceleration and voltage from the piezoelectric layers, which are used to compute the harvester's voltage

P.S. Varoto (✉)
Department of Mechanical Engineering, School of Engineering of São Carlos – USP, São Carlos, SP, Brazil
e-mail: varoto@sc.usp.br

A. Wicks (ed.), *Shock & Vibration, Aircraft/Aerospace, and Energy Harvesting, Volume 9*,
Conference Proceedings of the Society for Experimental Mechanics Series, DOI 10.1007/978-3-319-15233-2_2

FRF relating the output voltage to the input acceleration. The output voltage is usually measured across a load resistor connected in parallel or in series with the piezoelectric layers. By varying the value of the load resistance the performance of the harvesting circuit in terms of the output voltage generation can be assessed. Another common measurement consists of using a laser vibrometer to measure the harvester's tip velocity, that along with the input base acceleration are used to compute the tip velocity FRF, that can reveal additional important information on the performance of harvesting device. The goal of this paper is to review commonly employed vibration testing procedures used in obtaining experimental data that can be further used to validate analytical model of piezoelectric energy harvesters. Two different test setups, one for linear tests and the second for nonlinear tests are used in order to simulate different test scenarios when testing a specific harvesting device. Experimental results are shown and discussed, and major conclusions and recommendations are taken for future use.

2.2 Analytical Model Description

This section presents a very brief description of the analytical models used in the remaining parts of the paper. Two models are employed, as shown in Figs. 2.1 and 2.2. Figure 2.1 shows the linear cantilever harvester model, where the device is grounded at the left end and free at the opposite end of the cantilever. The free end carries a tip mass (Mt) and the substrate is partially covered on both sides by piezoelectric ceramic layers that are, in the present case connected in series, as seen from Fig. 2.1. Piezoelectric layers and the substrate are assumed to have the same width, and as it will be seen in the next section, they are also have the same length, or, the beam will be fully covered by piezoelectric layers. The model shown in Fig. 2.1 is well known and has been widely used in modeling the electromechanical energy conversion process.

When the piezoelectric layers are connected in series, it can be shown [7] that the FRF relating the output voltage from the harvester to the input base applied displacement can be written as

$$\frac{V}{wW}(\omega)=\sum_{r=1}^{\infty}\left\{\sigma_r-\chi_r^{ser}\frac{\displaystyle\sum_{r=1}^{\infty}\frac{j\omega\kappa_r\sigma_r}{(\omega_r^2-\omega^2+j2\zeta_r\omega\omega_r)}}{\frac{1}{R_l}+j\omega\frac{C_p}{2}+\displaystyle\sum_{r=1}^{\infty}\frac{j\omega\kappa_r\chi_r^{ser}}{(\omega_r^2-\omega^2+j2\zeta_r\omega\omega_r)}}\right\}\times\frac{U_r(x)}{(\omega_r^2-\omega^2+j2\zeta_r\omega\omega_r)} \tag{2.1}$$

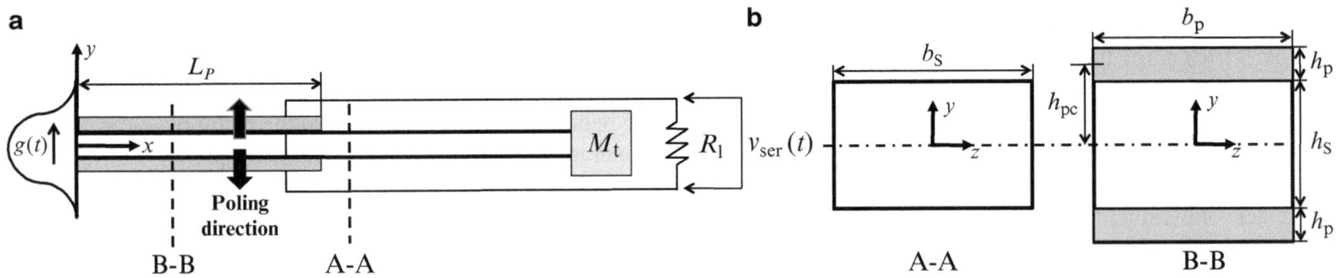

Fig. 2.1 Beam type cantilever piezoelectric energy harvesting model

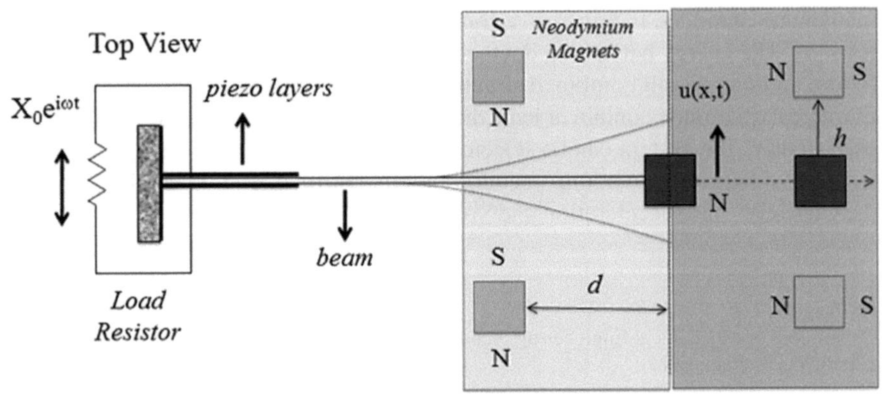

Fig. 2.2 Nonlinear cantilever energy harvesting model

Equation 2.1 was obtained by using the Euler-Bernoulli model for the cantilever energy harvester, and details on how to obtain this electromechanical FRF can be found in several references [1–3, 10].

Figure 2.2 shows the nonlinear energy harvesting model that is used in the experimental analysis to be further presented. As seen the harvesting device consists of a cantilever metallic beam usually denoted as the substructure and partially covered by piezoelectric ceramic on both sides and carrying a lumped mass at its free end. The piezoceramic layers are connected in series to a load resistor R_l and the input to the system consists of a base input harmonic acceleration that is applied at the harvester's grounded side. The tip mass consists of a neodymium magnet that is rigidly attached to the substructure's free end and that oscillates in the vicinity of a second magnet. A repulsive or attractive nonlinear force can be then generated at the beam's free end, depending on the magnetic polarity of these magnets. This contactless interaction between the magnetic fields of the tip magnets introduces a nonlinear force on the bimorph beam what in turn can generate nonlinear oscillations of the harvesting device.

The governing equations of motion of the model shown in Fig. 2.2 can be obtained by applying the Lagrange equations and can be shown to be expressed as [11]

$$\ddot{\eta} + \tilde{c}_f \dot{\eta} + \tilde{\kappa}\eta + \tilde{b}\eta^3 - \tilde{c}V - \tilde{d}V\eta^2 = -\frac{F}{m_t}\cos(\omega t) \tag{2.2}$$

$$\dot{V} + \frac{V}{2Rle} = \frac{c}{4e}\dot{\eta} - \frac{3d}{4e}\eta^2\dot{\eta} \tag{2.3}$$

Equations (2.2) and (2.3) constitute a system of coupled and nonlinear differential equations and correspond to the electromechanical model for the harvesting system shown in Fig. 2.2. Solution of this system of coupled equations for the harvester transverse displacement and output voltage can be achieved by employing perturbation methods [11].

2.3 Test Apparatus

Two experimental setups are used in order to assess the dynamic behavior of the harvesting systems shown in Figs. 2.1 and 2.2. Figure 2.3 shows a bimorph cantilever energy harvester that is mounted on the vibration exciter's table through an impedance head (PCB 288D01, 22.4 mV/N, 100 mV/g). The impedance head is used in order to monitor the input signals (force and acceleration) at the interface between the vibration exciter and the system to be tested. As pointed in [9], knowledge of interface forces and motions in base driven tests can be very useful in the model validation. The harvester used in this case is a commercially available bending sensor (Piezo Systems T226-H4-203X) measuring 26 × 6.4 × 0.66 mm made of reinforced brass (substructure) and fully covered by piezoelectric layers. The bimorph harvester is mounted on a test fixture [3] designed to proper simulate the fixed-free boundary condition. A tear-drop ICP miniature accelerometer (PCB 352A24, 100 mV/g) is used to measure the input acceleration signal on the clamping fixture and this signal will be further used to estimate the voltage FRF of the harvester.

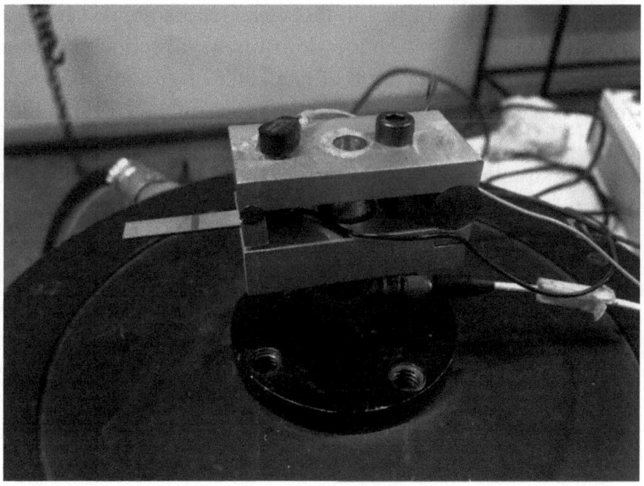

Fig. 2.3 Linear cantilever energy harvesting setup

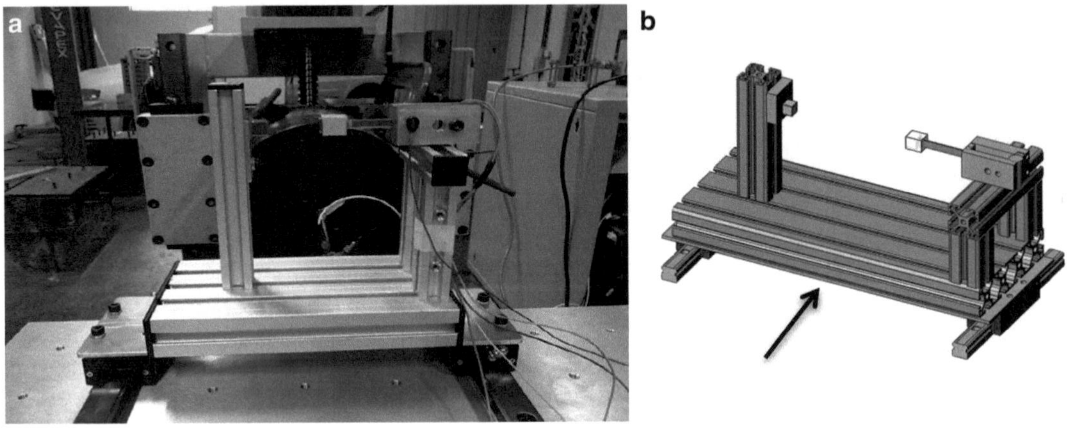

Fig. 2.4 Nonlinear cantilever energy harvesting setup: (**a**) actual test setup; (**b**) illustration of the test setup showing excitation direction (*arrow*)

The second experimental setup used in this work is shown in Fig. 2.4. A special fixture was constructed in order to test the harvesting device in both linear and nonlinear testing configurations. Figure 2.4a shows an illustration of the testing apparatus and Fig. 2.4b shows an illustrative description of the test setup. A MIDÉ Volture V22BL energy harvester was used in the tests. The sensor is attached to the test fixture through a clamping device that is used to simulate the cantilever boundary condition. A neodymium magnet is attached to the harvester's free end and a second magnet is mounted on a vertical tower that part of the fixture as shown in Fig. 2.4. The magnetic gap between the tip and fixed magnet can be varied by moving the vertical tower along the longitudinal axis of the test fixture. By varying the linear distance between the two magnets an attractive or repulsive (depending on the relative position of the magnetic poles) contactless nonlinear restoring force can be induced on the transverse motion of the harvester. This nonlinear magnetic force is actually a nonlinear effect that is induced on the harvesting system in order to generate nonlinear dynamic behavior, which can present some benefits in terms of voltage generation and therefore improve the performance of the harvesting system. The test fixture assembly is mounted on the top of linear tracks and attached to the armature of the vibration exciter that in turn will provide the excitation signals to the harvesting system as indicated by the arrow in Fig. 2.4b.

An impedance head is also used to monitor the interface signals between the vibration exciter and the harvesting system, as shown in Fig. 2.4a. The miniature accelerometer is mounted on the clamping fixture in order to measure the transverse input acceleration and will be further used to estimate the harvester's voltage FRF.

2.4 Experimental Results

This section presents and discusses experimental results obtained from tests performed according to the setups described in the previous section. Figure 2.5 show results obtained in two base driven tests employing the test setup of Fig. 2.3. Two test scenarios were examined in this case, where the test item, composed of the bimorph harvester and the clamping fixture was mounted in two different locations on the vibration exciter table. First, the test item was positioned on the top of the exciter's table such that the location of its center of gravity is aligned with the vertical axis of symmetry of the vibration exciter. The importance of this alignment issue relies on the fact that the vibration exciter table moves vertically and significant sources of misalignments can induce rocking motions of the exciter's table [9], what in turn can affect the resulting measured FRF data.

The situation just mentioned can be clearly seen on the results presented in Fig. 2.5. Figure 2.5a shows the resulting transmissibility FRF relating the bimorph output voltage to the input acceleration measured by the miniature accelerometer. In this case the piezoelectric layers were connected in series with a 1 MΩ load resistor. Two curves are shown in Fig. 2.5a, the blue corresponding to what is called symmetric mounting, where the vertical axis passing through the center of gravity of the energy harvesting system and the main vertical axis of the exciter bare table are approximately coincident. This measured voltage FRF is very consistent, noise free and can be used to characterize the energy harvesting dynamic behavior on the 0–1,000 Hz frequency range used in the experiment. In this case, the system was drive by a pseudorandom excitation signal, that is very useful since it is a periodic signal in the time window, making the use of digital windows unnecessary, and in addition being very cost effective since a reduced number of averages are required in the data acquisition process. The blue curve shown in Fig. 2.5b corresponds to a measurement where the harvester along with its clamping fixture is fixed to an off-center

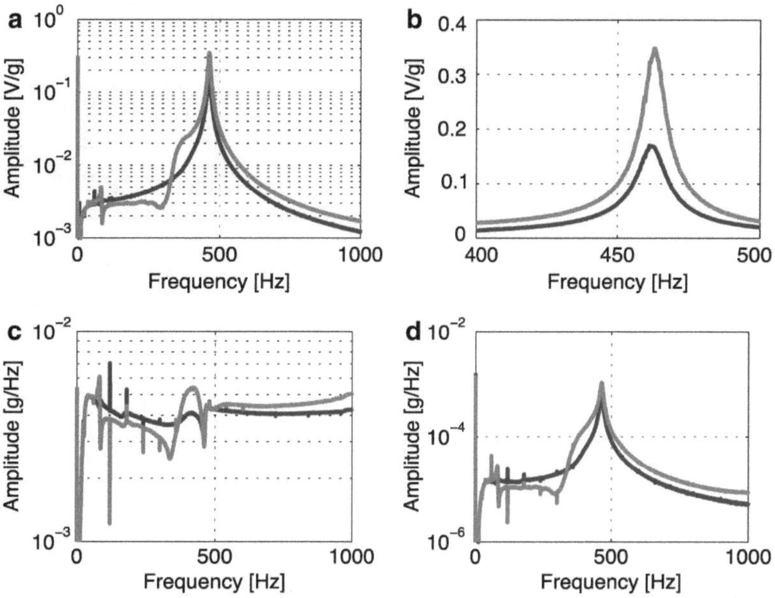

Fig. 2.5 Experimental results for symmetrical (*blue*) and unsymmetrical (*red*) assemblages: (**a**) output voltage FRF related to base acceleration; (**b**) zoomed FRF; (**c**) input acceleration frequency spectra; (**d**) output voltage frequency spectra

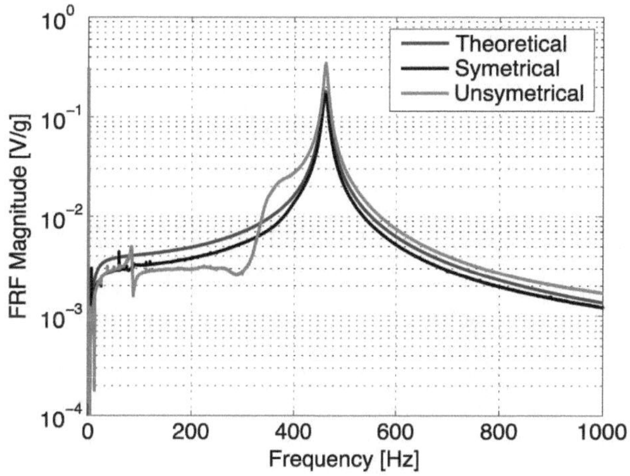

Fig. 2.6 Comparison between symmetrical (*blue*) and unsymmetrical (*red*) FRFs to theoretical FRF (*black*)

point on the vibration exciter table. In this case the mass moment of inertia of the harvesting system induces significant rocking motions on the exciter table due to the unsymmetrical mounting and the results of this inadequate assemblage significantly affects the resulting measured voltage FRF, as shown in Fig. 2.5a. Figure 2.5b represents a close up on both symmetrical and unsymmetrical measurements on a zoomed frequency range in the vicinity of the harvester's natural frequency. It can be seen that amplitude discrepancies occur when inappropriate assemblage is used to characterize the harvester's dynamics. Figure 2.5c, d show similar behavior on the measured input and output frequency spectra signals. Figure 2.6 shows a comparison between the symmetrical and unsymmetrical FRFs and the theoretical voltage FRF computed by Eq. (2.1), in this case used for a single mode representation. Although the natural frequency is sufficiently close to the theoretical value for both measurements the unsymmetrical result shows amplitude discrepancy in relation to the theoretical curve.

The experimental results obtained from sine sweep tests using the test setup of Fig. 2.3 are shown in Fig. 2.7. In this case three tests were performed with different input levels, as indicated, and the resulting output voltage to input acceleration transmissibility FRF was measured for each test. Results shown in Fig. 2.7b indicate that different peak amplitudes are obtained and this in principle contradicts the FRF concept, since it essentially reflects the amount of output to each unit of input applied to the test system. Hence, when linearity holds, increasing input levels should in principle lead to proportionally increasing output levels and the ratio between them should remain constant. On the contrary, results shown in Fig. 2.7 not only show different amplitude levels for the measured FRFs but also a slight shift of the harvester's natural frequency peak.

A possible explanation for the amplitude differences observed on the results shown in Fig. 2.7b is the fact that when passing through the resonance, the vibration exciter presents the force drop-off phenomenon [9], and this can be clearly seen on the results for the input acceleration frequency spectrum shown in Fig. 2.7a. A second possibility that can explain both the amplitude differences and the natural frequency deviation is that the system reveals nonlinear behavior, what is usually

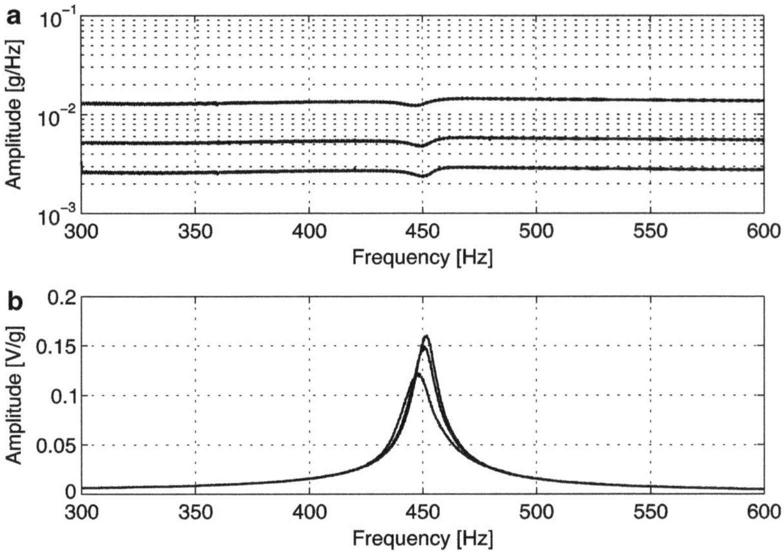

Fig. 2.7 Experimental results from sine sweep tests: (**a**) input acceleration frequency spectra; (**b**) FRF

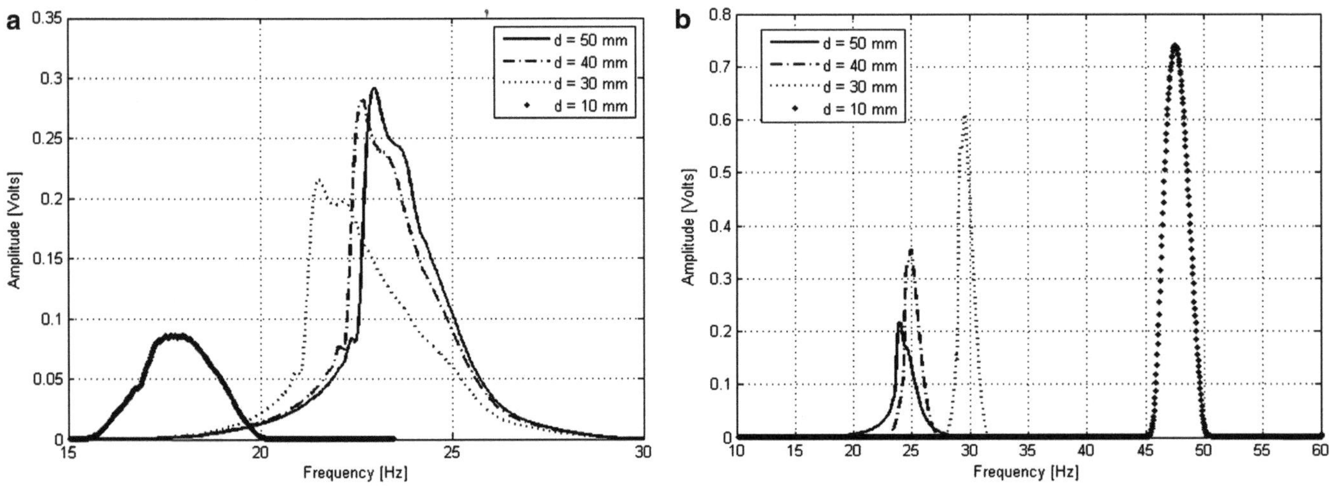

Fig. 2.8 Nonlinear measurements: (**a**) repulsive magnetic force; (**b**) attractive magnetic force

observed when sine sweep is used to drive the system under study. In this case, care should be taken in choosing appropriate excitation levels such that the system's response falls within linear limits. On the other hand, the use of sine sweep signal is suitable for studying nonlinear behavior, and should be preferred in this case in relation to random or pseudorandom excitation signals, since the later tends to smooth nonlinear effects mostly due to the nature of the signal (white noise) and the linearization effects of the Fourier transform used to compute the FRF.

Finally, the results shown in Fig. 2.8 were obtained with the test setup shown in Fig. 2.4. In this case the system was driven by a sinusoidal excitation signal that covered a reduced frequency range. The linear distance between the tip magnet attached to the free end of the harvester and the magnet fixed on the vertical tower shown in Fig. 2.4 was varied in order to generate different magnetic forces, thus inducing different nonlinear effects on the dynamics of the harvester. Figure 2.8a show results for a repulsive magnetic force while Fig. 2.8b show similar results but in this case the magnets were inverted such that an attractive magnetic restoring force could be achieved. Results clearly show differences in amplitude and on the location of the resonant peak, indicating that the use of the magnets significantly altered the dynamic behavior of the harvester.

2.5 Closing Remarks

This paper addressed important issues regarding experimental procedures used in the characterization of piezoelectric energy harvesters in laboratory testing. It was shown that adequate assemblage of the harvesting system on the vibration exciter table during base driven tests is very important in order to obtain meaningful experimental results. It was also discussed the application of different test setups as well as the use of different excitation signals in the process of characterization of the dynamic behavior of piezoelectric energy harvesters.

Acknowledgements The author would like to acknowledge the support received from CAPES and University of São Paulo, in Brazil in the development of this work.

References

1. Erturk A, Innam DJ (2008) A distributed parameter electromechanical model for cantilevered piezoelectric energy harvesters. J Vib Acoust 130:041002
2. Erturk A, Innam DJ (2008) On mechanical modeling of cantilevered piezoelectric vibration energy harvesters. J Intell Mater Syst Struct 19:1311–1325
3. Erturk A, Innam DJ (2009) An experimentally validated bimorph cantilever model for piezoelectric energy harvesting from base excitations. J Vib Acoust 18:025009
4. Erturk A, Hoffmann J, Inman DJ (2009) A piezomagnetoelastic structure for broadband vibration energy harvesting. Appl Phys Lett 94:254102
5. Stanton S, Mcgehee C, Mann B (2009) Reversible hysteresis for broadband magnetopiezoelastic energy harvesting. Appl Phys Lett 95:174103. 3pp
6. Stanton S, Mcgehee C, Mann B (2010) Nonlinear dynamics for broadband energy harvesting: investigation of a bistable piezoelectric inertial generator. Phys D Nonlinear Phenom 10:640–653
7. Abdelkefi A, Nayfeh AH, Hajj MR (2012) Global nonlinear distributed-parameter model of parametrically excited piezoelectric energy harvesters. Nonlinear Dyn 67:1147–1160
8. Abdelkefi A, Nayfeh AH, Hajj MR (2012) Effects of nonlinear piezoelectric coupling on energy harvesters under direct excitations. Nonlinear Dyn 67:1231–1232
9. McConnell KG, Varoto PS (2008) Vibration testing: theory and practice, 2nd edn. Wiley, New York
10. Franco VR Optimization techniques applied to piezoelectric energy harvesting. Ph.D. dissertation, University of Sao Paulo, Brazil (in Portuguese)
11. Mineto AT (2013) Energy harvesting from nonlinear structural vibration signals. Ph.D. thesis, University of Sao Paulo, Brazil (in Portuguese)
12. Erturk A, Innam DJ (2008) Issues in mathematical modeling of piezoelectric energy harvesters. Smart Mater Struct 17:065016

Chapter 3
Innovative Piezoelectric Cantilever Beam Shape for Improved Energy Harvesting

Iman Mehdipour and Francesco Braghin

Abstract Piezoelectric cantilevered beams have been used as a MEMS energy harvester for the last decade because of their less natural frequencies in comparison with other types of boundary conditions. Defining a new shape of cantilever beam to reduce the natural frequency in compared with conventional one is so worthwhile because it causes more flexible bending stiffness and a larger bonding area of piezoelectric layer. So, higher efficiency of the PZT energy harvester can be expected. In order to achieve this goal, a new S-shape PZT cantilever energy harvester is proposed. In this study, the software COMSOL Multiphysics is used to analysis and investigate the characteristics of the suggested model. Preliminary results from modal analysis confirm that in the same volume of mass and effective length of the straight and S-shape beam, the proposed model is more flexible and experiences several natural frequencies which are less than second natural frequency of the conventional cantilevered energy harvester. It is predicted, by mechanical and electrical analyzing, the proposed model produces much higher output voltage than the conventional flat vibration energy harvester, mainly because of lower resonance frequencies. In better word, the proposed model needs less amplitude of excitation force for its maximum efficiency.

Keywords S-shaped cantilevered beam • Piezoelectric energy harvester • Low frequency vibration • Maximum harvested voltage • COMSOL software

3.1 Introduction

Harvesting energy from wasted ambient vibration has received great interest during last decades because of high power density of vibrational sources [1]. This harvested energy, which is converted from mechanical energy to electrical one by means of electrostatic, electromagnetic, and piezoelectric techniques, can provide appropriate electrical energy for low power consumption electronic devices [2–4].

The conventional single-degree-of-freedom (SDOF) has the maximum efficiency in a very narrow bandwidth around its natural frequency. The ambient vibration frequency of structures like machine and civil structure are much lower than the natural frequency of traditional energy harvesters. One way to tune the single-degree-freedom energy harvester to achieve maximum efficiency is reducing the natural frequency of the energy harvester. For instance, for a straight beam energy harvester, adding mass at the tip of the harvester or increasing the length of the beam can help to reduce the natural frequency of the energy harvester. Therefore, for very low frequency of ambient vibration, the length of the harvester should be impractically long or excessively heavy [5].

To overcome these limitations, several researchers proposed different structures. Increasing bandwidth frequency for vibration energy harvester by means of (2 DOF) vibrating body is a way without changing the proof mass [6]. Tunable rotational energy harvester which consists of a rotational spring and a suspended weight is another way to tune for low ambient frequency [7]. A multi-resonant energy harvester [8] and the spiral geometry of the beam [9] are also two methods to reduce the natural frequency of the energy harvester. A new zigzag shape vibration energy harvester is proposed by Karami and Inman [10, 11] and established an analytical model of energy harvesting structure. A new horizontal S-shaped PZT cantilevered model of energy harvester is provided by Huicong Liu et al. [12]. This kind of energy harvesting model is appropriate to scavenge energy from vibrations at frequency less than 30 Hz. Recently, a corrugated cantilevered beam

I. Mehdipour • F. Braghin (✉)
Department of Mechanical Engineering, Politecnico Di Milano, Via La Masa 1, 20156 Milan, Italy
e-mail: francesco.braghin@polimi.it

A. Wicks (ed.), *Shock & Vibration, Aircraft/Aerospace, and Energy Harvesting, Volume 9*,
Conference Proceedings of the Society for Experimental Mechanics Series, DOI 10.1007/978-3-319-15233-2_3

model of a vibration energy harvester which had a sinusoidal length shape and trapezoidal cross section is suggested [5]. They could show that this model is more flexible in comparison with conventional shape (i.e. flat). PZT length, thickness, and cross section shape optimization of energy harvester is another way to reduce natural frequencies and increase the performance of energy harvesters [13, 14].

In this study, a new inversed S-shaped cantilevered PZT energy harvester is proposed. To verify the performance of this new model, first, a straight beam model with first natural frequency equals to 100 Hz is designed. Then, the inversed S-shaped model is optimized by changing the effective length and length of the middle beam of the model to reach the first Eigen-frequency equal to 100 Hz with this constraint that the volume, mass, and materials are the same with the mentioned straight energy harvester. To have general verification, both mechanical and electrical analysis has done. The results confirm that the proposed model is more flexible mechanically according to its other natural frequencies. Regarding electrical analysis, the extracted electric voltage from the proposed model is higher in comparison with the equivalent straight energy harvester.

3.2 Modeling of Energy Harvesting with Different Cantilever Structures

Two different geometrics of the bimorph piezoelectric energy harvester are designed and modeled us by helping the COMSOL Multiphysics software as shown in Fig. 3.1. In two structures, substrate material is structural steel and PZT-5A, which are defined in materials library of the COMSOL Multiphysics, is used as a piezoelectric layer on the top and bottom of the substrate.

The structure of a rectangular straight and inversed S-shaped beams with rectangular cross section are shown in Fig. 3.1.

The dimensions of the energy harvester for the straight and the inversed S-shaped beams are given in the following table (Table 3.1):

Regarding the dimensions of the beams, the length of the straight beam is defined to reach the first Eigen-frequency in 100 Hz. In order to compare the performance of straight beam with the proposed beam model in the same volume, mass and the first Eigen-frequency, the length and the thickness of the substrate and PZT layer is optimized to reach this goal.

3.3 Results and Discussion

Various analyses are carried out to investigate the effects of the inversed S-shaped beam model on mechanical and electrical performance of energy harvester. The analysis and their results are discussed in details in the following sections:

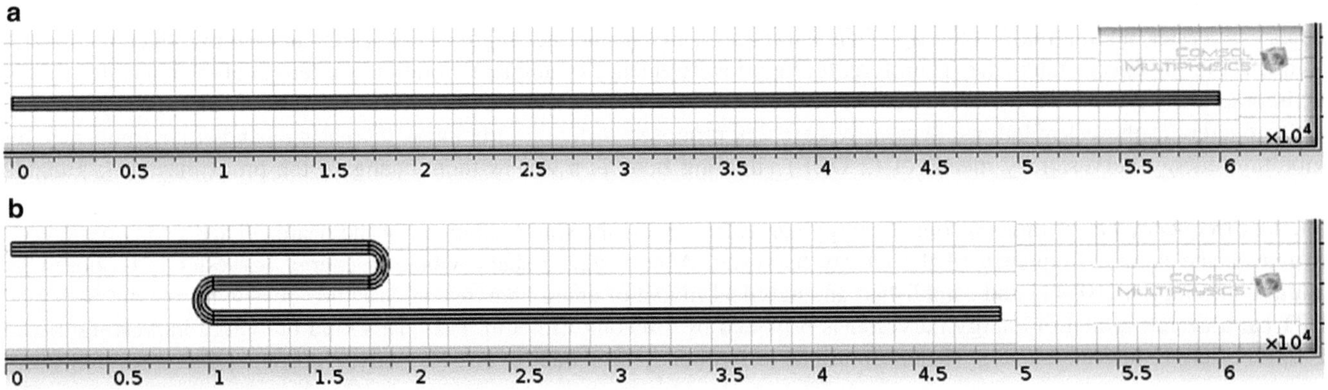

Fig. 3.1 Two different structures of energy harvester beams (**a**) straight beam (**b**) inversed S-shaped beam

Table 3.1 Dimensions of straight and inversed S-shaped beams

Beam type	Straight beam	Inversed S-shaped beam
Length of the harvester (μm)	60,000	Leff = 49,000 (Ltot = 70,000)
Thickness of the substrate layer (μm)	200	171
Thickness of the PZT layer (μm)	210	180

Table 3.2 First six natural frequencies of straight and S-shaped beams

Beam type/Eigen-frequency (Hz)	Straight beam	Inversed S-shaped beam
First	100	100
Second	626	547
Third	1,753	1,224
Fourth	3,341	1,538
Fifth	5,662	3,239
Sixth	8,440	4,778

3.3.1 Eigen-Frequency Analysis

As far as, calculating the Eigen-frequency of beams is the well-known primary mechanical analysis; therefore, in this section, the first six Eigen-frequencies of both straight and S-shaped beams are calculated and compared. Table 3.2 listed the first six natural frequencies:

As mentioned in the previous section, dimensions of two beams are proposed to achieve the same first Eigen-frequency for the same materials, volume, and mass. Table 3.2 confirms that two beams' first natural frequencies are the same. Regarding other five natural frequencies, the results verified that inversed S-shaped beam is more flexible in comparison with the straight beam. It means that to reach second, third and other natural frequencies of S-shaped beam, it needs less excitation force in comparison with straight beam.

3.3.2 Electrical Analysis

In this section, electrical analysis is carried out to find the maximum voltage produced by the energy harvester along the length of the beam. In primary step, open circuit voltage output along the length of beams at the goal natural frequency (first Eigen-frequency) is compared.

Figure 3.2 clearly depicts the open circuit voltage of upper piezoelectric layer versus the length of the energy harvester. Graphs confirms that maximum electric potential generates at the left side of the energy harvester beam where is clamped. In both graphs, by passing from left end to right end, the open circuit voltage decreases because the bonding area of piezoelectric layer decreases. Due to simple geometry of straight beams, this decreasing trend is smooth. However, second graph confirms that complicated geometries of energy harvester beams can seriously change this trend. According Fig. 3.2b, there are two sinusoidal curve which are in the same voltage amplitude. The positive voltage area means that one-half of the curves are under tension and another half is under pressure. In aspect of voltage generation, these curves can be neglected because their average generated voltage is zero. Regarding maximum generated voltage in the same natural frequency, the proposed model generates more voltage. In order to have more accurate judgment about the performance of the proposed model, some analyzes have done in frequency domain to obtain the maximum, average and electric potential per length of the harvester in a range of frequencies near the first natural frequency.

Maximum electric potential generated by upper piezoelectric layer for a range of normalized frequency is plotted in Fig. 3.3. The graph clearly verified that for a range of frequencies around the first natural frequency, the maximum electric potential generated by upper layer for S-shaped energy harvester is more than the generated voltage by straight one. However, this difference is more considerable in a range of frequencies higher than the first natural frequency. So, for frequencies less than the first natural frequency, the maximum generated voltage is almost the same in both two structures.

Another useful parameter to compare two different energy harvester structures is average generated electrical potential along the upper piezoelectric layer. Figure 3.4 obviously shows the trend of this parameter for a range of frequencies around the first natural frequency of two beams for upper layer. Average generated voltage for S-shaped beam for a range of frequencies less than the first natural frequency is higher than straight one. However, this trend is inversed for frequencies higher than first natural frequency of beams.

Electric potential per length of upper layer of energy harvester is illustrated for a range of frequencies in Fig. 3.5. At a first glance, it is obvious that totally, the performance of the proposed model is better than the straight one because the positive or negative generated voltage per length of the energy harvester is higher. However, this difference more considerable for frequencies less than the first natural frequency. Therefore, the extracted voltage from Inversed S-shaped beam is much more in comparison with straight one for the first natural frequency and frequencies less than it.

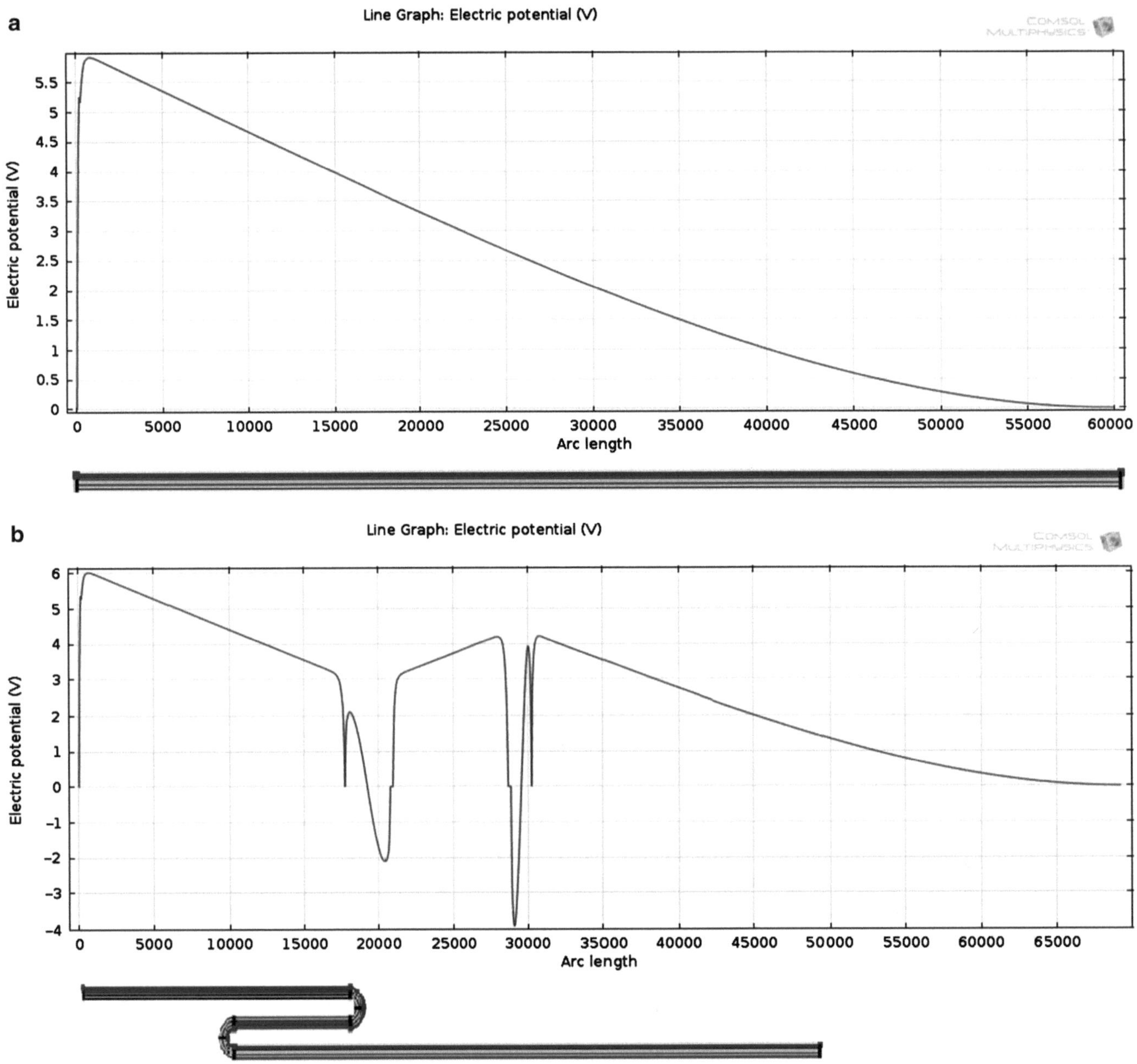

Fig. 3.2 Open circuit electric potential versus length of the energy harvester (**a**) straight beam (**b**) Inversed S-shaped beam

3.4 Conclusion

In this work, a new structure of vibration energy harvester is proposed and the performance of this model is investigated and compared with the equivalent straight beam in the first natural frequency, volume, mass, and materials. To verified this model, two types of analysis: mechanical and electrical are carried out to obtain natural frequencies, maximum, averaged and density of generated voltage for upper layer in a range of frequencies near the first natural frequency. Extracted results are listed in below:

1. The inversed S-shaped Cantilevered energy harvester is more flexible than the equivalent straight one.
2. The maximum generated electric potential of the inversed S-shape model is more than the straight energy harvester for a range of frequencies around the first natural frequency (100 Hz).
3. Generally, the performance of the inversed S-shaped energy harvester is more because the harvested voltage is much higher than straight one for a range of frequencies around the first natural frequency.

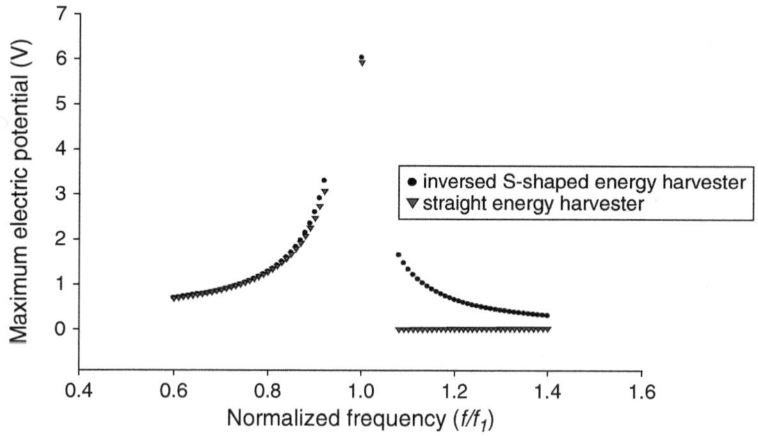

Fig. 3.3 Maximum electric potential versus normalized frequency in the first natural frequency (100 Hz)

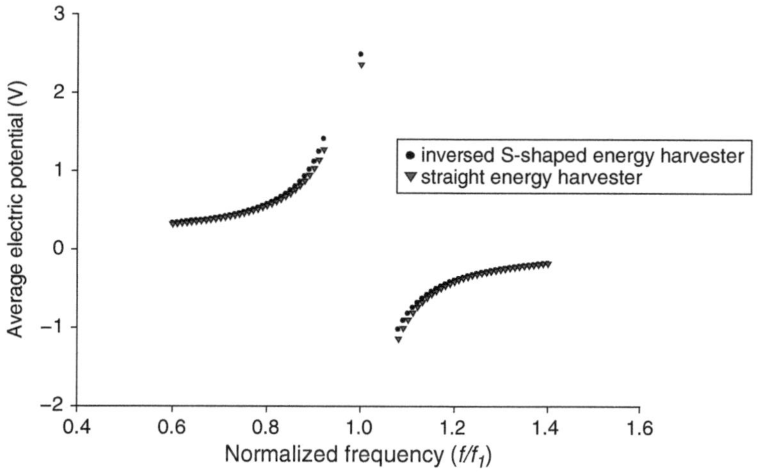

Fig. 3.4 Average electric potential along the upper piezoelectric layer length versus normalized frequency in the first natural frequency (100 Hz)

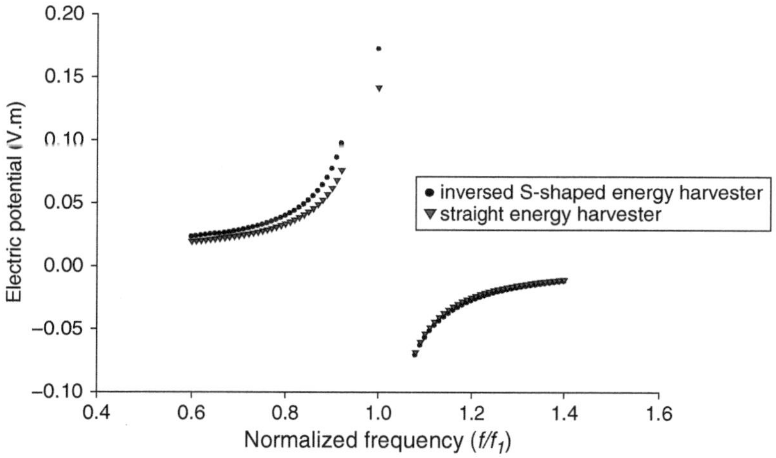

Fig. 3.5 Electric potential per length of upper piezoelectric layer versus normalized frequency in the first natural frequency (100 Hz)

Reference

1. Roundy S, Wright PK, Rabaey JM (2004) Energy scavenging for wireless sensor networks with special focus on vibrations. Kluwer, Boston
2. Sodano HA, Inman DJ, Park G (2005) Generation and storage of electricity from power harvesting devices. J Intell Mater Syst Struct 16:67–75
3. Jung HJ, Kim IH, Jang SJ (2011) An energy harvesting system using the wind-induced vibration of a stay cable for powering a wireless sensor node. Smart Mater Struct 20:075001
4. Jung HJ, Park J, Kim IH (2012) Investigation of applicability of electromagnetic energy harvesting system to inclined stay cable under wind load. IEEE Trans Magn 48:3478–3481
5. In-Ho Kim, SeungSeop Jin, Seon-Jun Jang, Hyung-Jo Jung (2014) A performance-enhanced energy harvester for low frequency vibration utilizing a corrugated cantilevered beam. Smart Mater Struct 23:037002. (7 pp)

6. Kim IH, Jung HJ, Lee BM, Jang SJ (2011) Broadband energy-harvesting using a two degree-of-freedom vibrating body. Appl Phys Lett 98:214102
7. Jang SJ, Kim IH, Jung HJ, Lee YP (2011) A tunable rotational energy harvester for low frequency vibration. Appl Phys Lett 99:134102
8. Seo MH, Choi DH, Kim IH, Jung HJ, Yoon JB (2012) Multi-resonant energy harvester exploiting high-mode resonances frequency down-shifted by a flexible body beam. Appl Phys Lett 101:179905
9. Hu HP, Xue H, Hu YT (2007) A spiral-shaped harvester with an improved harvesting element and an adaptive storage circuit. IEEE Trans Ultrason Ferroelectr Freq Control 54:1177–1187
10. Karami MA, Inman DJ (2011) Electromechanical modeling of the low-frequency zigzag micro-energy harvester. J Intell Mater Syst Struct 22:271–282
11. Karami MA, Inman DJ (2011) Analytical modeling and experimental verification of the vibrations of the zigzag microstructure for energy harvesting. J Vib Acoust Trans ASME 133:011002
12. Huicong Liu, Chengkuo Lee, Takeshi Kobayashi, Cho Jui Tay, Chenggen Quan (2012) A new S-shaped MEMS PZT cantilever for energy harvesting from low frequency vibrations below 30 Hz. Microsyst Technol 18:497–506
13. Sunithamani S, Lakshmi P (2014) Simulation study on performance of MEMS piezoelectric energy harvester with optimized substrate to piezoelectric thickness ratio. Microsyst Technol. doi:10.1007/s00542-014-2226-4
14. Sunithamani S, Lakshmi P, Eba Flora E (2013) PZT length optimization of MEMS piezoelectric energy harvester with a non-traditional cross section: simulation study. Microsyst Technol. doi:10.1007/s00542-013-1920-y

Chapter 4
Energy Harvesting from Piezoelectric Stacks Using Impacting Beam

Yiğit Özpak, Murat Aykan, and Mehmet Çalışkan

Abstract Piezoelectric materials can be used for energy harvesting from ambient vibration due to their high power density and ease of application. Two basic methods, namely, tuning the natural frequency to the operational frequency and increasing the operation bandwidth of the harvester are commonly employed to maximize the energy harvested from piezoelectric materials. Majority of the studies performed in recent years focus mostly on tuning the natural frequency of the harvester. However, small deviations in operating frequency from the natural frequency can cause excessive loss in the power output. It is then advantageous to design a harvester which is capable operating in a wide frequency band. This goal could be achieved both by expanding effective bands of natural frequencies and introducing a frequency-rich external input to the system. The main idea is to supply constant excitation energy into the harvester system to obtain high energy levels by changing system characteristics. In this study, in order to investigate the effects of impacts on energy harvested, an analytical model of an impacting beam with piezoelectric stack at its tip is developed. Experimental validation of analytical results is also performed.

Analytical expressions to obtain the response of harvester and impact forces during motion are formulated in MATLAB® and solved iteratively. Validation of the analytical model is performed by comparing with test results. Moreover, harvester efficiency for broadband frequency excitations is tested and its characteristic properties are investigated in detail.

Keywords Piezoelectric materials • Vibration energy harvesting • Experimental validation • Impacting beam • Structural dynamics

4.1 Introduction

Energy harvesting from environmental sources has gained great attention in recent years. Common environmental sources exploited in energy harvesting can be listed as mechanical energy, thermal energy, light energy, electromagnetic energy, natural energy, human body, chemical and biological energies. For these kinds of energy conversions, there is no need for additional input to supply into system. The basic motivation is the conversion of the natural movements into electric charge. Adoption of a proper method for working conditions is the starting point of energy harvesting. For high power output in scale of kilowatts; solar, wave energy and wind could be used [1]. On the other hand; sources like mechanical vibration, radio frequency, inductive and light are more feasible ways of energy harvesting for the milli watt scale. Key components of harvesting energy consist of a transducer which converts energy into another form, such as piezoelectric material, and a medium to store the converted energy. Energy is conveniently converted into electrical form for the ease of storage.

In this study, mechanical vibration to electricity conversion mechanism is studied. Electromagnetic [2], electrostatic [3] and piezoelectric [4] means are the three main techniques of converting the available vibration energy to electricity.

Piezoelectric materials in mechanical-to-electrical energy conversion are very efficient means to supply energy needs for small electronic equipment. Great majority of the machinery are exposed to vibration in working conditions. Consequently,

Y. Özpak (✉)
Department of Mechanical Engineering, Middle East Technical University, Çankaya, Ankara 06800, Turkey

Defense Systems Technologies Division, ASELSAN Inc., Çankaya, Ankara 06172, Turkey
e-mail: yozpak@aselsan.com.tr

M. Aykan
Defense Systems Technologies Division, ASELSAN Inc., Çankaya, Ankara 06172, Turkey

M. Çalışkan
Department of Mechanical Engineering, Middle East Technical University, Çankaya, Ankara 06800, Turkey

A. Wicks (ed.), *Shock & Vibration, Aircraft/Aerospace, and Energy Harvesting, Volume 9*,
Conference Proceedings of the Society for Experimental Mechanics Series, DOI 10.1007/978-3-319-15233-2_4

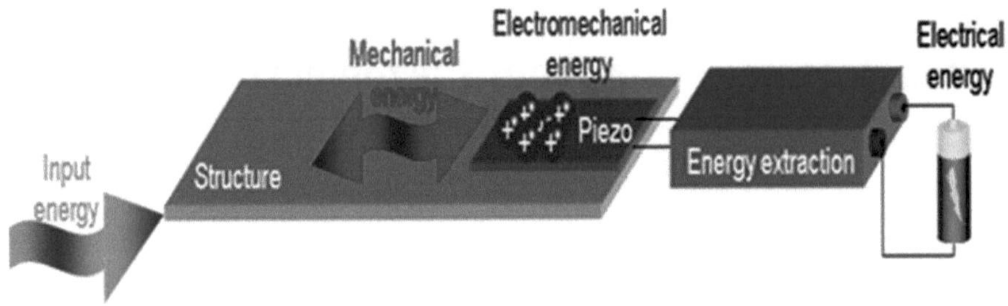

Fig. 4.1 Schematic view of mechanical energy conversion into electric energy using piezo [1]

with decreasing power needs of electronic cards used in such machinery, vibration based energy harvesting using piezoelectric materials becomes popular. Figure 4.1 is an explanatory visual to summarize this procedure.

The main goal of this study is to maximize the power output of a harvester for the same amount of environmental vibration energy. Two basic methods, namely, tuning the natural frequency to the operational frequency and increasing the operation bandwidth of the harvester are commonly employed to maximize the energy harvested from piezoelectric materials. In literature, there is different ways of frequency tuning methods such as mechanical [5], magnetic [6], electrical methods [7] and autonomous frequency tuning methods [8]. However, deviation from the resonance causes significant decrease in power output and makes the harvester effective band smaller. Therefore, it is advantage to design harvester working on broadband loading environment. Achievement of this goal is anticipated to expand the areas of use for piezoelectric materials, due to the increase in power output. In this study energy harvesting from piezoelectric stacks is investigated using an impacting beam.

In the literature, there are two main approaches to model cantilever beam impact. These are Newtonian coefficient of restitution and numerical prediction of the impact force [9]. Advantages of Newtonian coefficient of restitution method are that it is simple to apply in solutions, does not require complex equations and has a wide range of applications. However, coefficient of restitution can change over short impact durations and it should be determined experimentally [9]. Moreover, to find the effective mass of a cantilever impacting beam against stopper is another challenging drawback of this method. The second method involves prediction of impacting force between two bodies. This method requires a complex numerical procedure and difficult mathematical equations to solve come up since the responses of stopper and impacting beam should be solved simultaneously.

In this study, an analytical model of an impacting beam with piezoelectric stack at its tip is developed, in order to investigate the effect of impacts on energy harvested. Experimental validation of analytical results is also performed. Analytical expressions to obtain the response of harvester and impact forces during motion are formulated in MATLAB® and solved iteratively. Validation of the analytical model is performed by comparing with test results. Moreover, harvester efficiency for broadband frequency excitations is tested and its characteristic properties are investigated in detail.

4.2 Methodology

In this study, impacting beam problem given in Fig. 4.2 is investigated. In order to find impact force, steady state response of the impacting cantilever beam and stopper under base excitation is analyzed. Lumped-parameter solutions are used to model each part individually. Coupled response equations are solved simultaneously to find impact force between impacting beam and stopper. In this study, a numerical solution procedure is adopted. This procedure comes with coupled equations and at each step of the solution, responses should be monitored to see whether the impacting process still continues or not. After the impact force is obtained, electrical outputs of piezoelectric material are calculated. The quasi-static solution procedure is used and equivalent circuit models are investigated to simplify the solution. Results of experiments and their comparison with their analytical counterparts are discussed and evaluated comparatively. Finally, experiments are performed to obtain the effective frequency band of the harvester.

4.3 Analytical Expressions for Cantilever and Stopper Via Impact

This section discusses analytical and numerical methods to model dynamic behaviour of a cantilever beam, its stopper and piezoelectric material placed on the beam. Impact force between cantilever and stopper can be calculated by solving coupled steady state responses of the both system. Figure 4.3 shows the schematic view of cantilever and stopper system under base excitation.

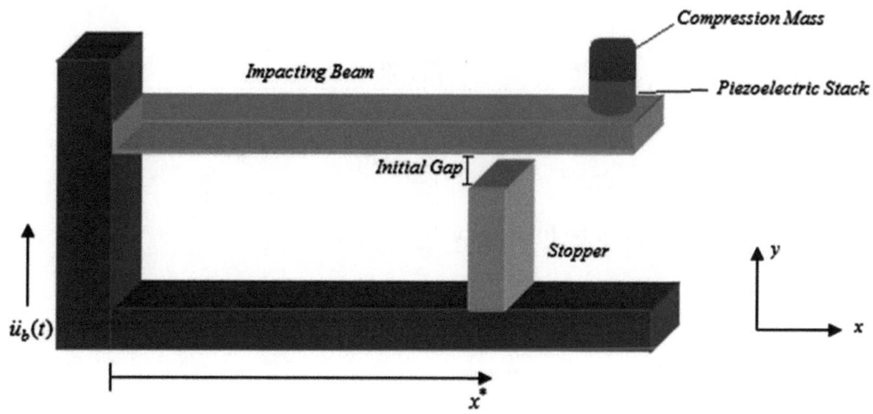

Fig. 4.2 Impacting beam problem

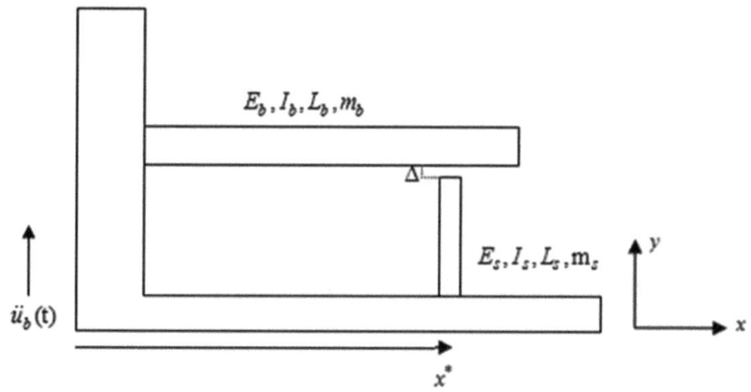

Fig. 4.3 Schematic drawing of impacting beam system

Beam motion during impact can be expressed as

$$w(x,t) = w_b(x,t) + w_{imp}(x,t) \tag{4.1}$$

$$w(x,t) = w_b(x,t) + \int_0^t g_b(x, t-\tau) F(\tau) d\tau \tag{4.2}$$

where $w(x,t)$ total response of the beam, $w_b(x,t)$ is the response coming from base acceleration, $g_b(x,t)$ is unit impulse response function and $F(t)$ is the impact force. The stopper motion during the impact can be given as

$$u(y,t) = u_b(y,t) - u_{imp}(y,t) \tag{4.3}$$

$$u(y,t) = u_b(y,t) - \int_0^t g_s(y, t-\tau) F(\tau) d\tau \tag{4.4}$$

where $u(y,t)$ is total response of the stopper, $u_b(y,t)$ is the response coming from base acceleration, $g_s(y,t)$ is unit impulse response function and $F(t)$ is the impact force.

The contact positions and contact instant can be calculated from the solution of the Eq. (4.5) as

$$w(x^*, t) = u(L_s, t) - \Delta \tag{4.5}$$

where Δ is a positive value for this coordinate system convention adopted.

Unit impulse response functions can be defined for the nth mode when the force is applied to beam at $x = x^*$ and to the stopper at $y = L_s$ as follows.

$$g_b(x,t)=\sum_{r=1}^{n_b}\frac{\phi_{br}(x)}{m_b w_{db}}\phi_{br}\left(x^*\right)\mathrm{e}^{-\zeta_{br}w_{nbr}t}\sin\left(w_{db}t\right)\quad r=1,2,\ldots \tag{4.6}$$

$$g_s(y,t)=\sum_{r=1}^{n_b}\frac{\phi_{sr}(y)}{m_s w_{ds}}\phi_{sr}\left(L_s\right)\mathrm{e}^{-\zeta_{sr}w_{nsr}t}\sin\left(w_{ds}t\right)\quad r=1,2,\ldots \tag{4.7}$$

The equilibrium point can be found from Eq. (4.8) as

$$w_b\left(x^*,t\right)+\int_0^t g_b\left(x^*,t-\tau\right)F(\tau)\,d\tau=u_b\left(\mathrm{L}_s,t\right)-\int_0^t g_s\left(\mathrm{L}_s,t-\tau\right)F(\tau)\,d\tau-\Delta \tag{4.8}$$

However, in Eq. (4.8) impact force *F(t)* is not known. Rearranging Eq. (4.8), the integration can be divided in two parts.

$$\int_0^t g_b\left(x^*,t-\tau\right)F(\tau)\,d\tau=\int_0^{t-\Delta t} g_b\left(x^*,t-\tau\right)F(\tau)\,d\tau+\int_{t-\Delta t}^{t} g_b\left(x^*,t-\tau\right)F(\tau)\,d\tau \tag{4.9}$$

$$\int_0^t g_s\left(\mathrm{L}_s,t-\tau\right)F(\tau)\,d\tau=\int_0^{t-\Delta t} g_s\left(\mathrm{L}_s,t-\tau\right)F(\tau)\,d\tau+\int_{t-\Delta t}^{t} g_s\left(\mathrm{L}_s,t-\tau\right)F(\tau)\,d\tau \tag{4.10}$$

Δt is defined as the very small fixed time step to solve one part of integration with area calculation. Rearranging Eqs. (4.9) and (4.10) as

$$\int_0^t g_b\left(x^*,t-\tau\right)F(\tau)\,d\tau=\int_0^{t-\Delta t} g_b\left(x^*,t-\tau\right)F(\tau)\,d\tau+g_b\left(x^*,t-\tau\right)F(\tau)\,\Delta t \tag{4.11}$$

$$\int_0^t g_s\left(\mathrm{L}_s,t-\tau\right)F(\tau)\,d\tau=\int_0^{t-\Delta t} g_s\left(\mathrm{L}_s,t-\tau\right)F(\tau)\,d\tau+g_s\left(\mathrm{L}_s,t-\tau\right)F(\tau)\,\Delta t \tag{4.12}$$

Inserting Eqs. (4.11) and (4.12) into the Eq. (4.8); Eq. (4.13) is obtained which is the equilibrium of motion of beam and stopper. The reason of these modifications is to be able to solve the impact force by numeric iterations.

$$\begin{aligned}&\mathrm{w}_b\left(x^*,t\right)+\int_0^{t-\Delta t} g_b\left(x^*,t-\tau\right)F(\tau)\,d\tau+g_b\left(x^*,t-\tau\right)F(\tau)\,\Delta t=u_b\left(\mathrm{L}_s,t\right)-\\&\int_0^{t-\Delta t} g_s\left(\mathrm{L}_s,t-\tau\right)F(\tau)\,d\tau-g_s\left(\mathrm{L}_s,t-\tau\right)F(\tau)\,\Delta t-\Delta\end{aligned} \tag{4.13}$$

Using Eq. (4.13) and introducing S1, S2 and S3 the following relations are obtained as

$$S_1=\mathrm{w}_b\left(x^*,t\right)+\int_0^{t-\Delta t} g_b\left(x^*,t-\tau\right)F(\tau)\,d\tau \tag{4.14}$$

$$S_2=u_b\left(\mathrm{L}_s,t\right)-\int_0^{t-\Delta t} g_s\left(\mathrm{L}_s,t-\tau\right)F(\tau)\,d\tau \tag{4.15}$$

$$S_3=g_b\left(x^*,t-\tau\right)\Delta t+g_s\left(L_s,t-\tau\right)\Delta t \tag{4.16}$$

Total impact force could be defined as by Eq. (4.17)

$$F(t) = -\frac{S_1 - S_2 + \Delta}{S_3} \tag{4.17}$$

when there is contact, displacement and force function can be represented as:

$$w\left(x^*, t\right) \leq u\left(L_s, t\right) - \Delta \tag{4.18}$$

$$F(t) \geq 0 \tag{4.19}$$

When there is no contact,

$$w\left(x^*, t\right) \geq u\left(L_s, t\right) - \Delta \tag{4.20}$$

$$F(t) \leq 0 \tag{4.21}$$

In numerical solution, contact force should be controlled in every step and if it is positive, it means that two bodies are still in contact and otherwise separated. In this problem, time step should be chosen carefully to ensure the convergence of the solution. If it is too small, numerical solution time becomes very long and solution may diverge. If it is too large, contact may not occur and high frequency harmonics cannot be monitored. The whole procedure is drawn in Fig. 4.4 as a flowchart.

4.4 Electromechanical Modeling of Piezoelectric Stacks

In this study piezoelectric material and compression mass are bolted to the cantilever beam from its end; therefore, tip mass of the beam consists of piezoelectric stack, compression mass, bolt and washer. Figure 4.5 shows the cross-section of this configuration.

During the motion of the base where the piezoelectric stack and compression mass are inserted, the total force on the piezo electric stack could be calculated from Newton second law as

$$F_{pz} = m_c \ddot{b}(t) \tag{4.22}$$

where F_{pz} is the total force, m_c is compression mass, $\ddot{b}$ (t) is base acceleration. The strain formed on each layer of piezoelectric stack should be calculated by distributed parameter modelling. Same methodology is used during modelling the cantilever and stopper. Total response can be written as

$$u(x, t) = \sum_{r=1}^{\infty} \frac{U_r}{w_r} \int_0^t F(\tau) \sin w_r (t - \tau)\, d\tau \quad \mathrm{r} = 1, 2, \ldots \tag{4.23}$$

Strain at any point and time in force direction can be calculated as

$$S_j(x, t) = \frac{\partial u(x, t)}{\partial x} \tag{4.24}$$

Stress and force in force direction can also be written as

$$T_j = S_j Y_{ij} \tag{4.25}$$

$$F_{pz} = T_j A_{pz} \tag{4.26}$$

where Y_{ij} is the young's modulus, T_j is stress vector and S_j is strain vector of the piezoelectric material, i is the direction of electric field, j is the direction of mechanical strain or stress, A_p is the cross section.

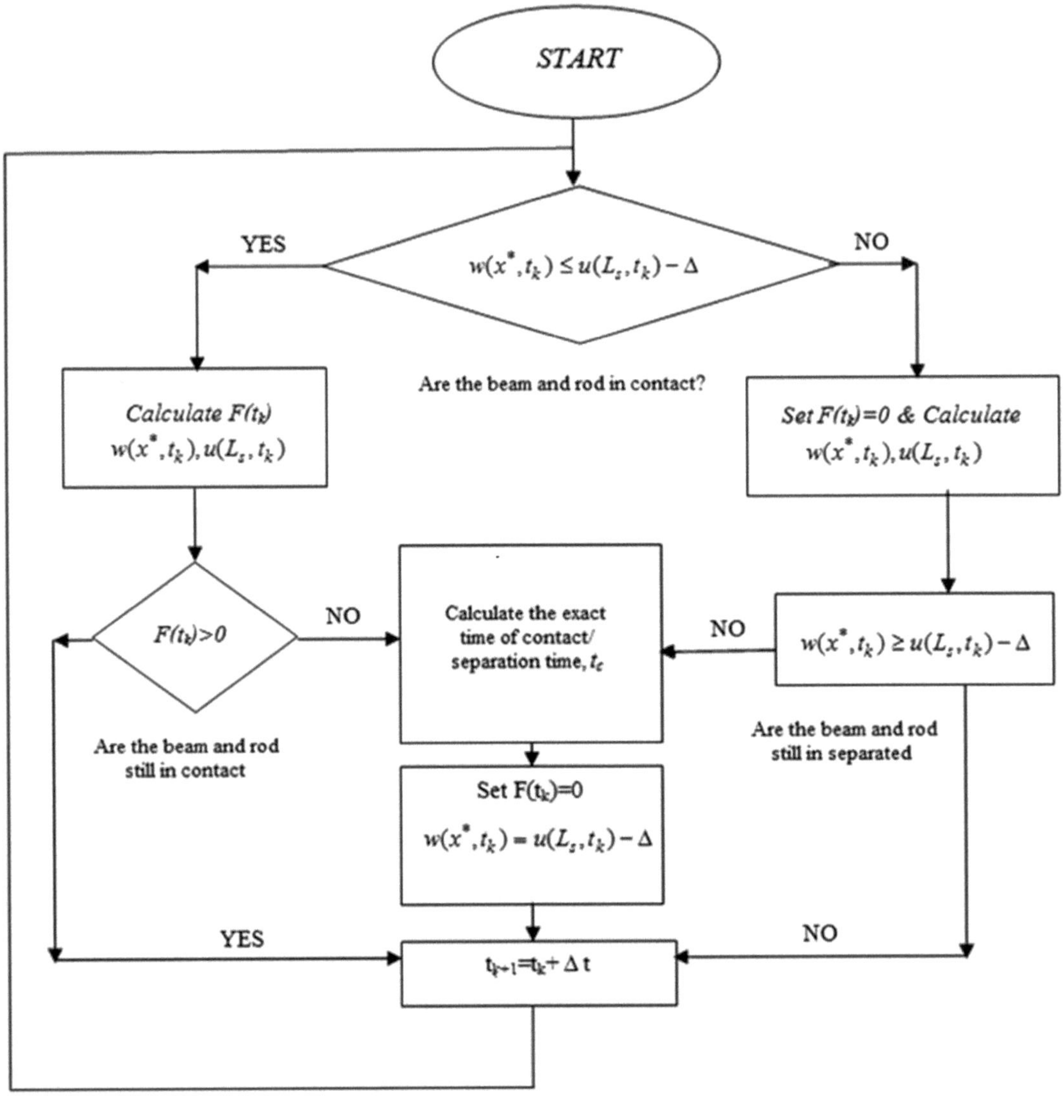

Fig. 4.4 Schematic flowchart of numerical implementation of impact model

Fig. 4.5 Compression mass model of piezoelectric material

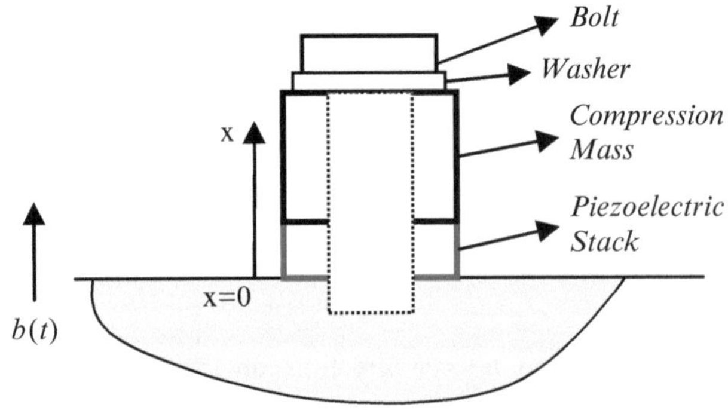

4.5 Experimental Validation of the Analytical Model

In this part of the study, results of experiments and their comparison with their analytical counterparts are discussed and evaluated comparatively. The section starts with validation of analytical solutions in reference to experimental results. Later in the section, frequency dependence of harvester voltage output is investigated.

In the experiments CMAR03 piezoelectric stacks are employed to measure voltage output. Piezoelectric stacks have 25 layers made up of NCE57 ceramics. Isolation material is used between the beam and the piezoelectric material layers as well as under the compression mass resting on the top of piezoelectric layers during the mounting process.

The harvester is designed and built in this study is shown in Fig. 4.6. In this case, the stopper is used to investigate effects of impacts on energy harvesting. Piezoelectric material and tip mass on the beam is screwed to the beam and the accelerometer is glued on the mounting screw as shown.

In the experiments, first natural frequency of the cantilever beam is observed as 62.5 Hz which is almost the same with theoretical case. The experimental burst random test result is shared in Fig. 4.7. The measurements are performed by LMS SCADAS III and PCB 356A16 and 352A24 accelerometers. Accelerometers are glued on the clamped end of the beam and tip mass using Loctite 401. Frequency resolution is 0.25 Hz and a sampling frequency 2,048 Hz is assigned. Number of averages in spectral analysis is 40. Block size is 8,192.

The validation between experimental and analytical methods is performed for tip displacement, acceleration and voltage output for both with and without impact cases. These responses are shown in Figs. 4.8 and 4.9. Time response measurement is performed using Dewesoft software and Dewetron data acquisition system (DEWE 510). The gap size between stopper and beam is measured as $3 \times 10{-}4$ m.

Some differences between experimental results and analytical solutions could be seen. However, in analytical modelling, mode expansion theorem is used with responses to both harmonic base excitation and impulsive excitation at the location of the stopper. Therefore, in analytical solution, effects of only first five natural frequencies are seen. Though, in experimental response, contributions at all frequencies take part in the total response.

Displacements with and without impact conditions are almost same when experimental and theoretical results are analysed. Their oscillation frequencies are observed as 62.5 Hz (experimental) and 63 Hz (theoretical), respectively. In acceleration time histories, it is discovered that base or first natural frequency motion still keeps the main harmonic form, as the base excitation comes to the beam is the same as with that of due to the first natural frequency of the cantilever beam. However, during this motion high frequency terms are also included in acceleration because, impact force excites higher natural frequencies as well. The zoomed view given in Fig. 4.9c shows one oscillation at 62.5 Hz. Small oscillations at the

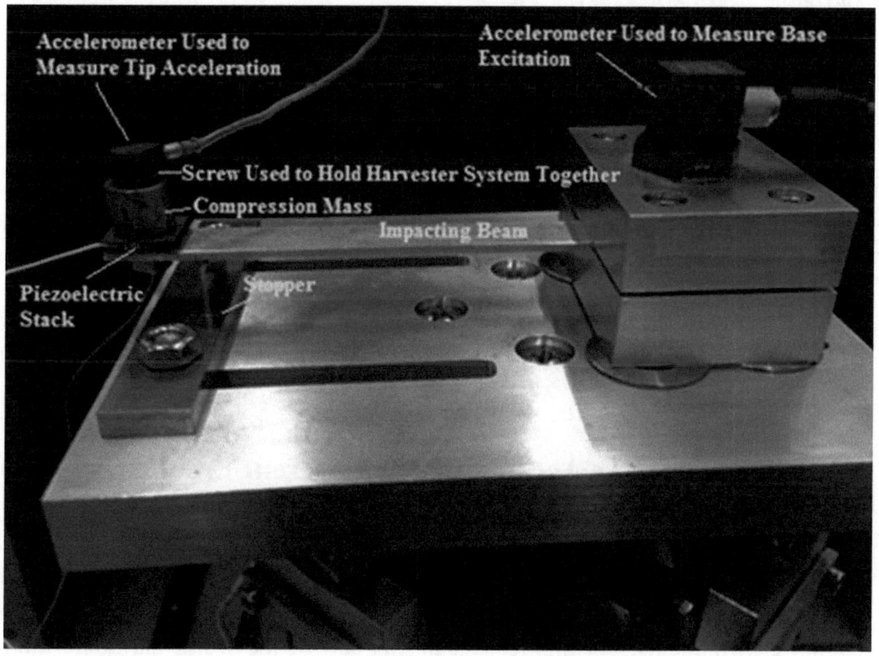

Fig. 4.6 Experimental setup used for impacting beam

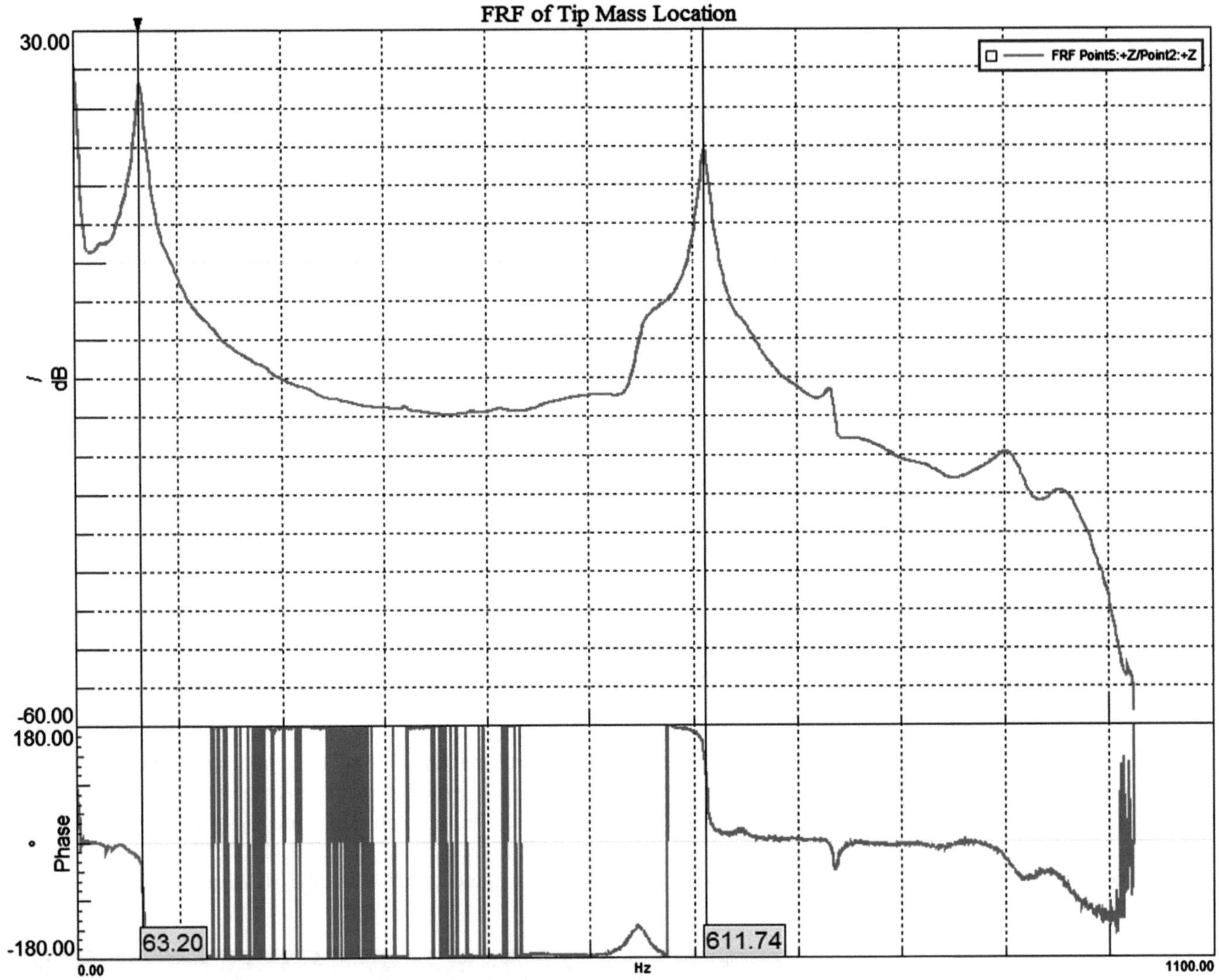

Fig. 4.7 Experimental frequency response function of tip mass location

first part of the motion comes from the fifth natural frequency of the beam at 5,962 Hz and the other oscillation within the cycle is associated with the third natural frequency at 1,612 Hz. Moreover, in the experiments involving impacts, high frequency components or chattering are observed on the whole response. High frequency components are expected in the response since impacts generate sub and super harmonics of the excitation frequency due to its nonlinear nature. Effects are found to be present in zoomed acceleration time histories. Figure 4.10 shows the rms power spectrum of tip acceleration in impacting process up to 3 kHz. Inserting a stopper to the system converts linear system to a nonlinear one, introducing a backlash effect. This non-linear term cause harmonics of excitation within a frequency bandwidth.

4.6 Harmonic Base Acceleration at Different Frequencies with Impact and Without Impact Conditions

Small deviations in the input excitation frequency from any natural frequency are expected to cause excessive decrease in the power output. Therefore, increasing the frequency bandwidth of the harvester enables the harvester to be used in broadband loading environments. For this reason, in this part of the work, harmonic motions of differing frequency are introduced

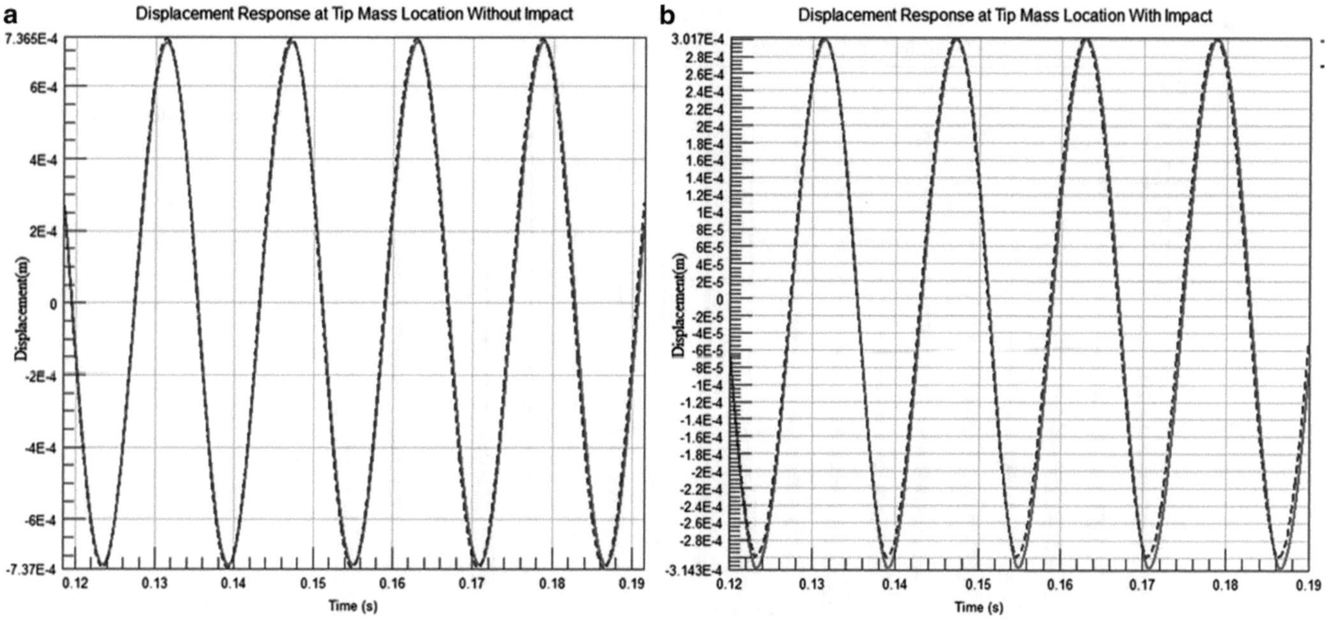

Fig. 4.8 Steady-state displacements of tip mass location (**a**) without impact conditions (**b**) with impact conditions

as input by using PCB 2100 E 11 shaker to understand the effects of stopper on the effective frequency bandwidth of the harvester. Experience with the accelerometers in the tests has disclosed insufficiency of amplitude range. Natural frequency of the beam is decreased to 56.2 Hz by increasing the tip mass to extend the working frequency bandwidth. At higher frequency excitations such as 90 Hz, the accelerometer located on the tip mass location is found to be saturated. Hence, drop in natural frequency equips the harvester with a wider band of operation. For this purpose, white noise shown in Fig. 4.11 is given as an input to realize effects of impact on frequency response functions of the harvester.

For the control of shaker and measurement of acceleration levels, LMS SCADAS III data acquisition system is used. In the measurements, frequency resolution is specified as 0.195313 Hz, maximum frequency of interest is 1,000 Hz, sampling frequency is 3,200 Hz and number of averages is 40. The frequency response characteristic is obtained by the closed looped controlled white noise given in Fig. 4.11 is shown in Fig. 4.12. Between 35 and 58 Hz, amplitudes in non-impacting conditions are found to be larger than those of impacting conditions. In this particular case, this range of frequencies are very close the cantilever natural frequency which is 56.2 Hz. However, during the impact, natural frequency shifts to the right of 56.2 Hz and cover a wider range such as from 55 to 66 Hz.

In the white noise experiment, voltage measurement is also performed in both impact and no impact conditions. Figure 4.13 shows the voltage production in piezoelectric material in each case. When the time history of the voltage produced is investigated, measured voltage levels are discovered to be very high compared to the non-impacting conditions. These indicate the impact with stopper and excitation of the wider frequency band.

Impacts lead to decrease in amplitude of response at natural frequency. On the other hand, natural frequency band expands to the right of natural frequency and this yields an increase in other frequency response outputs as shown in Fig. 4.12. Therefore, to validate these extractions, cantilever beam is vibrated with difference frequencies both with impact and without impact conditions for the same base excitation levels. Compared frequencies during the harmonic loading are 30, 40, 45, 50, 56, 60, 65, 70, 75 and 80 Hz. For these frequencies, constant base excitation (for each frequency itself) is given to the harvester system and time histories are collected for with impact and without impact conditions. Table 4.1 is the summary of this test result. Test results given in Table 4.1 supports the results obtained from Fig. 4.12. For the time history measurement of voltages, Dewesoft software is used and sample frequency is taken as 100 kHz. A fourth order, analogue low pass filter with a cut off frequency of 20 kHz is also used in the measurement. The gap size between stopper and beam is measured as $3 \times 10{-}4$ m.

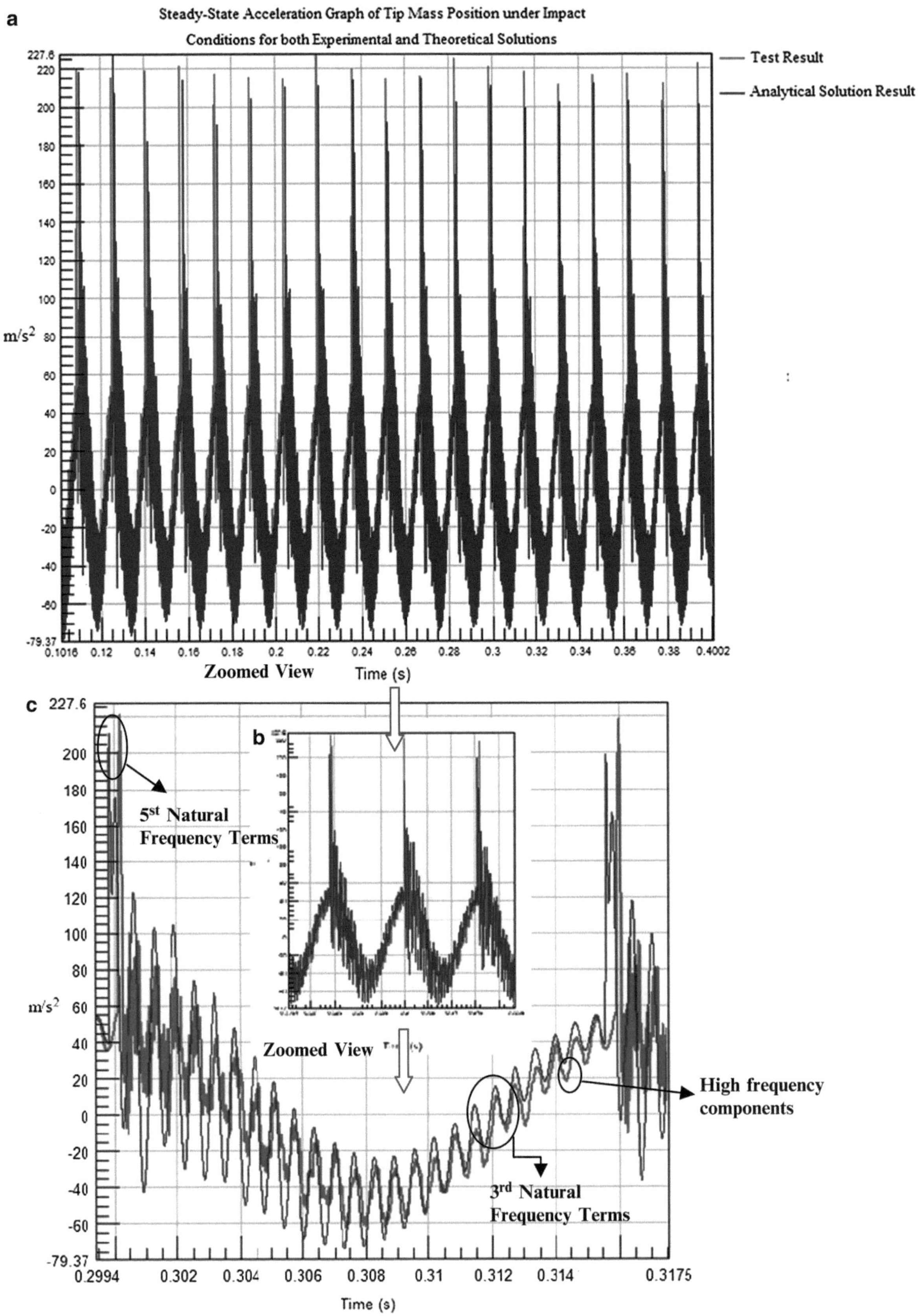

Fig. 4.9 Steady-state acceleration history of tip mass position under impact conditions for both experimental and theoretical solutions (**a**) total response, (**b**) zoomed view of graph a, (**c**) shows one oscillation at 62.5 Hz

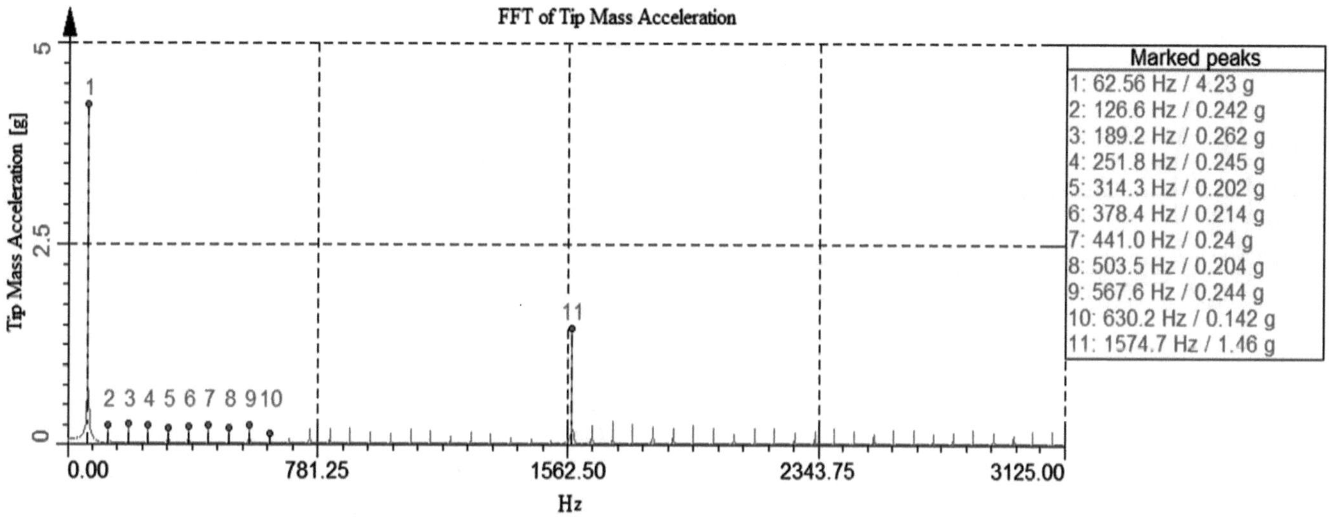

Fig. 4.10 Experimental rms power spectrum of impacting cantilever condition

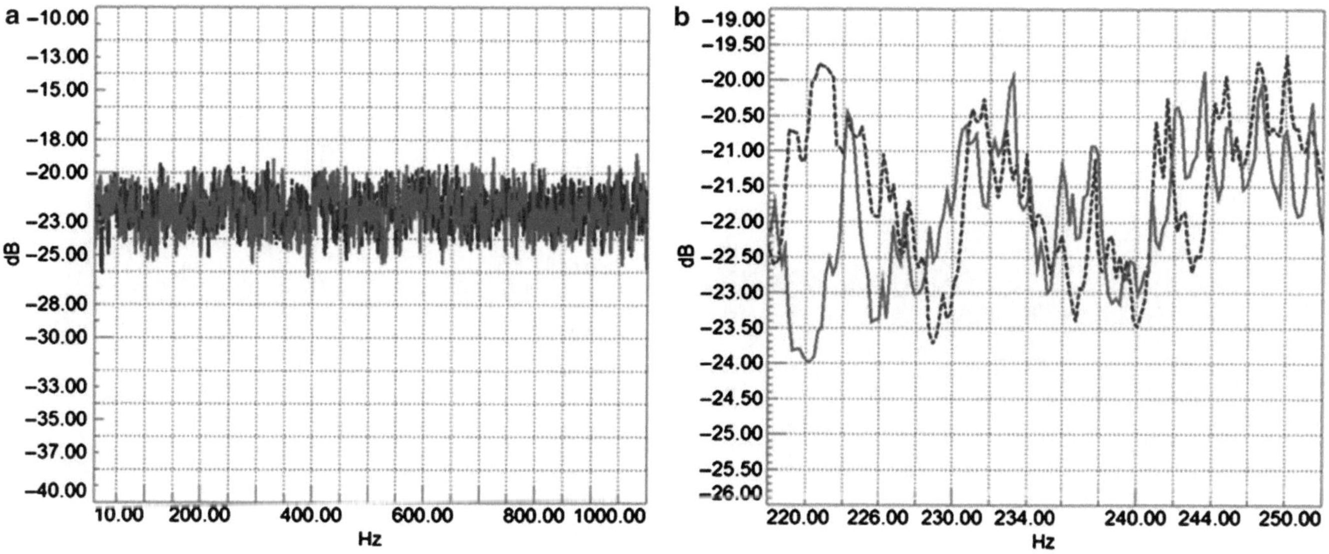

Fig. 4.11 (**a**) Power spectral density (PSD) of input white noise to the harvester (**b**) zoomed view

4.7 Conclusion

The cantilever beam naturally amplifies the input at its resonance frequencies. However, as the damping is very small the bandwidth of energy generation is fairly limited. The motion stopper included in the design eliminates this handicap in two ways. Firstly, the impacts generate various harmonics which are instrumental in enrichment the frequency content. Secondly, as the motion stopper transforms the linear system to a nonlinear one, frequency shifts occur as discussed in Sect. 4.5. However, this modification has also a disadvantage, which is the loss of response amplitude at the resonant frequencies since the motion is disturbed by the stopper.

In the experiments, different harmonic excitation at different frequencies and random excitation are applied to the test setup at its based. Firstly, white noise is applied to the harvester for impact and not impact case at the same level and frequency response characteristics of the harvester is obtained. The frequency response characteristic plots show that,

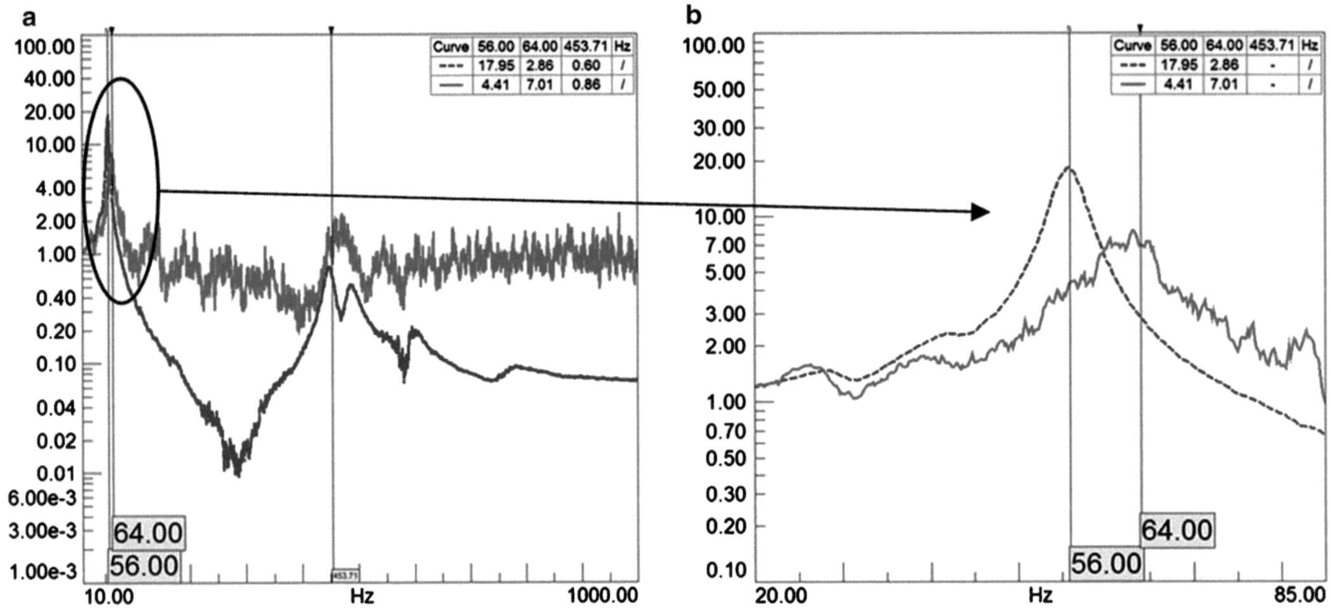

Fig. 4.12 (**a**) Frequency response characteristic of tip mass location (**b**) zoomed view

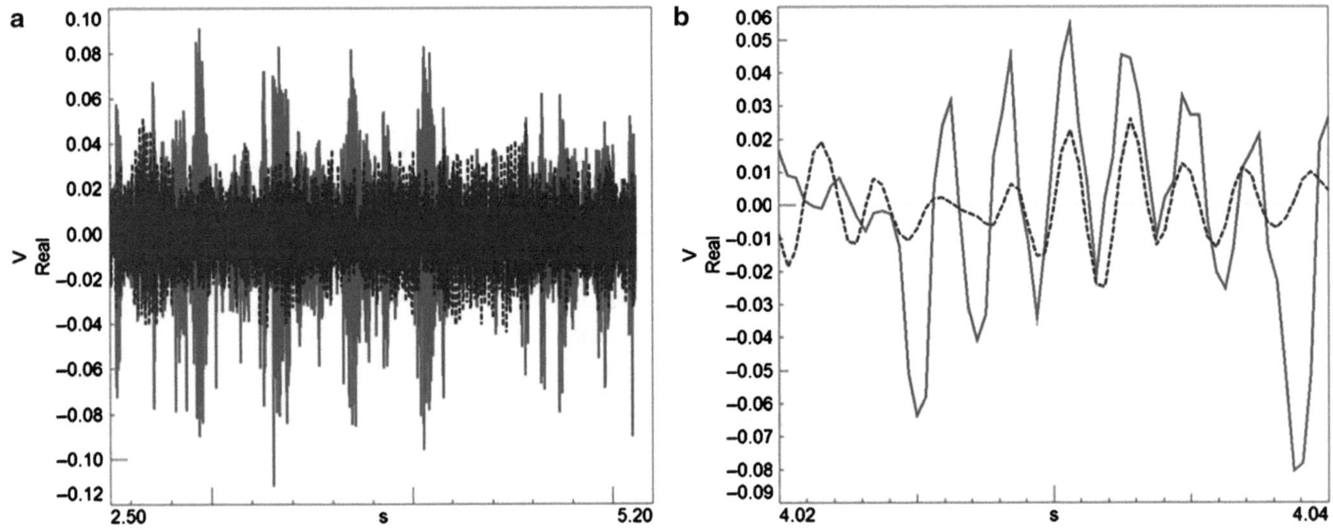

Fig. 4.13 (**a**) Voltage measurement in white noise experiment (**b**) zoomed view

impacting beam has higher amplitude up to 1 kHz apart from 35 to 58 Hz band. For validation study, harmonic excitation is given to the harvester from 30 to 80 Hz with 5 Hz steps. Results show that, harvested voltage increases with impact for the frequency ranges of 30–35 Hz and 58–80 Hz while decreases for the frequency range of 40–58 Hz. The frequency response characteristic is verified by this method adopted. The impact restricts the motion of the beam which causes a decrease in amplitude at the natural frequency, accompanied by an increase in the bandwidth of the natural frequency. Therefore, this leads to a decrease in harvester efficiency at the first natural frequency. However, the efficiency is shown to increases for the rest of the frequency bandwidth.

Acknowledgements The author would like to thank ASELSAN Inc. for useful support during the work.

Table 4.1 Voltage production capability of harvester at different frequencies

Frequency (Hz)	Voltage change
30	160 % increase
40	No increase
45	30 % decrease
50	53 % decrease
56	75 % decrease
60	No increase
65	50 % increase
70	100 % increase
75	200 % increase
80	200 % increase

References

1. Ibrahim SW, Ali WG (2012) A review on frequency tuning methods for piezoelectric energy harvesting systems. J Renew Sustain Energ 4:062703
2. Williams CB, Yates RB (1996) Analysis of a micro-electric generator for microsystems. Sensors Actuators A 52:8–11
3. Mitcheson P, Miao P, Start B, Yeatman E, Holmes A, Green T (2004) MEMS electrostatic micro-power generator for low frequency operation. Sensors Actuators A 115:523–529
4. Roundy S, Wright PK, Rabaey JM (2003) A study of low level vibrations as a power source for wireless sensor nodes. Comput Commun 26:1131–1144
5. WU X, Lin J, Kato S, Zhang K, Ren T, Liu L (2008) A frequency adjustable vibration energy harvester. In: Proceedings of PowerMEMS, Sendai
6. Challa VR, Prasad MG, Shi Y, Fisher FT (2008) A vibration energy harvesting device with bidirectional resonance frequency tunability. Smart Mater Struct 17(1):015035
7. Peters C, Maurath D, Schock W, Manoli Y (2008) Novel electrically tunable mechanical resonator for energy harvesting. In: Proceedings of PowerMEMS, Sendai
8. Lallart M, Anton SR, Inman DJ (2010) Frequency Self-tuning Scheme for Broadband Vibration Energy Harvesting. J Intell Mater Syst Struct 21(9):897–906
9. Mak KH, McWilliam S, Popov AA, Fox CHJ (2011) Performance of a cantilever piezoelectric energy harvester impacting a bump stop. J Sound Vib 330:6184–6202

Chapter 5
Nonlinear 2-DOFs Vibration Energy Harvester Based on Magnetic Levitation

I. Abed, N. Kacem, M.L. Bouazizi, and N. Bouhaddi

Abstract The nonlinear dynamics of a two-degree-of-freedom (2-DOFs) vibrating energy harvester (VEH) based on magnetic levitation is modeled and investigated. The equations of motion have been derived while taking into account the magnetic nonlinearity and the electro-magnetic damping. The associated linear eigenvalue problem has been analyzed and optimality conditions have been expressed in term of distance minimization between the two eigenfrequencies of the considered system. The resulting optimal design parameters have been substituted into the coupled nonlinear equations of motion which have been numerically solved. It is shown that the performances of a classical single degree of freedom VEH can be significantly enhanced up to 270 % in term of power density, up to 34 % in term of frequency bandwidth and up to 10 % in term of resonance frequency attenuation.

Keywords Energy harvesting • Nonlinear dynamics • Magnetic levitation • Electro-magnetic damping • Frequency bandwidth

5.1 Introduction

Over recent years, many research activities are oriented to the development of some harvester devices to produce inexhaustible electric energy in their local environment. The purpose in using Vibration energy harvesters (VEHs) is related to the reduction of power requirement as well as the communication in an autonomous way and to the disposition of the monitor, the measure, and the process data in a hostile environment. On this concept, VEHs can be used in many fields such as environmental monitors, wireless sensors and medical implants [1–3]. In order to make VEHs usable, there are several types of transduction, where the most common transduction modes are piezoelectric [4], electrostatic, magnetostrictive and electromagnetic [5] and each of them presents advantages and disadvantages.

The linear VEHs has a low bandwidth and designed to collect power in the ambient dominant frequency. When the excitation and resonance frequency of the system do not coincide, linear VEHs are known to under perform. So, to increase the collected power and the bandwidth frequency, researches are oriented towards the study of nonlinear systems. For instance, Daqaq et al. [6] reported softening frequency response characteristics in a parametrically forced piezoelectric device with structural nonlinearities. Mann et al. [7] showed analytically and experimentally how magnetic levitation could be used to extend device bandwidth through a hardening response. Nevertheless, this extension is limited by dry friction dissipation phenomenon. Mahmoudi et al. [8] proposed an alternative to overcome this issue by guiding the moving magnet vertically

I. Abed
Applied Mechanics Department, FEMTO-ST Institute, UMR 6174, University of Franche-Comté, 24 chemin de l'Epitaphe, 25000 Besançon, France

Preparatory Engineering Institute of Nabeul (IPEIN), 8000 M'Rezgua, Nabeul, Tunisia

Tunis El Manar University (UTM), El Manar, Tunisia

N. Kacem (✉) • N. Bouhaddi
Applied Mechanics Department, FEMTO-ST Institute, UMR 6174, University of Franche-Comté, 24 chemin de l'Epitaphe, 25000 Besançon, France
e-mail: najib.kacem@femto-st.fr

M.L. Bouazizi
Preparatory Engineering Institute of Nabeul (IPEIN), 8000 M'Rezgua, Nabeul, Tunisia

Tunis El Manar University (UTM), El Manar, Tunisia

Mechanical Department, College of Engineering Salman Bin Abdulaziz University, Al Kharj, Saudi Arabia

A. Wicks (ed.), *Shock & Vibration, Aircraft/Aerospace, and Energy Harvesting, Volume 9*,
Conference Proceedings of the Society for Experimental Mechanics Series, DOI 10.1007/978-3-319-15233-2_5

in an elastic way by means of sandwich beams, combined with an hybrid piezoelectric and electromagnetic transductions. Although their numerical results demonstrated high performances, the electronic circuit may become very complex.

In this paper, we propose a design of a VEH based on the magnetic levitation of two coupled magnets. The equations of motion have been derived and include the magnetic nonlinearity and the electro-magnetic damping. While respecting the optimality conditions expressed in term of distance minimization between the two eigenfrequencies of the associated linear problem, the coupled nonlinear equations of motion have been solved numerically using the harmonic balance method (HBM) [9] coupled with the asymptotic numerical continuation technique (ANM) [10]. We prove numerically that the nonlinear coupling between the magnets permits the performance enhancement of a classical single degree of freedom VEH in terms of frequency attenuation, power density and frequency bandwidth.

5.2 Design and System Modeling

5.2.1 *Proposed Device*

Inspired by [5, 7], an extension of magnetic levitation for a multi-degree of freedom vibration energy harvester is proposed. As a proof of concept, the considered device shown in Fig. 5.1 is composed of four magnets M_i where $i \in [1,4]$. M_1 and M_4 are fixed with respect to a Teflon tube inside which M_2 and M_3 are subjected to magnetic levitation forces. All magnets are placed vertically in such a way that all opposed surfaces have the same pole and wire-wound copper coils are wrapped horizontally around the separation distance between each two adjacent magnets.

Three reference frames have been applied in order to describe the electromagnetic induction model represented in Fig. 5.1. The first reference frame is fixed in space and is used to describe the motion amplitude Y_0 and excitation frequency Ω of the outer housing. The second and third reference frames (designed as x_2 and x_3) describe the motion of the moving magnets M_2 and M_3 which are subjected to restoring forces expressed as follows:

$$\mathbf{F}^{m}_{(2)} = \left(-\frac{M_2 g {d_0}^2}{(M_1Q_1 - M_3Q_3)}\left(\frac{M_1Q_1}{(d_0-v_2)^2} - \frac{M_3Q_3}{(d_0+v_2-v_3)^2}\right) + M_2 g\right)\overrightarrow{\mathbf{x}_2} \tag{5.1}$$

$$\mathbf{F}^{m}_{(3)} = \left(-\frac{M_3 g {d_0}^2}{(M_2Q_2 - M_4Q_4)}\left(\frac{M_3Q_3}{(d_0+v_2-v_3)^2} - \frac{M_4Q_4}{(d_0+v_3)^2}\right) + M_3 g\right)\overrightarrow{\mathbf{x}_3} \tag{5.2}$$

When the device is subjected to an external mechanical vibration, the moving magnets oscillate around their equilibrium positions and a current is induced in each coil as shown in Fig. 5.2, resulting in the following electrical damping forces:

$$F^{e}_{(2)} = ce_1\dot{v}_2 - ce_2\left(\dot{v}_3 - \dot{v}_2\right) \tag{5.3}$$

$$F^{e}_{(3)} = ce_3\dot{v}_3 - ce_2\left(\dot{v}_2 - \dot{v}_3\right) \tag{5.4}$$

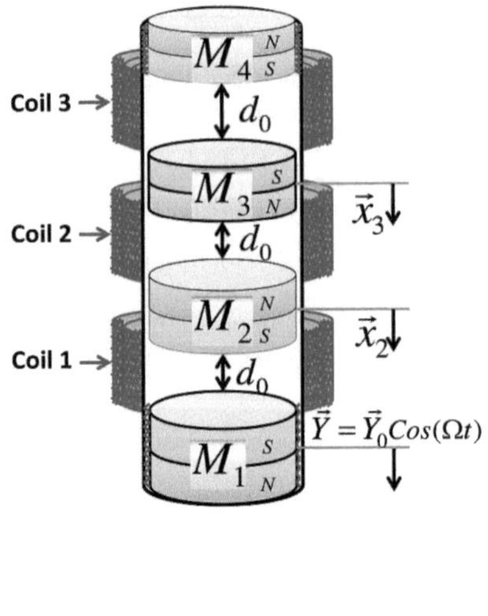

Fig. 5.1 A schematic diagram of the two degree of freedom vibration energy harvester based on magnetic levitation

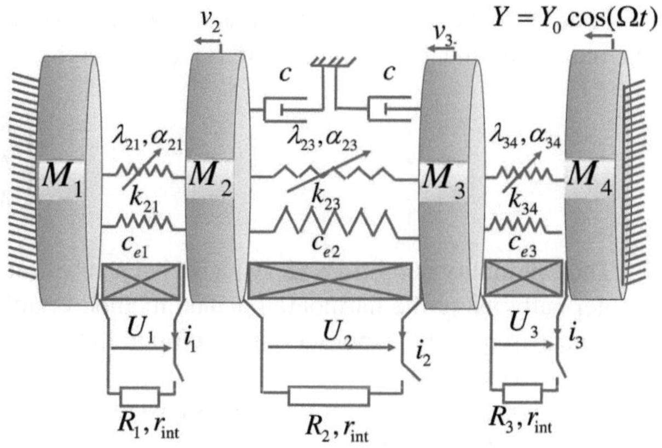

Fig. 5.2 Equivalent electro-mechanical system of the two degree of freedom vibration energy harvester based on magnetic levitation

The electrical damping can be expressed as a function of the internal resistance of the coil, the resistance of external load and the electromechanical coupling coefficient.

$$ce_j = \frac{{\alpha_j}^2}{R_j + r_{int}} \tag{5.5}$$

where $\alpha_j = N_j B l$, with B the average magnetic field strength, N_j is the number of coil turns and l is the coil length.

Expanding the non-linear magnetic forces written in Eq. (5.2) in Taylor series up to the third order and summing all forces applied to each moving magnet in the vertical direction, we obtain the following governing equations of motion for the mechanical system:

$$\begin{cases} \begin{pmatrix} M_2\ddot{v}_2 + (c + ce_1 + ce_2)\dot{v}_2 - ce_2\dot{v}_3 + k_{23}(v_2 - v_3) + k_{21}(v_2) \\ +\alpha_{23}(v_2 - v_3)^2 - \alpha_{21}(v_2)^2 + \lambda_{23}(v_2 - v_3)^3 + \lambda_{21}(v_2)^3 \end{pmatrix} = -M_2\ddot{Y} \\ \begin{pmatrix} M_3\ddot{v}_3 + (c + ce_2 + ce_3)\dot{v}_3 - ce_2\dot{v}_2 \\ +k_{34}(v_3) + k_{32}(v_3 - v_2) + \alpha_{34}(v_3)^2 \\ -\alpha_{32}(v_3 - v_2)^2 + \lambda_{34}(v_3)^3 + \lambda_{32}(v_3 - v_2)^3 \end{pmatrix} = -M_3\ddot{Y} \end{cases} \tag{5.6}$$

If the magnetic intensities (Q) are equal, $\forall j \in [2, 3]$ the linear and nonlinear stiffnesses can be written as:

$$\begin{aligned} k_{j(j+1)} &= \frac{2gM_jM_{j+1}}{(M_{j-1}-M_{j+1})d_0}; k_{j(j-1)} = \frac{2gM_jM_{j-1}}{(M_{j-1}-M_{j+1})d_0} \\ \alpha_{j(j+1)} &= \frac{3}{2d_0}k_{j(j+1)}; \alpha_{j(j-1)} = \frac{3}{2d_0}k_{j(j-1)} \\ \lambda_{j(j+1)} &= \frac{2}{{d_0}^2}k_{j(j+1)}; \lambda_{j(j-1)} = \frac{2}{{d_0}^2}k_{j(j-1)} \end{aligned} \tag{5.7}$$

5.3 Optimality Conditions and Average Power

Firstly, the associated linear problem is written in its matrix form. Then, in order to guaranty the symmetry of the rigidity matrix, we assume that $k_{23} = k_{32}$ which results in the following relation between the magnets in term of mass:

$$M_4 + M_1 = M_2 + M_3 \tag{5.8}$$

While respecting Eq. (5.8), the eigen frequencies can be written as follows:

$$ {\Omega_1}^2, {\Omega_2}^2 = \left(\frac{1}{2}\left\{\frac{(k_{23}+k_{21})M_3+(k_{34}+k_{32})M_2}{M_2M_3}\right\} \pm \frac{1}{2}\sqrt{\begin{bmatrix} \left\{\frac{(k_{23}+k_{21})M_3+(k_{34}+k_{32})M_2}{M_2M_3}\right\}^2 \\ -4\left\{\frac{(k_{23}+k_{21})(k_{34}+k_{32})-k_{23}k_{32}}{M_2M_3}\right\} \end{bmatrix}} \right) \tag{5.9}$$

In order to enhance the performances of the considered VEH in term of bandwidth, we minimize the distance separating the naturel frequencies expressed in Eq. (5.9). Doing so, we obtain the following optimality conditions:

$$\begin{cases} M_2 = M_1 - M_4 \\ M_3 = 2M_4 \\ M_1 > 2M_4 \end{cases} \tag{5.10}$$

The optimality conditions (5.10) have been substituted in the system of nonlinear equations (5.6) which have been solved numerically using the harmonic balance method coupled with the asymptotic numerical continuation technique. Thus, the nonlinear frequency responses can be plotted in term of velocity that will be used to determine the average power delivered to the electrical load.

Indeed, the magnetic transduction is ensured by three coils. The oscillations of the movable magnets cause magnetic field variations in the separation zones, which provides an induced current (Lenz's Law). The induced current can be expressed as a vibration velocity function $\dot{v}_j(t) = V_j \Omega \sin(\Omega t); \forall j \in [2, 3]$. Then, the average power delivered to the electrical load can be written as follows:

$$P_m = \frac{1}{T}\int_0^T P(t)dt = \frac{\Omega^2}{2}\left(\sum_{j=1}^{3}\left(ce_j\left(V_{j+1} - V_j\right)^2\right)\right) \tag{5.11}$$

5.4 Results and Discussion

5.4.1 *Optimal Design Parameters*

In order to enhance the bandwidth of the proposed device, the distance between the eigenfrequencies written in Eq. (5.9) is plotted with respect to M_1 and M_4 as shown in Fig. 5.3. Remarkably, Eq. (5.10) are not verified for regions 1 and 2 which are separated by an optimal zone for which $f_2 - f_1 < 4.3\,Hz$. Inside this region, a set of design parameters, listed in Table 5.1, is chosen.

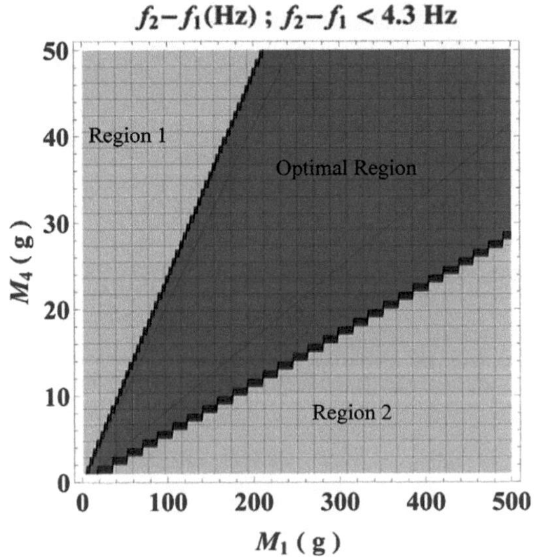

Fig. 5.3 Variation of the distance separating the eigenfrequencies of the proposed VEH with respect to M_1 and M_4. Inside the optimal zone, $f_2 - f_1 < 4.3\,Hz$

Table 5.1 Physical and geometric properties of the proposed 2-DOFs vibration energy harvester

r_{int}	Internal resistance (Ω)	188
l	Total coil length (m)	0.015
d_0	Separation distance (m)	0.015
B	Residual magnetic flux density (T)	1.18
c	Mechanical damping [7] (Ns/m)	0.116
ρ_m	Magnet density (kg/m^2)	7800
$R_j, j \in [1,3]$	External load resistance (Ω)	10^4
$N_j, j \in [1,3]$	The coil number (turns)	2000
M_1	Mass (kg)	0.125
M_2	Mass (kg)	0.0195
M_3	Mass (kg)	0.02
M_4	Mass (kg)	0.01

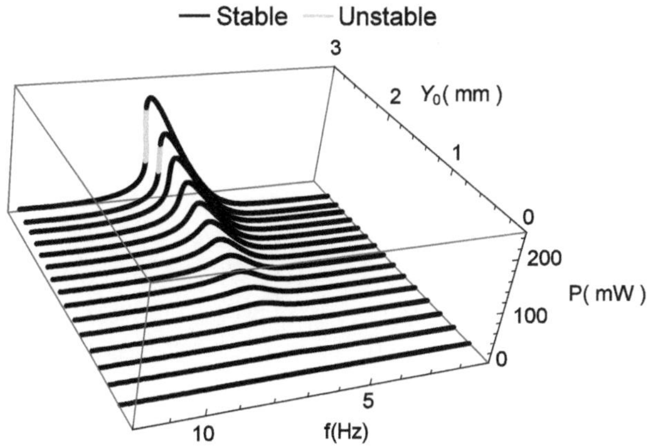

Fig. 5.4 Variation of the frequency responses in term of harvested power with respect to the excitation amplitude for the proposed device. *Black and gray lines* denote respectively stable and unstable branches

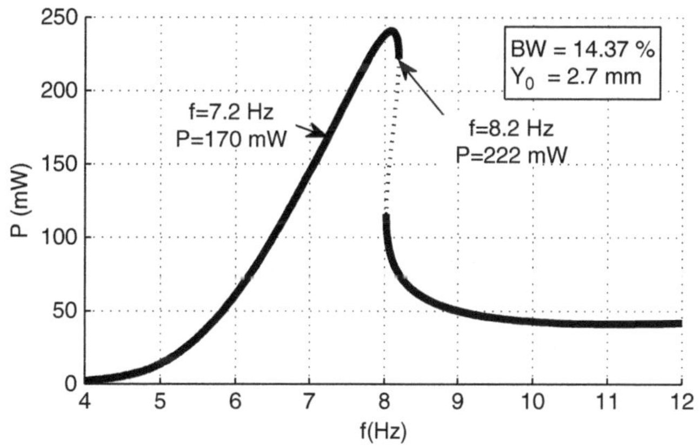

Fig. 5.5 A typical frequency response of the proposed device showing a hardening behavior for an excitation amplitude $Y_0 = 2.7\,mm$, the frequency bandwidth is $BW = 14.4\,\%$. *Solid and dashed lines* denote respectively stable and unstable branches

5.4.2 *Bandwidth Enhancement*

Several numerical simulations have been performed for the set of design parameters listed in Table 5.1. Figure 5.4 displays the evolution of the frequency response in term of harvested power for different values of amplitude excitation Y_0 up to $3\,mm$. It is shown that bistability takes place for large excitation amplitudes ($Y_0 > 2\,mm$) for which the dynamic response has two solutions for a given frequency inside a range separated by two bifurcation points.

Hence, one can take advantage of the nonlinear spring hardening effect in order to enlarge the frequency bandwidth of the VEH. Interestingly, Fig. 5.5 shows that the frequency bandwidth of the proposed device can reach $14.4\,\%$ for an excitation amplitude $Y_0 = 2.7\,mm$.

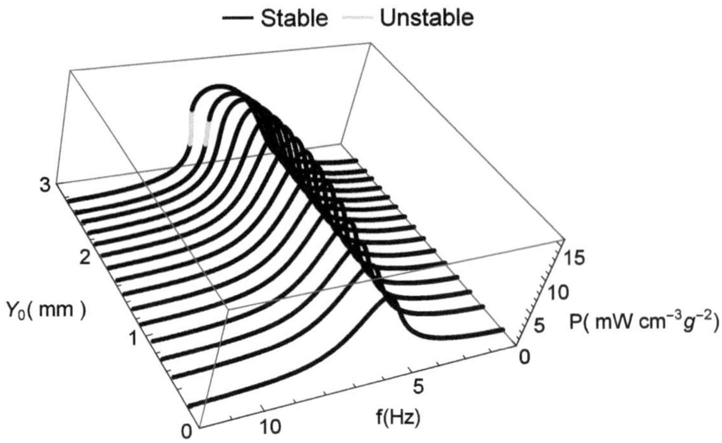

Fig. 5.6 Variation of the frequency responses in term of normalized power density with respect to the excitation amplitude for the proposed device. *Black and gray lines* denote respectively stable and unstable branches

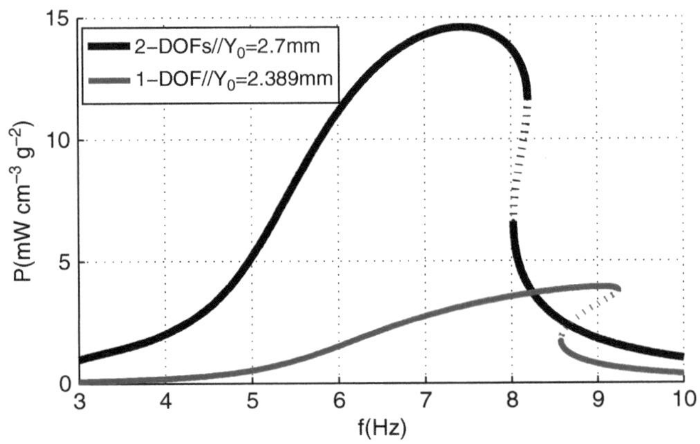

Fig. 5.7 Comparison of the frequency responses in term of normalized power density between the proposed device (2-DOFs) and the single degree of freedom VEH presented in [7]

Table 5.2 Performances of 1-DOF and 2-DOFs vibration energy harvesters based on magnetic levitation

Model	Resonance frequency (Hz)	Frequency bandwidth (%)	Normalized power ($mWcm^{-3}g^{-2}$)	Power (mW)
1-DOF	6.96	10.5	3.93	35.38
2-DOFs	6.60	14.4	14.57	240.50

5.4.3 *Normalized Power Density Enhancement*

Since the harvested power varies with respect to the excitation amplitude and in order to compare the proposed device with the one investigated in [7], in terms of performances, we propose the use of the normalized power density as a kind of energetic efficiency. The latter varies slightly with respect to Y_0 which is proved in Fig. 5.6 showing that the maximum of frequency curves is almost constant independently of the excitation amplitude.

Figure 5.7 shows that the energetic efficiency of a single degree of freedom VEH based on magnetic levitation can be enhanced up to 270 % by coupling two moving magnets. Moreover, the bandwidth can be boosted up to 34 % and the frequency attenuation is about 10 % as listed in Table 5.2.

5.5 Conclusion

The non-linear dynamics of a two-degree-of-freedom vibrating energy harvester (VEH) was modeled including the main sources of non-linearities. The associated linear eigen problem was analytically solved and optimality conditions were derived in term of distance minimization between the two eigenfrequencies. The resulting coupled nonlinear equations of motion were solved numerically for the optimal design, using the harmonic balance method coupled with the asymptotic numerical continuation technique.

Several numerical simulations have been performed to highlight the performance of the proposed device. Particularly, the power density, the bandwidth and the frequency attenuation can be boosted up to 270 %, 34 % and 10 % respectively compared to the case of a single degree of freedom magnetic levitation based VEH, thanks to the nonlinear coupling between the magnets. Future work will include the extension of the proposed concept to large arrays of coupled levitated magnets.

References

1. Ding Z, Perlaza SM, Esnaola I, Poor HV (2013) Power allocation strategies in energy harvesting wireless cooperative networks. IEEE Trans Wireless Commun 13:846–860
2. Donelan JM, Li Q, Naing V, Hoffer JA, Weber DJ, Kuo AD (2008) Biomechanical energy harvesting: generating electricity during walking with minimal user effort. Science 319:807–810
3. Wang W, Wang N, Jafer E, Hayes M, O'Flynn B, O'Mathuna C (2010) Autonomous wireless sensor network based building energy and environment monitoring system design. In: Environmental Science and Information Application Technology (ESIAT), International Conference
4. Zhou L, Sun J, Zheng XJ, Deng SF, Zhao JH, Peng ST, Zhang Y, Wang XY, Cheng HB (2012) A model for the energy harvesting performance of shear mode piezoelectric cantilever. Sens Actuators A Phys 179:185–192
5. Foisal ARM, Hong C, Chung GS (2012) Multi-frequency electromagnetic energy harvester using a magnetic spring cantilever. Sens Actuators A Phys 182:106–113
6. Daqaq MF, Stabler C, Qaroush Y, Seuaciuc-Osório T (2009) Investigation of power harvesting via parametric excitations. J Intel Mater Syst Struct 20(5):545–557
7. Mann BP, Sims ND (2009) Energy harvesting form the nonlinear oscillations of magnetic levitation. J Sound Vib 15(319):515–530
8. Mahmoudi S, Kacem N, Bouhaddi N (2014) Enhancement of the performance of a hybrid nonlinear vibration energy harvester based on piezoelectric and electromagnetic transductions. Smart Mater Struct 23(7):075024
9. Cochelin B, Vergez C (2009) A high order purely frequency-based harmonic bal-ance formulation for continuation of periodic solutions. J Sound Vib 324:243–262
10. Arquier R, Karkar S, Lazarus A, Thomas O, Vergez C, Cochelin B (2010) Manlab:an interactive path-following and bifurcation analysis software. http://manlab.lma.cnrs-mrs.fr/s

Chapter 6
Parameter Identification of Riveted Joints Using Vibration Methods

Elif Altuntop, Murat Aykan, and Melin Şahin

Abstract Rivets are widely used in several industries including aerospace, shipbuilding and construction. Aircraft components such as wings and fuselages are some examples of riveted structures. Accurate parameter identification of these joints is critical since excessive number of rivets is present in such structures. Furthermore, modeling structures with fasteners has always been a challenge since these members might show nonlinear behavior. In this study, the FEM of a continuous plate is constructed and modal tests are performed in order to have a valid modeling strategy. After a good correlation between finite element analyses (FEA) and tests is obtained, a finite element model with riveted joints is constructed and parameters of these fasteners are identified by means of vibration measurements and optimization.

Keywords Finite element method • Parameter identification • Modal testing • Riveted joint • Optimization

6.1 Introduction

Engineering structures usually consist of substructures which are assembled with various fasteners such as bolts, screws, welds and rivets. Identification of the dynamic behavior of joints is of great importance since most of the deformation and damping in assembled structures arise from these connectors [1]. Aerospace substructures such as wing and fuselage contain excessive number of rivets which affect the dynamic response. Therefore, defining an accurate numerical model for these fasteners is an important task which involves validation and updating procedures. In this study, a numerical method is used to model the rivets where the dynamic parameters of the rivets are obtained from experimental modal analyses.

Numerous approaches have been developed to identify joint parameters (i.e., stiffness and damping properties) which use experimental and numerical methods [2–8]. Experimental methods focus on measurements of frequency response functions (FRFs) or identified modal parameters. Former approach bases on the fact that joint parameters can be obtained using the FRFs of the system with and without joints. Using substructure FRF synthesis method, Tsai and Chou [2] identified the stiffness and damping values of a single bolt. Wang and Liou [3] developed a method to obtain the parameters of joint using contaminated FRFs of the main structure and substructures.

Modal parameters such as natural frequencies, damping ratios and mode shapes were also utilized to estimate the stiffness and damping characteristics of joints. For instance, using complete mode shapes and eigenvalues, Inamura and Sata [4] identified joint parameters. Kim et al. [5] proposed an approach based on the use of a condensed finite element model with incomplete mode shapes. However, in case of closely spaced modes and high damping, modal parameters may not be extracted accurately [6] which lead to inaccurate joint parameters.

Furthermore, model updating techniques that involve Finite Element (FE) modeling and experimental data are also common in the literature for joint identification [6]. These techniques can broadly be classified into two categories: the direct methods and the penalty methods. The direct methods use the mass and stiffness matrices taken from a FEM and joint parameters are determined from the solution of characteristic equations which require solving both those matrices and

E. Altuntop (✉)
Department of Aerospace Engineering, Middle East Technical University, Çankaya, Ankara 06800, Turkey

Defense Systems Technologies Division, ASELSAN Inc., Çankaya, Ankara 06172, Turkey
e-mail: elifaltuntop@gmail.com

M. Aykan
Defense Systems Technologies Division, ASELSAN Inc.,Çankaya, Ankara 06172, Turkey

M. Şahin
Department of Aerospace Engineering, Middle East Technical University, Çankaya, Ankara 06800, Turkey

A. Wicks (ed.), *Shock & Vibration, Aircraft/Aerospace, and Energy Harvesting, Volume 9*,
Conference Proceedings of the Society for Experimental Mechanics Series, DOI 10.1007/978-3-319-15233-2_6

measured frequency response data [7]. Alternatively, in penalty methods, the error between the FE analyses and experimental results are minimized by means of sensitivity and Jacobian matrices which involve the derivatives of the stiffness, damping and mass matrices with respect to each of the modal parameters to be updated. As an alternative to model updating methods, Park and Kim [8] solved joint parameters using an optimization method by devising a cost function as the difference between experimental and analytical results.

In order to update the model accurately, the FEM of the joint also has to be accurate. Several approaches can be found in the literature for modeling riveted connections using FEM and analysis software [9–12]. One of the methods for modeling rivets in FEM is to represent rivets as rigid links of infinite stiffness. In the study of Cabell [9], the riveted joints are modeled with RBE2 elements which couple six degrees of freedom between neighboring nodes using MSC Nastran. Another method is to use spring and beam elements as connectors. Xiong and Bedair [10] represented the riveted joint by a circular beam and two disks that are connected rigidly at the end of the beam. The rivet itself is modeled by a single beam element where spring elements are used to connect the nodes in the contact region along the rivet/plate boundary. Fung and Smart [11] represented the riveted joint with three dimensional solid elements which is well suited for introducing friction, preload and tolerance. In the study of Körük and Şanlıtürk each rivet is assumed to be connecting the clamping parts through an effective diameter where all the nodes inside this diameter are coupled [12].

In this study, two aluminum plates connected by rivets are used as a sample structure in the identification of the parameters of joints via vibration methods. Vibration measurements are conducted in order to obtain the first five resonance frequencies of the sample plate in fixed-free conditions. Each rivet is modeled using the bushing joint in ANSYS® which has six degrees of freedom where each degree of freedom has an associated stiffness. Then, stiffness values of the rivets are optimized according to resonance frequencies obtained from the modal tests.

6.2 Methodology

First of all, the FEM of a continuous plate is constructed and modal tests are performed in order to have a valid modeling strategy. In this step, the material properties of the plate and mass values of accelerometers are optimized. After a good correlation between finite element analyses (FEA) and tests is obtained, another plate having six rivets is manufactured. Modal tests are also repeated on the riveted plate and resonance frequencies are obtained. In this plate, rivets are modeled as bushing elements in ANSYS® in which stiffness values of the joints are to be determined using the Optimization Toolbox in MATLAB. Having obtained a good agreement between the resonance frequencies coming from the modal test and FEA is established, stiffness values are then recorded. Flow chart of the study can be seen in Fig. 6.1.

6.3 Experimental Studies

In order to obtain the stiffness values of the riveted joint, two aluminum plates having identical dimensions were manufactured as shown in Fig. 6.2. The dimensions of the plates are 600 mm in length, 300 mm in height and 2 mm in width. One of the plates having no rivets is called as the continuous plate while the other plate having six identical rivets is called as riveted plate. The shank diameter of the rivet is 4.75 mm, the head diameter is 9.5 mm and the length of the shank is 9.5 mm. The mass of each rivet is measured as 0.750 g. As mentioned before, the continuous plate in this study is used to validate FEM which is further used in model updating and optimization steps.

Experiments were performed in fixed-free conditions using a shaker with a push-rod. Excitation was given from the right corner of the plates so as to excite mode shapes properly. LMS Scadas was used to record the acceleration and force data in order to measure FRFs for further analysis of resonance frequencies and corresponding mode shapes. Step sine input was given from the signal generator between 1 and 128 Hz with 0.25 Hz resolution covering the first five modes of the plate. Twenty accelerometers were attached to the same points on the plates which were determined according to the mode shapes coming from FEA. FRFs measured from one of the points corresponding to riveted joints are presented in Fig. 6.3. Resonance frequencies are recorded in Table 6.1 for both of the experiment setups. Note that the mode numbers (i.e. 1, 2, 3, 4 and 5) in the table correspond to the mode shapes in the order of 1st out-of-plane bending, 1st torsion, 2nd out-of-plane bending, 2nd torsion bending and 3rd out-of-plane bending.

Five natural frequencies of the continuous and riveted plates are measured with high coherence. Mode shapes corresponding to first five resonance frequencies of continuous and riveted plate are presented in Figs. 6.4 and 6.5.

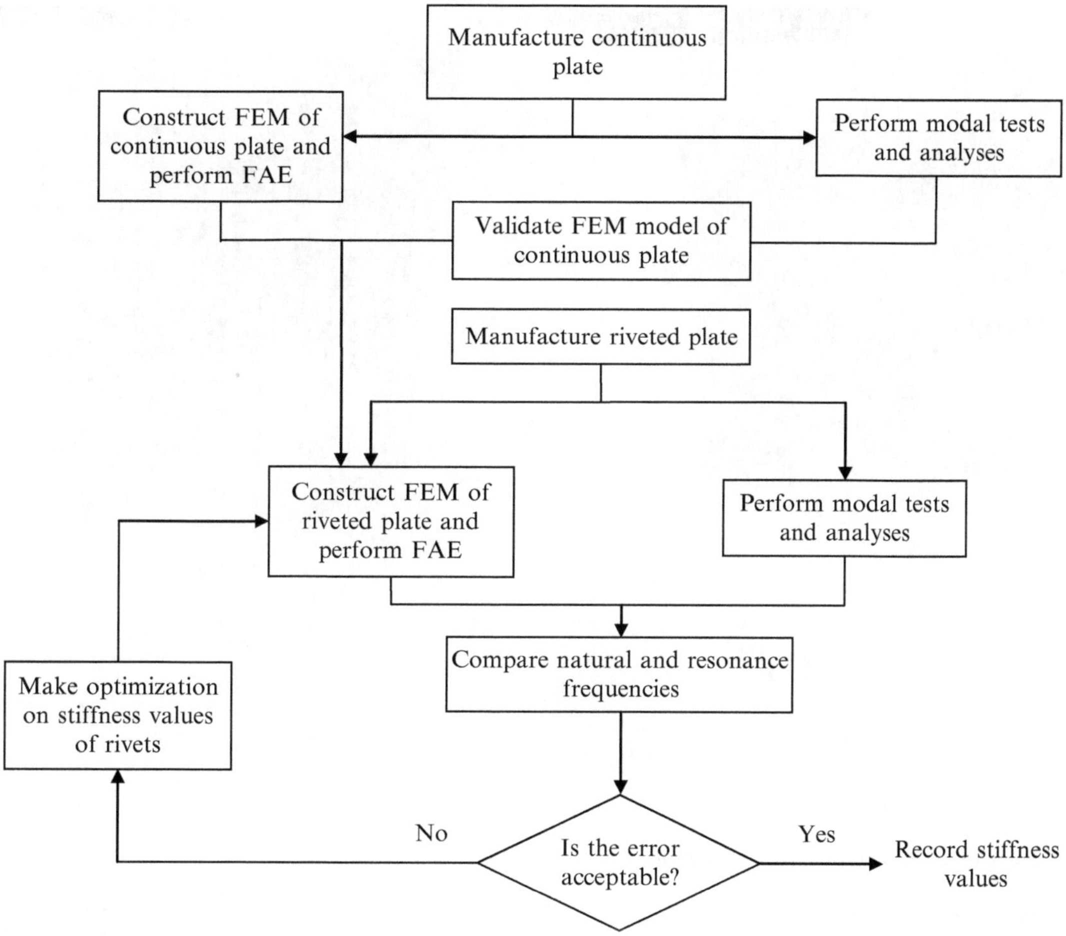

Fig. 6.1 Flow chart of the parameter identification of the rivets

6.4 Model Validation of the Continuous Plate

The FEM of the continuous plate is prepared in ANSYS® with 5,850 elements. Continuous plate is modeled with mixture of triangle and quadrilateral shell elements and thickness of 2 mm. Accelerometers and force transducers are modeled as point masses where one edge of the plate is fixed. Five natural frequencies are obtained and corresponding mode shapes can be seen in Fig. 6.6.

Table 6.2 shows the natural frequencies coming from finite element model and experiments. Absolute errors between first five natural frequencies show that there is a good correlation between finite element model and the test. For the identification part of this study, same material properties and point masses will be used for modeling in ANSYS®.

6.5 Parameter Identification of Riveted Joints

Identification of axial and rotational stiffness values of the riveted joints is performed using the Optimization Toolbox implemented in MATLAB and ANSYS®. Genetic algorithm is chosen as the solver where upper and lower bounds are defined for stiffness values. This toolbox calls a MATLAB function developed by the authors in which the input file created using ANSYS for the riveted plate is processed. Furthermore, ANSYS® is called within the program and modal analysis is performed, resonance frequencies in the output file are taken and least square error is calculated using experimental and numerical values. Fitness function is set as the least square error and depending upon the value of it, new stiffness values are generated by the algorithm. Note that rivets are assumed to be identical, i.e. values assigned for each degree of freedom are same for all six of them. This program runs until the function tolerance or maximum iterations are reached.

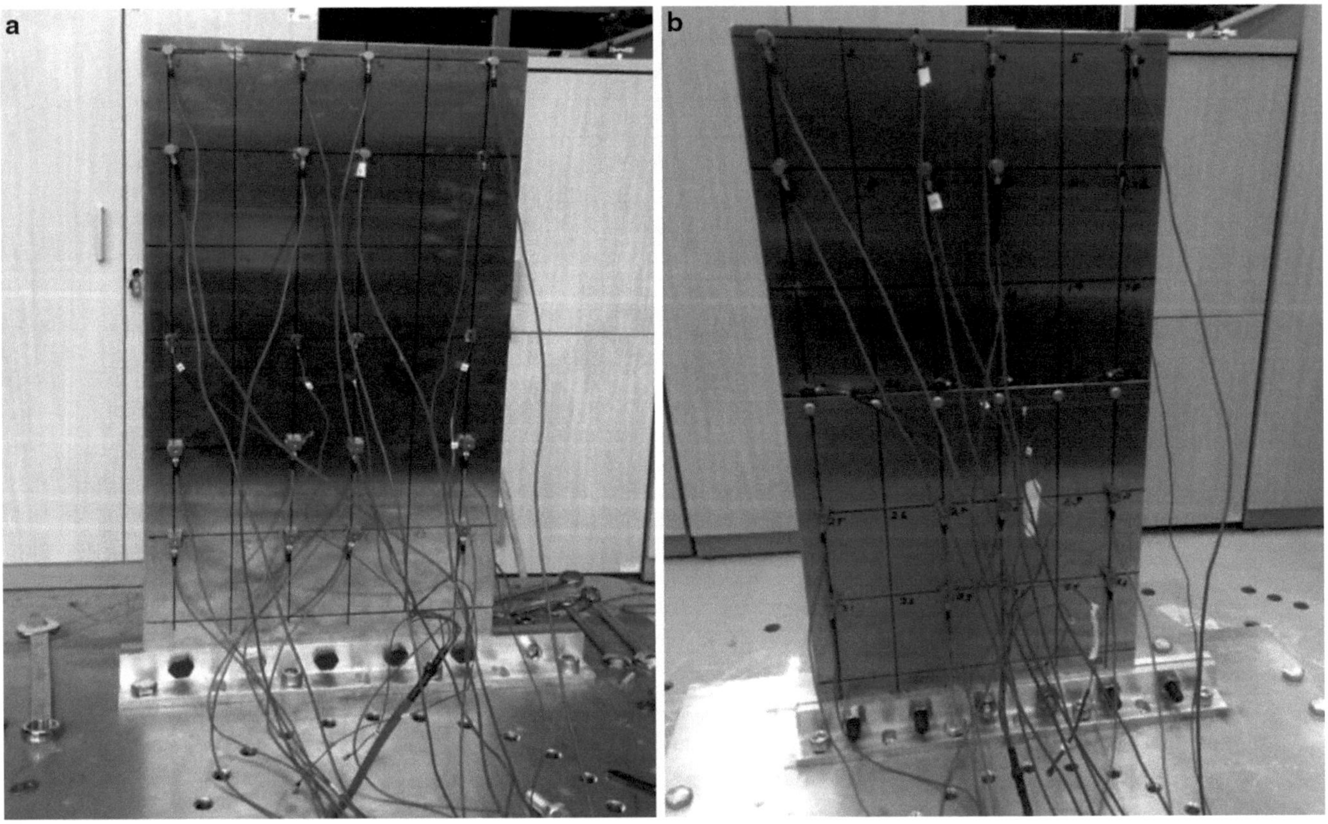

Fig. 6.2 Test structure (a) continuous plate (b) riveted plate

Fig. 6.3 FRFs measured from one of the points corresponding to riveted joints (*red*: continuous plate, *blue*: riveted plate)

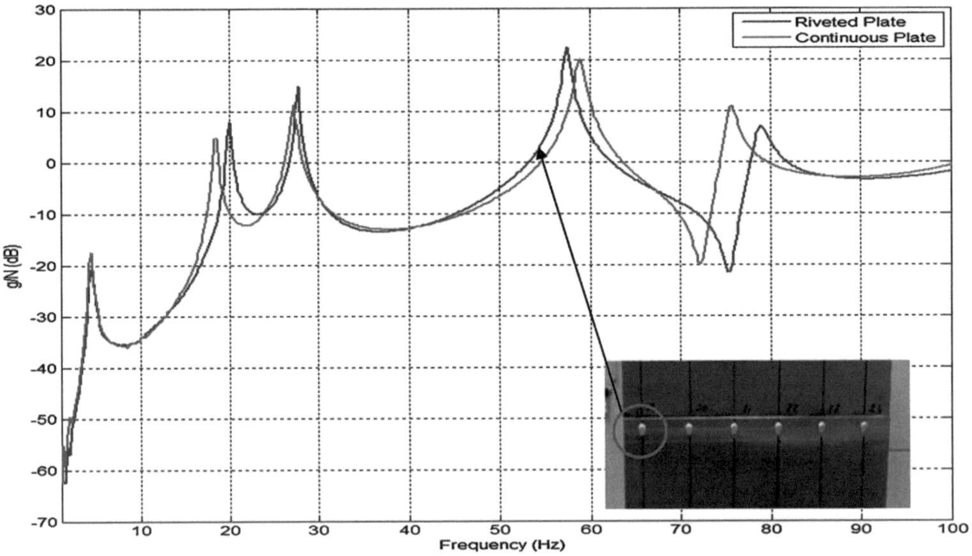

Lower and upper bounds of stiffness values are set as 1 and 1,000,000 respectively. Axial stiffness values in vertical and transverse directions and corresponding bending stiffness values are assumed to be identical. Natural frequencies obtained from optimization and the errors between resonance frequencies taken from modal test and optimization are presented in Table 6.3. Note that final least square error is found to be 4.15 (Table 6.4).

Table 6.1 Experimental natural frequencies of continuous and riveted plate

	Experimental natural frequencies [Hz]	
Mode number	Continuous plate	Riveted plate
1	4.41	4.43
2	18.42	19.92
3	27.28	27.74
4	58.87	57.44
5	75.57	78.82

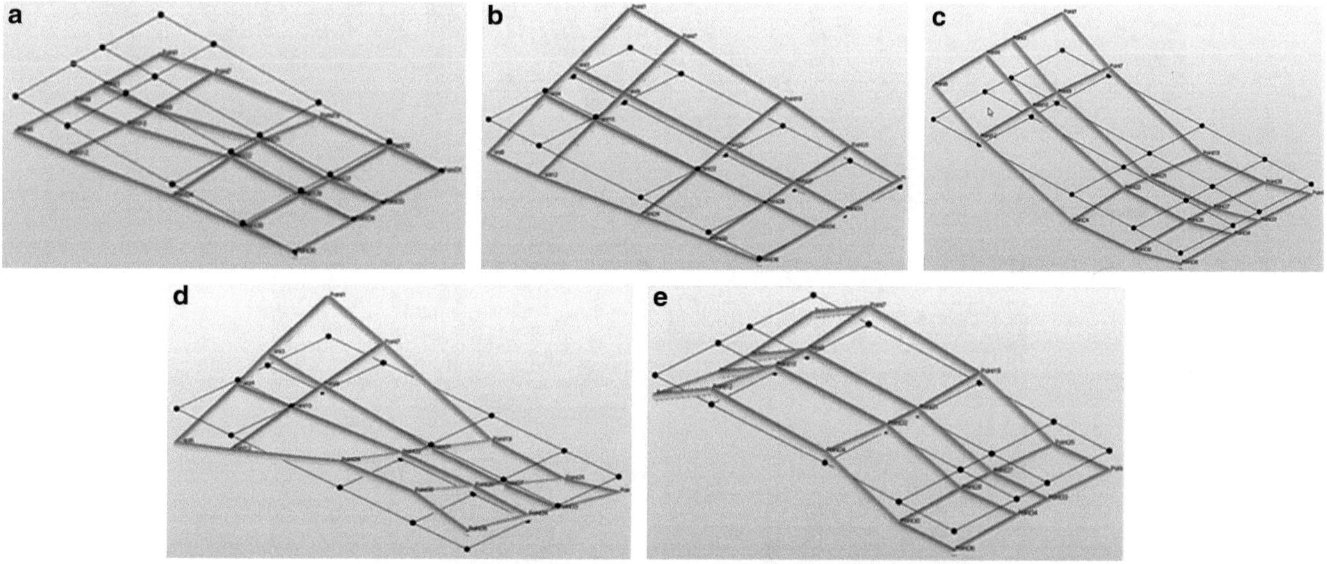

Fig. 6.4 First five mode shapes of the continuous plate from modal tests (**a**) mode 1, (**b**) mode 2, (**c**) mode 3, (**d**) mode 4, (**e**) mode 5

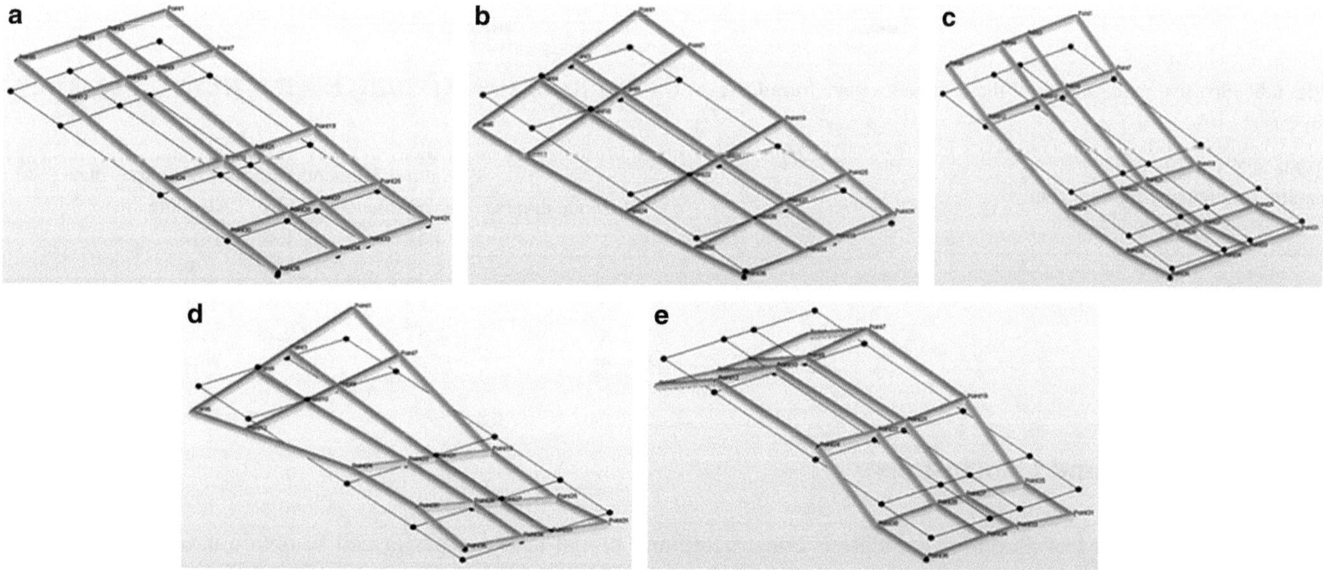

Fig. 6.5 First five mode shapes of the riveted plate from modal tests (**a**) mode 1, (**b**) mode 2, (**c**) mode 3, (**d**) mode 4, (**e**) mode 5

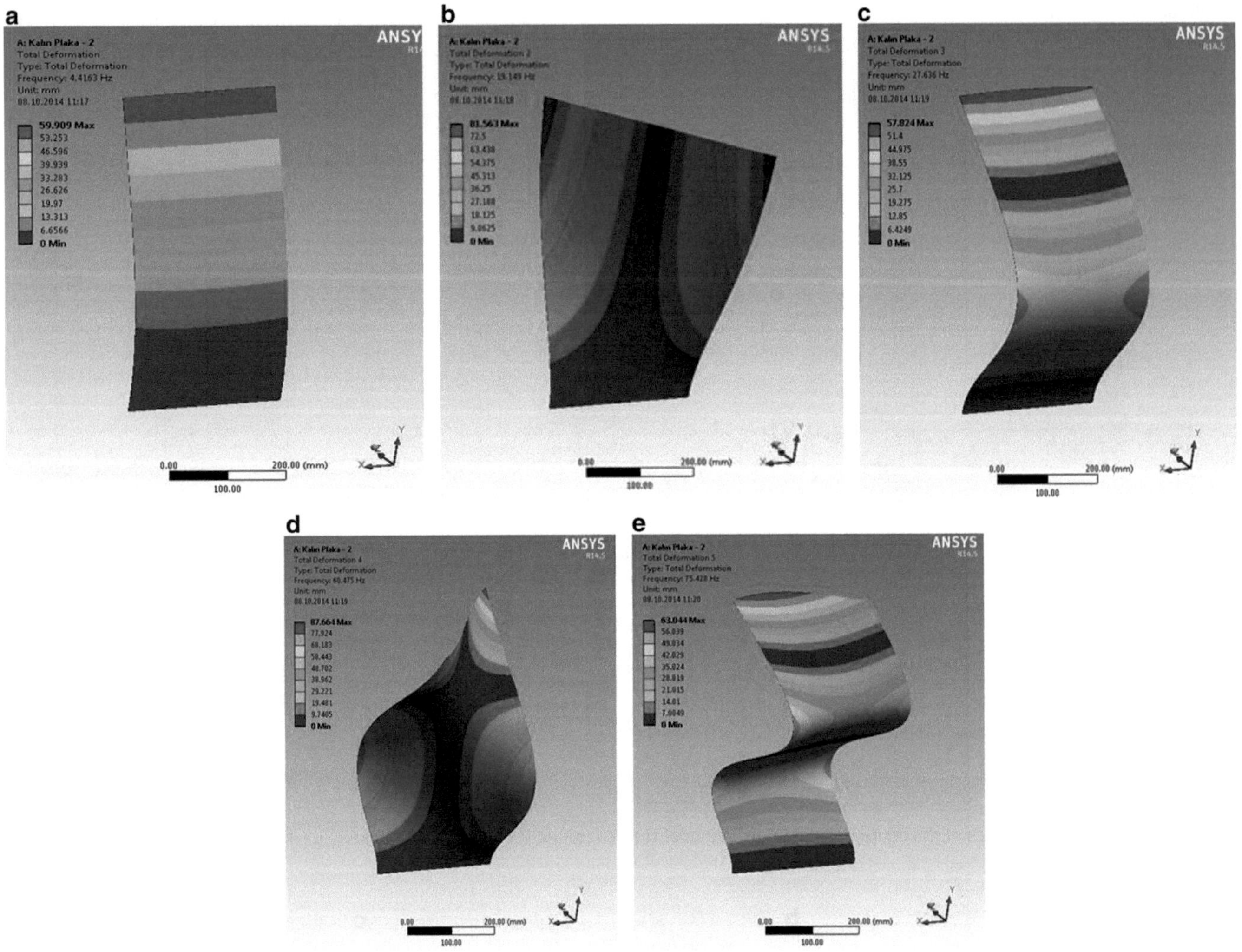

Fig. 6.6 First five mode shapes of the continuous plate from FEA (**a**) $f_1 = 4.41$ Hz, (**b**) $f_2 = 19.15$ Hz, (**c**) $f_3 = 27.64$ Hz, (**d**) $f_4 = 60.48$ Hz, (**e**) $f_5 = 75.43$ Hz

Table 6.2 Comparison of natural frequencies coming from experiment and finite element model

	Natural frequencies of the continuous plate [Hz]		
Mode number	Experimental	FEM	Absolute error [%]
1	4.41	4.42	0.14
2	18.42	19.15	3.96
3	27.28	27.64	1.30
4	58.87	60.48	2.73
5	75.57	75.43	0.19

6.6 Discussion and Conclusions

In this study, the FEM of a continuous plate is constructed and modal tests are performed to obtain a valid model. After the material properties of the plate and mass values of accelerometers are optimized, another plate having six rivets is manufactured. Experiments are also repeated on the riveted plate to get resonance frequencies. Rivets are modeled as bushing elements in ANSYS® having six stiffness values correspond to six degrees of freedom and these values are determined using the Optimization Toolbox in MATLAB. Having obtained a good agreement between the resonance frequencies coming from the modal test and FEA is established, stiffness values are then recorded.

Table 6.3 Natural frequencies of continuous plate obtained from optimization and corresponding errors

Mode number	Natural frequencies obtained from optimization	Resonance frequencies obtained from experiment	Absolute error [%]
1	4.41	4.43	0.33
2	19.22	19.92	3.49
3	26.94	27.74	2.89
4	58.78	57.44	2.33
5	75.04	78.82	4.80

Table 6.4 Stiffness values obtained from optimization and common formulae

	Experimental values
Longitudinal stiffness [N/mm]	785,348.7
Lateral stiffness [N/mm]	849,643.2
Transverse stiffness [N/mm]	849,643.2
Stiffness under lateral bending [Nmm/rad]	879,055.8
Stiffness under transverse bending [Nmm/rad]	879,055.8
Stiffness under torsion [Nmm/rad]	782,024.5

The parameter identification method developed in this study is not limited to rivets but also can be used for any kind of joint such as bolts, spot welds, bearings, etc. There may be complicated joints where parameters cannot be obtained using theoretical approaches. In such cases, complicated joints that are difficult to model can easily be modeled by this method as it works as a black box method.

The residual errors in frequencies are mostly due to not being able to model the sheet-to-sheet interface. This interface requires frictional contact modeling which is unsupported in modal analysis of commercial FEA packages. However, the errors are acceptable for engineering purposes. The method can also easily be adapted to include damping values and eigenvectors.

Currently, the proposed method is being modified to be used in damage localization where missing or damaged rivet locations will be identified with a similar procedure.

References

1. Beards CF (1983) The damping of structural vibration by controlled interface slip in joints. Am Soc Mech Eng J Vib Acoust Stress Anal Reliab Des 105:369–373
2. Tsai JS, Chou YF (1988) The identification of dynamic characteristics of a single bolt joint. J Sound Vib 125:487–502
3. Wang JH, Liou CM (1991) Experimental identification of mechanical joint parameters. J Vib Acoust 113:28–36
4. Inamura T, Sata T (1979) Stiffness and damping properties of the elements of a machine tool structure. Annals CIRP 28:235–239
5. Kim TR, Wu XM, Eman KF (1989) Identification of the joint parameters for a taper joint. Am Soc Mech Eng J Eng Ind 111:282–287
6. Ibrahim RA, Pettit CL (2005) Uncertainties and dynamic problems of bolted joints and other fasteners. J Sound Vib 279:857–936
7. Arruda JRF, Santos JMC (1993) Mechanical joint parameter estimation using frequency response functions and component mode synthesis. Mech Syst Signal Process 7:493–508
8. Kim HK, Park YS (1993) An efficient response analysis method for a non-linear/parameter changing system using sub-structure modes. Proc IME C J Mech Eng Sci 207:41–52
9. Cabell RH (2010) Vibration response models of a stiffened aluminum plate excited by a shaker. Noise Control Eng J 58:389–402
10. Xiong Y, Bedair OK (1999) Analytical and finite element modeling of riveted lap joints in aircraft structure. AIAA J 37:93–99
11. Fung CP, Smart J (1994) An experimental and numerical analysis of riveted single lap joints. J Aerosp Eng 208:79–90
12. Körük H, Şanlıtürk KY (2007) Validation of finite element models of structures with riveted joints using vibration data. Inter-Noise 2007, Istanbul, Turkey

Chapter 7
Dynamic Ground Testing: Ground Vibration Tests Through Control Surface Excitation

G. Osmond, A. Azzat, S. Leroy, and O. Delverdier

Abstract Flutter flight tests are a crucial step in aircraft certification. The typical approach to flutter flight testing is to excite the aircraft using its control surfaces at several stabilized test points and to measure the corresponding structural response. Before beginning the flight test campaign, however, a preliminary ground vibration test (GVT) performed on the prototype is necessary in order to ensure the reliability of the flutter predictions.

These tests are currently costly in terms of preparation time, hardware, resources and time spent in the final assembly line. Due to the high development costs, each day that can be saved in the test program is crucial. As AIRBUS' current strategy is to develop more derivative aircraft from its flagship projects as opposed to completely new aircraft, a reduction in the number of measurements can be considered. A lighter GVT (or DGT) can therefore be used to save costs.

The DGT approach proposed in this paper also relies on the adaptation of in-flight excitation techniques in order to reduce the need for external exciters. These tests would be achieved through sine sweep excitations performed with the control surfaces on ground. However several challenges have to be addressed.

Keywords Structure • Vibration • Dynamics • Aircraft • Certification

Abbreviations

DGT	Dynamic Ground Tests
FAL	Final Assembly Line
OGT	Optimized Ground Tests
ADIS	Aircraft systems in charge of generating c/s orders
TLM	Telemetry tools
C/S	control Surfaces
OMA	Operational modal analysis

7.1 Introduction

In the past few years, AIRBUS has launched several derivative developments based on existing programs. Economic pressure has led to a reduction of the lead time dedicated to get the type certification for these new derivative programs. These efforts are primarily made possible by the fact that the development does not start from a blank page, meaning that knowledge gained from the legacy product can be used to efficiently update the theoretical models. Even if the time slot devoted to ground vibrations tests have been greatly reduced on the A350-900 compared to the previous programs (9 days vs. 14 days), the challenge has been raised to reduce these tests to only one day for the derivative programs. In addition to the aforementioned challenges, a consolidated modal base has to be transmitted quickly to the design office after the end of the test. This point requires a drastic optimization of the modal filtering and validation process.

Thus, this test would be used to validate the theoretical model when we already have a good level of confidence in it. The tests would no longer be performed during the FAL period but under the flight test team responsibility, only a couple of days before the first flight. The requested reduction is so significant that it could not be addressed by an evolution of the

G. Osmond • A. Azzat • S. Leroy (✉) • O. Delverdier
Airbus, Flight and Integration Test Centre, EVAAA – Vibrations Tests, 31060 Toulouse Cedex, France
e-mail: stephane.leroy@airbus.com

A. Wicks (ed.), *Shock & Vibration, Aircraft/Aerospace, and Energy Harvesting, Volume 9*,
Conference Proceedings of the Society for Experimental Mechanics Series, DOI 10.1007/978-3-319-15233-2_7

current GVT techniques. We had to reconsider the entire testing process. The DGT concept relies on the idea that we can do on-ground what is done in-flight during the flight domain opening envelope, i.e. excite the aircraft via the control surface and measure the response with the embedded accelerometers, thus eliminating the for time-consuming shaker placement or sensor installation

7.2 A340-600 MSN360 Research Ground Vibration Test (2011)

In 2011, a test was performed which involved Airbus, Onera and DLR on the A340-600. Its main goal was to improve the GVT techniques. During these tests, the A340-600 was fully equipped with the FTI (Flight Test Instrumentation – a term used to describe the semi-permanent sensors and recorders installed on a development aircraft) and the GVT instrumentation installed externally to the aircraft specifically for this test. One full day was dedicated to control surface excitation. This FTI acquisition chain will be used for the whole flight test campaign, including Flutter tests. The resulting data had not been fully analysed, so it has been done in the frame of the DGT project, thus allowing the validation of the DGT principle as a replacement for a standard GVT.

This set of data provided a very useful first look at DGT feasibility because:

- The tested a/c (Airbus A340-600 MSN0360) was fully characterised using the GVT process in a comparable weight configuration. A robust base is thusly available for comparison with DGT results (frequency, damping and mode shapes).
- The a/c was equipped with both GVT autonomous instrumentation and FTI sensors.

This allows decomposing the gap between the standard GVT and the DGT in two main questions:

- Are control surface excitations capable of properly exciting target a/c modes?
- Are FTI sensors capable of properly capturing FRFs?

The first question will be answered using the GVT sensor set, knowing that they are functional and validated.

Because the second question will be answered using the FTI sensor set since it was acquired during all control surface excitations (Fig. 7.1).

In total, 22 runs were post-processed from the OGT. Operational Modal Analysis (OMA) was used to extract modal parameters. This method is considered acceptable because modal mass is needed model correlation purposes and OMA is considered to be sufficiently robust algorithm, especially when a sufficient level of response is measured in the whole frequency bandwidth.

Modes were extracted from various runs, compared and the best candidate was selected for comparison with GVT results.

Not all modes were extracted; we focused our attention on 17 modes, regarded as the most important.

These modes are compared with GVT results provided by ONERA-DLR.

CS location	parity	amplitude (degrees)	fmin	fmax
Outer Ailerons	SYM	10	0,5	3,5
elevators	SYM	10	0,5	3,5
Outer Ailerons	ANTI	10	1	3,5
Inner ailerons	ANTI	10	1	3,5
elevators	ANTI	10	1	3,5
Outer Ailerons	SYM	10	3	6
Inner ailerons	SYM	10	3	6
rudder		10	1	3,5
elevators	SYM	2,5	3	6
Outer Ailerons	ANTI	10	3	6
Inner ailerons	ANTI	10	3	6
elevators	ANTI	2,5	3	6
rudder		2,5	3	6

Fig. 7.1 Several runs with c/s excitations performed during A340-600 MSN360 OGT

Fig. 7.2 mismatch between DGT-GVT on A340-600

Requested modes from DGT	frequency mismatch DGT-Onera (%)	damping mismatch DGT-Onera (%)	C/S Excitation
2NZ wing bending	N/A	N/A	N/A
3NZ wing bending	4,2	45,3	Rudder
4NZ wing bending	1,5	31,5	I_Ail_Sym
Lateral engine SYM	1,3	37,2	O_Ail_Anti
Lateral engine ANTI	0,4	12,5	O_Ail_Sym
Vertical engine SYM	0,7	10,9	Elev_Sym
Vertical engine ANTI	0,6	4,1	Rudder
Rear Lateral Fuselage	0,0	42,9	Rudder
Front lateral Fuselage	1,3	18,4	Elev_Anti
2N vertical fuselage	0,8	12,3	Elev_Sym
2N lateral fuselage	0,8	12,3	Elev_Sym
Engine Roll and yaw	N/A	N/A	N/A
Fin Bending	4,6	3,3	Elev_Anti
HTTP yaw	16,9	51,4	Elev_Anti
HTTP roll	2,6	10,6	Elev_Anti
2NZ HTP bending	0,7	22,1	Elev_Sym
Wing torsion sym	0,2	11,3	I_Ail_Sym

Numerical values for frequency and damping are compared but not displayed in this paper. Only mismatch between GVT-DGT results is displayed in the table hereafter. A color code is given to assess extraction quality: green is acceptable, while red is too far from requested result quality (Fig. 7.2).

Some of the requested modes could not be identified. For example, the 2NZ wing bending mode could not be properly acquired, because the a/c was on its tires which resulted in a heavy coupling between this mode and the heave rigid body mode. Pitch and yaw at engine location was not captured. An inappropriate excitation is one explanation for this un-identified mode.

7.3 Instrumentation

Usually, a GVT is performed by means of hundreds of accelerometers glued on the external skins of the aircraft structure. The installation process can take several days of installation. On top of that, the instrumentation requires a scaffolding to be installed around the aircraft which is really a time consuming task (Fig. 7.3).

In order to avoid the sensors and scaffolding installation, the first idea is to take advantage of the FTI instrumentation normally dedicated to the flight domain opening and control law tuning which must be installed and validated for the first flight anyway.

On the A350-900, the flutter instrumentation is comprised of approximately 100 single-axis accelerometers dispatched within the structure. A similar number is foreseen for the A350-1000. Nevertheless, there is a drawback. The instrumentation is imprisoned in the airframe and cannot be easily accessed in case of failure. The analog to digital conversion is done locally, making any possible repair sensitive as the associated electronics is similarly distributed throughout the aircraft. On the other hand, the instrumentation is installed several months before the tests and can be validated in advance during the aircraft's stay in the FAL. This set of accelerometers is capacitive, which fully complies with the GVT requirements for low frequency measurement (Fig. 7.4).

The target assigned by the design office is to identify all the modes up to 15 Hz. Using 120 accelerometers instead of 500–800 sensors for a classical GVT, modal validation becomes tricky due to reduced fidelity of the MAC matrix. This risk has been mitigated by means of the A350-900 GVT results. The full modal basis was recalculated while retaining only the GVT sensors which have a corresponding FTI sensor.

A study has been performed on modal extraction algorithm robustness: even with very few sensors working on aircraft (studies were carried out with as little as 25 sensors), the frequency and damping of modes can still be extracted. Nevertheless, mode naming is not possible and mathematical models cannot be sorted out from physical ones. More sensors are necessary for a proper modal identification.

Fig. 7.3 A380 scaffolding installation

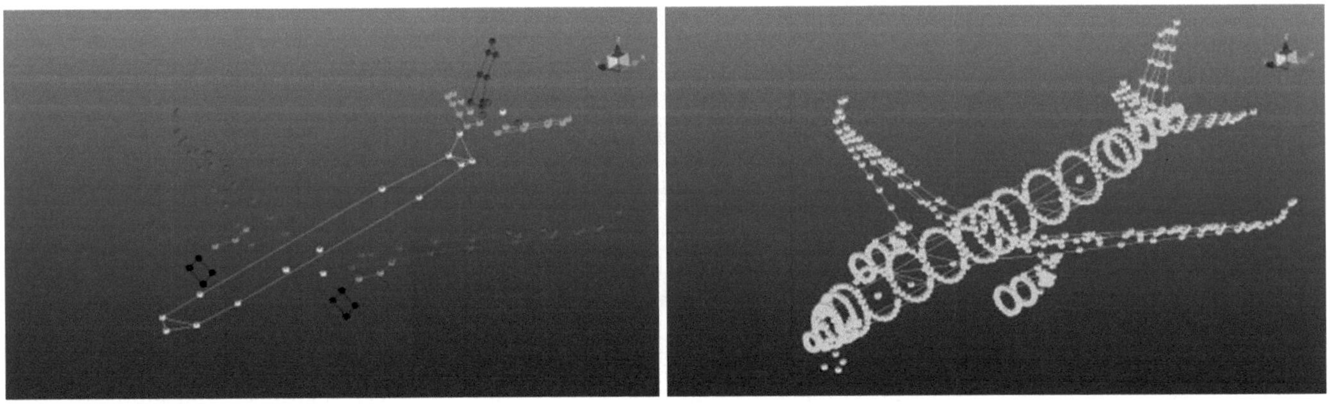

Fig. 7.4 A350-900 flight test instrumentation vs GVT

Some additional considerations have been identified. For instance, only one accelerometer is usually installed on each control surface for the flight, making it tricky to validate control surface modes. At last, it appears that the FTI is not fully sufficient to address all the modes we are interested in. There are definitely some additional measurements that must be added. In addition, there is a necessity to make the set-up more flexible. The idea was to mix the FTI with a reduced GVT installation. This would require adapting the GVT acquisition system so that it is able to merge data originating from the analog sensors and the numeric IENA data flow originating from the FTI.

A client–server architecture is designed to allow these exchanges. The synchronization between the GVT front-end and the FTI bus is ensured by a common IRIG-B source (Fig. 7.5).

This mixed technology between FTI sensors and autonomous sensors will allow us to:

- replace in short notice non-working FTI sensors without jeopardizing the test
- get better mode shape definition by adding sensors as needed

7.4 Excitation Signals

The proposed DGT method would use the control surfaces to artificially excite the aircraft. In flight, the excitation is obtained via the aerodynamics loads in flight, but on ground we must rely solely on the inertia of the control surface. As they are not designed for such a purpose, several limitations have to be taken into account, including temperature and amplitude. The efficiency of a control surface excitation that relies, on-ground, purely on inertia is questionable. One of the major challenges is to provide a solicitation which is sufficient to identify the different engine modes. It makes the "heavy" control surfaces, i.e. rudder and elevators, more interesting than the inner or outer ailerons which have lower inertia. Unfortunately neither the horizontal tail plane (HTP) nor the vertical tail plane (VTP) plays a role in the engine modes (except for engine vertical symmetric mode). So it must be concluded that the rudder or the elevators are not well adapted for extracting all engine modes.

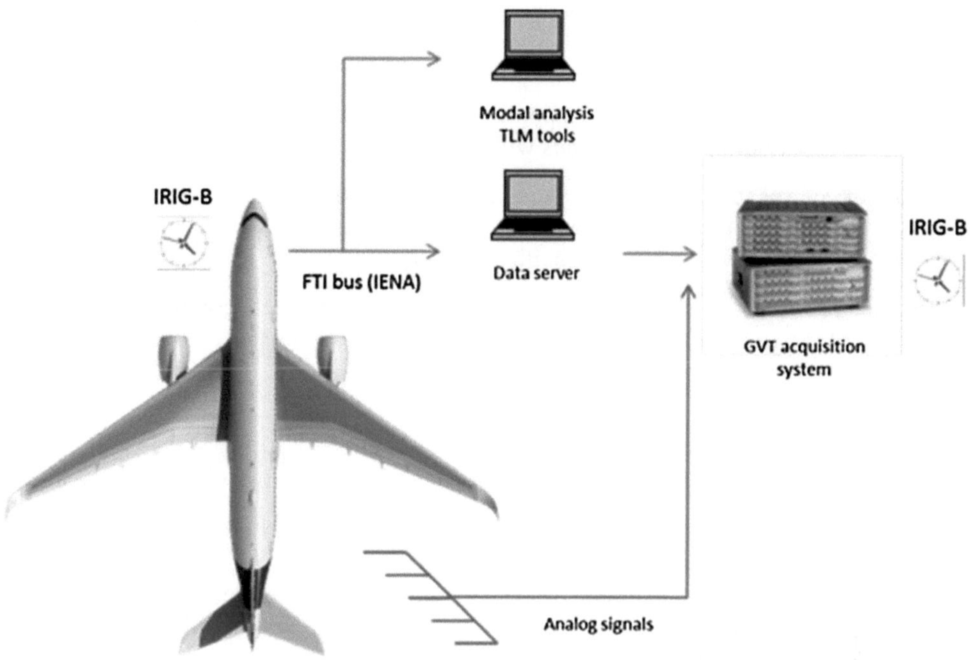

Fig. 7.5 Architecture synchronization between the GVT front-end and the FTI bus

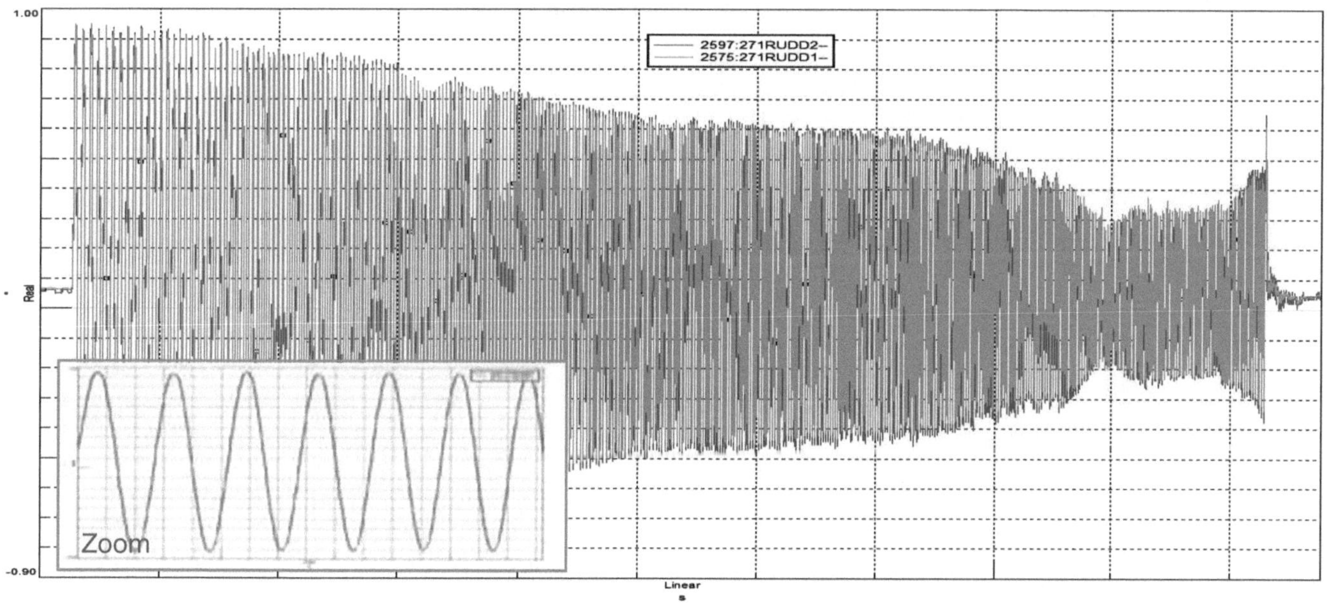

Fig. 7.6 Example of rudder sine-sweep excitation from 1 to 6 Hz

Since a linearity assessment is usually done for the engine modes in order to get rid of this issue, several options are being explored:

- Ensure the possibility to carry-out a regular GVT excitation via shakers located on the engines without impacting the tight schedule
- Identify the engine modes by performing additional excitations in-flight at low speed
- Modify the actuator to allow higher displacement

During a sine-sweep the actuator oil temperature increased and could even trigger the overheat protection when run for too long. Thus the oil trapped near the actuator needs to be renewed after each sine sweep by performing a large displacement of the control surface. Another option will be to install cooling systems dedicated to this test (Fig. 7.6).

There is a drawback. The ADIS system (AFDX and Digital Injection Signals) is used to inject the sine sweeps into the flight control system and thus move the control surface actuators. Basically, all injection signals need to be validated in a simulator before being used on aircraft. This makes the procedure to change the settings (sweep rate and type) inflexable. In order to save time, sine sweeps xill be split into three parts to cover the full bandwidth with different amplitudes. Each sweep will last 1 min.

In order to validate all of these open points concerning control surface excitation, a dry-run on the A350-900 is scheduled. Points to check during those dry-runs will be:

- ability of control surface excitations to excite engine modes
- need for additional sensor to properly capture mode shapes
- lowest sweep rate available to cope with heating issues.

This dry-run will be the final test to make a choice between purely control surface excitations and a mixed test with shaker excitations of the engines.

7.5 Streamlined Analysis

The proposed DGT extends the flight test opening techniques to ground testing. During the flight, the modal analysis is performed "online" (between test points) in order to follow the frequency and damping trends as the aircraft speed increases. The same tools will be used to quickly assess which modes can be identified after each run.

In addition, validated results are requested to be available within a 48 h period after the end of the test. The baseline scenario for this test is built on 100–150 runs per aircraft.

To cope with that large amount of data, special tools will need to be developed:

- Synchronization tool between FTI and autonomous sensors
- Automatic identification of relevant data based on ADIS states
- Semi-automatic sensor level extraction based on the Hilbert transform, and
- A post-processing modal data base tool to deal with the large number of modes generated.

Chapter 8
Adaptive Support of an Aircraft Panel

Manuel Baschke and Delf Sachau

Abstract The vibro-acoustic behavior of large lightweight structures such as aircraft fuselage has to be tested, especially to improve active and passive noise control means. Acoustic transmission laboratories with reverberant and anechoic rooms supply reliable test conditions for aircraft panels of a size up to several square meters. Shock mounts support these panels to realize free or pinned boundary conditions with minor damping, which is sufficient at high frequencies. At low frequencies, the vibration of the whole fuselage has to be considered. Therefore the support of the panel under test should provide the same dynamic impedance as the fuselage to which the panel would be connected in the aircraft. This frequency-dependent boundary condition can be realized by an adaptive support. The paper describes a numerical investigation of this supported panel which should behave like integrated into an aircraft fuselage.

Keywords Adaptive control • Numerical investigation • Boundary condition • Substructure • Aircraft panel

8.1 Introduction of Adaptive Boundary Conditions

The realization of adaptive boundary conditions have been simulated and tested on elementary physical experiments before it is investigated for an aircraft panel. The operating principle is explained with a hinged beam [1].

The basic idea of this strategy is to simulate parts of a complex structure like an aircraft fuselage by an adaptive algorithm, the Filtered x Least Mean Square Algorithm as explained in [2]. This algorithm sets impedances at the boundaries of the substructure to realize the same dynamical behavior like in the complex structure.

This method is realized for a hinged beam excited by the so called primary force F_p, Fig. 8.1 (top). In the middle of this figure the shearing forces Q_1, Q_2 and the bending moments M_1 and M_2 are shown. The bottom of Fig. 8.1 shows the principle of the adaptive boundary.

The cutting forces and bending moments are provided by controlled actuators to realize the same vertical displacements w and bending angles w' at the cut points.

With this approach it is possible, to realize almost the same dynamic behavior in the substructure like in the reference structure near the first three resonance frequencies. The relative errors for the vertical displacements and the bending angels are below 10 %. More details about the results can be found in [1].

In a next step the adaptive boundary conditions are investigated for a hinged thin plate with a finite element model [3]. The forced vibrations of the reference plate are calculated. The boundary conditions are implemented with the complex displacements for chosen nodes. These displacements are calculated in the reference structure. In this investigation the assumption of an ideal control success is made. In reality the adaptive filter are not able to fit the displacements of the individual nodes exactly. But the differences between the desired and measured displacements are minimized with the adaptive filters. In difference to [1] the actuators and sensors are collocated in this study. In [3] A. Schulz has shown, that it is possible, to realize nearly the same dynamic behavior in a 30×30 cm plate with 12 actuators and sensors like in the reference structure (size 1×1 m). He has used a triple of actuators and sensors in every corner. Due to his good results the investigation with a finite element model for an aircraft panel is done. The results are presented with this paper.

M. Baschke (✉) • D. Sachau
Helmut-Schmidt-University, University of the Federal Armed Forces Hamburg, Holstenhofweg 85, 22043 Hamburg, Germany
e-mail: baschkem@hsu-hh.de

A. Wicks (ed.), *Shock & Vibration, Aircraft/Aerospace, and Energy Harvesting, Volume 9*,
Conference Proceedings of the Society for Experimental Mechanics Series, DOI 10.1007/978-3-319-15233-2_8

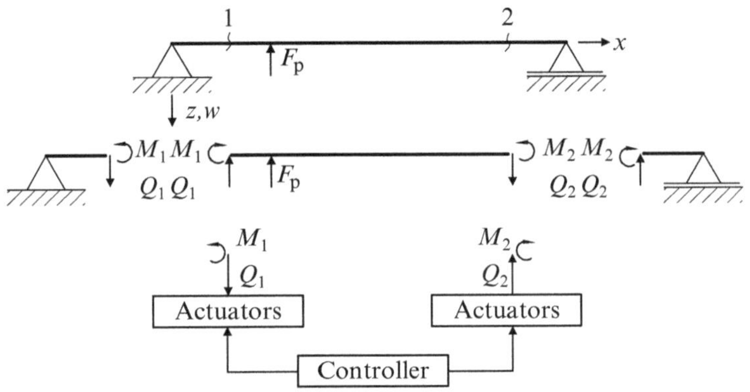

Fig. 8.1 Reference structure (*top*), free cut substructure (*middle*), adaptive boundary conditions (*bottom*) [1]

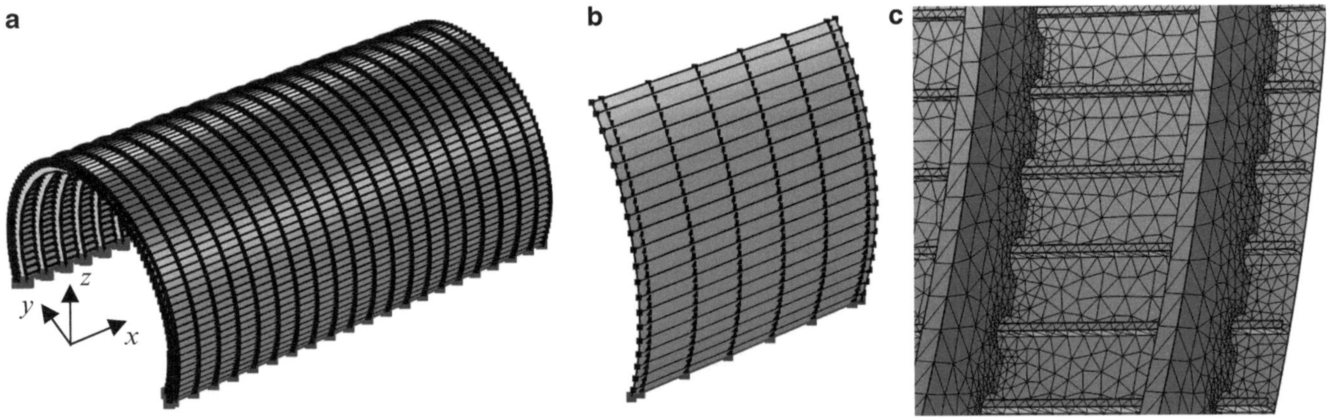

Fig. 8.2 Finite element models. (**a**) reference structure with fixed boundary conditions, (**b**) substructure with fixed boundary conditions, (**c**) detailed view of the mesh in the substructure

8.2 Finite Element Models

For the finite element calculations the software Comsol Multiphysics is used. This software is designed for the coupling between different physics. At the moment just the structural dynamics are relevant, but in further studies the coupling to the acoustics might become interesting.

The reference structure is a generic model of an aircraft fuselage with the following parameters. The length of the model is 10 m, the radius 3 m. Just the upper circle sector within 220° is considered due to the fact, that below this sector the aircraft floor is a rigid structure in comparison to the upper sector. Because of this the model is realized with fixed boundary conditions at the lower edges, Fig. 8.2a. The model includes frames and stringers. The stringers are even placed in 3° steps. The frames are implemented in distances of 0.5 m. The material of the model is aluminium. The damping is included with an isotropic structural loss factor of 0.01.

The frames and stringers are implemented with an L-profile. The ribs of both are connected to the outer skin. The lengths of the ribs are 30 mm for the stringers and 230 mm for the frames. The flanges are 15 mm (stringers) and 40 mm (frames) long. The thicknesses of the used components are: 3 mm for the stringers, 3.5 mm for the outer skin and 5 mm for the frames. Shell elements are used to model the whole structure.

The substructure is built with the same parameters as the reference structure. It is the lower left part of the foreground of the reference structure (Fig. 8.2a). It consists of six frames and 19 stringers with the same fixed boundary conditions on the ground (Fig. 8.2b). This leads to a size of approximate 3 m height and 2.7 m length.

Both models are meshed with the default mesher of the software Comsol. Triangles are used with linear shape functions. This leads to 931,000 degrees of freedom (DOF) for the reference structure and 140,000 for the substructure. A detailed view of the mesh is shown in Fig. 8.2c.

8.3 Calculations

The proceeding in the calculations is in analogy to [3]. In a first step the natural frequencies of the reference structure are calculated. With the above mentioned parameters four global mode shapes can be found. The natural frequencies of these global mode shapes are 17.69, 36.02, 40.13 and 49.86 Hz. The following analysis concentrates on the forced vibrations near these resonances (integer frequencies). These four forced vibrations of the reference structure should be imitated in the substructure. The local modes are not investigated, because they can be reproduced in the substructure with fixed boundary conditions. Figure 8.3 shows the four global mode shapes of the reference structure. The deformation is shown and the colours represent the total displacement from. The amplitudes of the displacements are not shown within these plots, because their absolute values are not important for these studies.

In Fig. 8.4 the force transmission point is shown for the unwrapped substructure. This point is identical in the reference structure and a force of 100 kN in radial direction is implemented. Furthermore the set boundary conditions in the

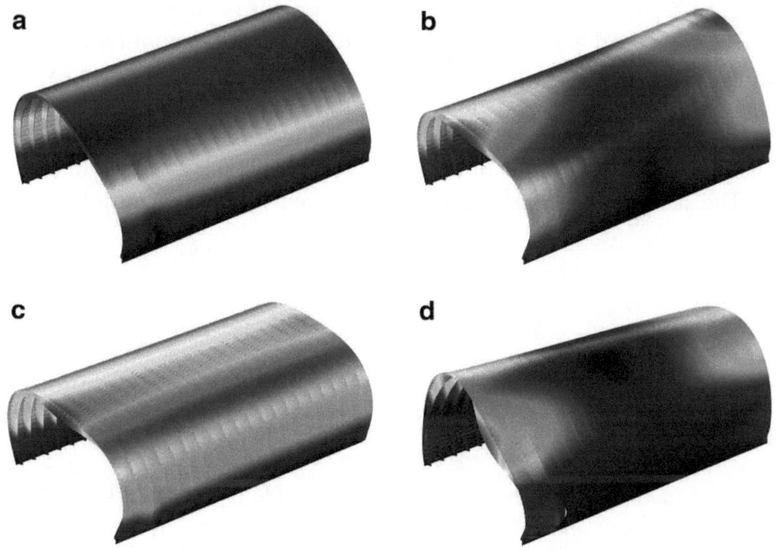

Fig. 8.3 Total displacement of forced vibrations of the reference structure at excitation frequencies. (**a**) 18 Hz, (**b**) 36 Hz, (**c**) 40 Hz, (**d**) 50 Hz

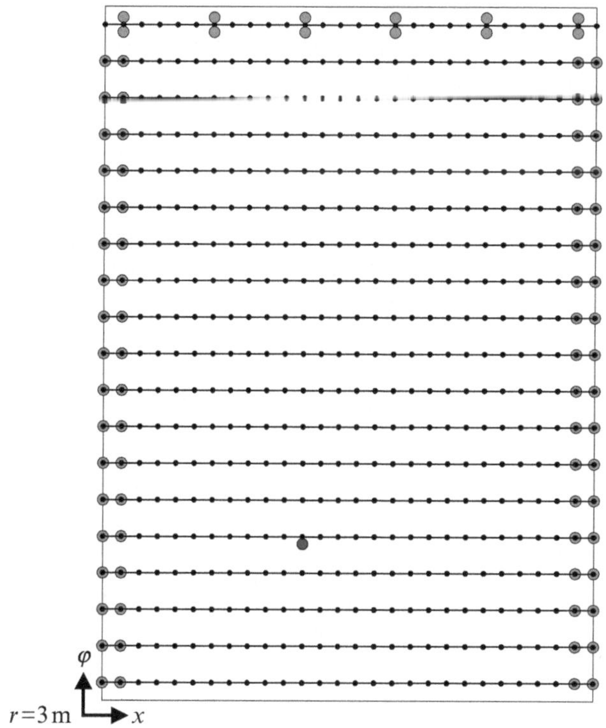

Fig. 8.4 Unwrapped substructure with primary excitation, set boundary conditions and evaluation points on the stringers in distances of 0.1 m (*black dots*)

substructure are shown. These boundary conditions are calculated in the reference structure. They are the radial displacements of these nodes. These are set in the substructure. This approach is identical to the already in the introduction mentioned ideal control success in analogy to [3]. The nodes for the implementation are chosen to the connection between the frames and the stringers on the one hand and the outer skin on the other hand, because the structure is stiffer at these points than in the skin fields between them. This makes it easier to excite the global modes with electrodynamic shakers. If these were put in the softer skin fields the local vibration of these would be excited.

In Fig. 8.4 it is shown, that the connection between the stringers and the outer skin is chosen for the left and right boundaries and the connection between the frames and the outer skin at the top of the structure. With this approach both the shearing forces and the bending moments are induced in the cut frames and stringers.

8.4 Results

Figure 8.5 shows exemplary the forced vibrations of the reference structure and the substructure for an excitation frequency of 36 Hz (second global mode shape) and an excitation amplitude of 100 kN. The colors represent the amplitudes of the total displacements from zero up to 3e-2 m. The deformation is scaled with a factor of 25. It can be determined, that the mode shape of the reference structure can be reproduced in the substructure. The forced vibrations of the other frequencies have a comparable correlation. Within this paper just the behavior for this frequency is presented graphical. For a more detailed analysis of the correlation of the structures, the displacements will be evaluated in the following.

The success of the approach with adaptive boundary condition is evaluated with the square error norm:

$$EN_{\mathrm{k}} = \frac{\| \boldsymbol{r}_{\mathrm{sub,k}} - \boldsymbol{r}_{\mathrm{ref,k}} \|}{\| \boldsymbol{r}_{\mathrm{ref,k}} \|} = \frac{\sqrt{\left(\sum_{i=1}^{532} |r_{\mathrm{sub,k}}(i) - r_{\mathrm{ref,k}}(i)|^2 \right)}}{\sqrt{\left(\sum_{i=1}^{532} |r_{\mathrm{ref,k}}(i)|^2 \right)}} \tag{8.1}$$

In this equation the vector $\boldsymbol{r}$ represents the radial displacements of all evaluated nodes (black dots in Fig. 8.4) in both models. The indexes sub and ref indicate in which model they are calculated. The index k represents the resonance frequency. So the values for k are from one to four. The sums are calculated for all $i = 532$ evaluated nodes, 28 nodes on each of the 19 stringers.

The values EN_{k} vary between 3.3 % (18 Hz) and 6.2 % (40 Hz). To localize the errors more detailed, matrixes with the relative square errors are calculated:

$$EN_{\mathrm{k}}(l, m) = \frac{\| R_{\mathrm{sub,k}}(l, m) - R_{\mathrm{ref,k}}(l, m) \|}{\| R_{\mathrm{ref,k}}(l, m) \|} \tag{8.2}$$

These matrixes show the relative square errors of every evaluated node. The parameter l represents the 19 evaluated nodes in a column of the $m = 28$ columns, Fig. 8.4. In contrast to (8.1) $\boldsymbol{R}$ is a matrix with the same entries as the vector $\boldsymbol{r}$ in (8.1), just sorted in matrix structure.

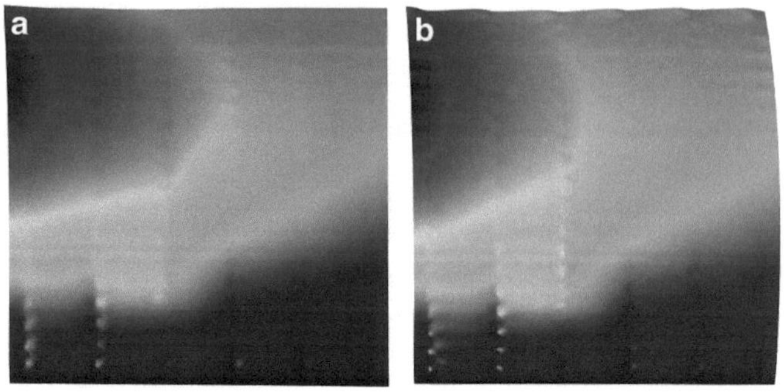

Fig. 8.5 Forced vibrations of (**a**) the reference structure and (**b**) the substructure for excitation frequency of 36 Hz

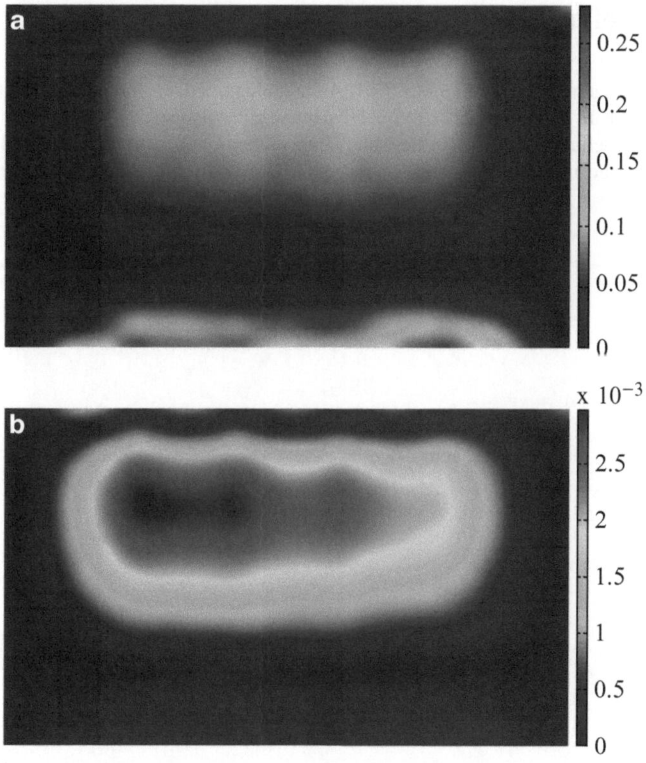

Fig. 8.6 (**a**) Relative square errors for every evaluated node for excitation frequency of 36 Hz – $\boldsymbol{EN}_2$ (**b**) absolute value of absolute error in radial direction for excitation frequency of 36 Hz in m

The relative errors are zero at the boundaries due to the fact, that the displacements are set equal. Figure 8.6a shows exemplary the matrix $\boldsymbol{EN}_2$. The other three matrixes look nearly the same. In general a tendency can be determined, that the relative errors increase, when the absolute displacements decrease. The largest relative errors occur at the nodes, which nearly do not move. To make this clear in Fig. 8.6b the absolute values of the absolute errors are shown for the same excitation frequency.

To compare the success of the presented approach of adaptive boundary conditions the square error norm EN_k is calculated for other bearings. The nodes, on which the displacements are set before, are fixed, hinged and free. For the fixed and the hinged bearing the values of the EN_k are between 97 % and 100 %, for the free bearing even between 93 % and 208 %. The matrixes $\boldsymbol{EN}_k$ are nearly complete red, if the same colour scale as in Fig. 8.6 is used.

During the calculations also boundary conditions for the displacements in flight direction have been investigated. It turned out, that it is not necessary, to set these boundary conditions. This is due to the fact, that the displacements in flight direction are very small for the investigated global mode shapes.

8.5 Conclusions

The above mentioned investigations show, that the concept of adaptive boundary conditions lead to a much more similar vibration behavior in the substructure for low frequencies in comparison to other boundary conditions.

Also the implementation in an experiment is possible. Our chair is in possession of an Airbus A400M fuselage and can borrow a substructure with nearly the same measurements as used in the presented calculations from Airbus. The adaptive boundary conditions could be realized with 84 small electrodynamic shakers from Kendrion Kuhnke Automotive GmbH, which can be mounted directly onto the substructure. These shakers cost round about 40€ each. So the total cost for these are below 3,500€. Furthermore all the other necessary equipment (filters, amplifiers, sensors, rapid control prototyping systems) is available at our chair. Herewith the implementation will probably be tested in future.

Acknowledgements This work is financially supported by the city of Hamburg in the frame of the “Hamburger Luftfahrtforschungsprogramm”.

References

1. Sachau D, Baschke M (2014) Adjustable impedance boundary for structure under test. In: Proceedings EURODYN 2014, Porto
2. Kuo S, Morgan D (1996) Active noise control systems. Wiley, New York
3. Schulz A (2013) Numerische Untersuchung von aktiven Randabschlüssen – Numerical investigation of active boundary conditions. Bachelor thesis, Helmut- Schmidt-Universität/University of the Federal Armed Forces Hamburg, Hamburg

Chapter 9
Calculating the Impact Force of Supersonic Hail Stones Using SWAT-TEEM

Tyler F. Schoenherr

Abstract In the aerospace industry, hail strikes on a structure are an environment that must be considered when qualifying a product. Performing a physical test on a product would require a test setup that would launch a fabricated hail stone at an expensive prototype. This test may be difficult or impossible to execute and destructive to the product. Instead of testing, a finite element model (FEM) may be used to simulate the damage and consequences of a hail strike. In order to use a FEM in this way, an accurate representation of the input force from a hail stone must be known. The purpose of this paper is to calculate the force that a hail stone imparts on an object using the inverse method SWAT-TEEM. This paper discusses the advantages of using SWAT-TEEM over other force identification methods and exercises the algorithm for a test series of hail strikes that include multiple angles of attack and multiple velocities which include speeds that are supersonic.

Keywords Hail • Force • Impact • SWAT • Ice

9.1 Introduction

Aerospace systems experience a variety of unique inputs ranging from aerodynamic loading to turbine induced vibration. The possibility of hail strikes on the system is a viable mechanical input to the system and must be analyzed when qualifying the system for its intended function. To determine if the system will survive the hail strike, a test is run either analytically or experimentally. If the system is qualified experimentally, then a test is run that fires a hail stone at the system and hits it in the location that causes the most damage to determine if the system will survive. However, firing a hail stone accurately at the known worst case location is difficult and expensive to execute. Also, the test may be destructive and most hardware is too costly or is one of a kind. The test would still not determine if the system could survive in a hail storm with multiple consecutive hail strikes.

Because of the difficulties of an experimental test, an analytical or finite element model is typically utilized. In order to properly use a calibrated finite element model, an accurate representation of the force is needed as an input to the model. To determine the force that a hail strike imparts on a system, characterization tests have to be performed that measure or calculate the force that a hail stone imparts on an object.

This characterization has been attempted in the past using different methods. Force gauges have directly been used to measure the force that the hail stone imparts on a plate [5, 7]. This method works for normal impacts and low frequency strikes, but force piezo electric gauges typically have low internal resonances and have issues measuring shear forces and moments if the strike is not perfectly normal. This would make measuring the force of high speed or glancing hail strikes hard to measure with this method.

Another method to measure the force a hail stone imparts on a surface has been proposed by Tippmann et al. [8]. They proposed to shoot the hail stone into a long cylindrical rod instrumented with strain gauges. The strain measured could be related to stress and the force with knowledge of the geometry of the rod. This process is very similar to a traditional Hopkinson bar test.

Sandia National Laboratories is a multi-program laboratory managed and operated by Sandia Corporation, a wholly owned subsidiary of Lockheed Martin Corporation, for the U.S. Department of Energy's National Nuclear Security Administration under contract DE-AC04-94AL85000.

T.F. Schoenherr (✉)
Experimental Mechanics, NDE and Model Validation Department, Sandia National Laboratories,
P.O. Box 5800, MS0557, Albuquerque, NM 87185, USA
e-mail: schoenherr.tyler@gmail.com

A. Wicks (ed.), *Shock & Vibration, Aircraft/Aerospace, and Energy Harvesting, Volume 9*,
Conference Proceedings of the Society for Experimental Mechanics Series, DOI 10.1007/978-3-319-15233-2_9

This paper presents another alternative to calculating the force imparted by a hail stone. The SWAT-TEEM (Sum of Weighted Accelerations Technique-Time Eliminated Elastic Motion) is the algorithm in this paper that is used to calculate the hail force. The SWAT algorithm developed by Gregory et al. [3] was later used to calculate the input forces on a system by Carne et al. [2]. Later, the algorithm was modified to use the time domain responses for the inversion instead of the mode shape and was renamed to SWAT-TEEM [6]. Both of these algorithms calculate the sum of the forces and moments from responses at the center of gravity.

Both SWAT and SWAT-TEEM algorithms are considered in this paper to calculate the hail impact forces. To determine which algorithm would be used, a characterization test is run. This test measures the mode shapes of the system for the SWAT algorithm. Because the characterization test involves hitting the object with a force gauge on a hammer, it provides a truth test to compare the results from SWAT and SWAT-TEEM to a measured hammer force. This truth test provides data to determine which algorithm to use for the hail impact test and provides evidence for the determination of the useful bandwidth of the calculated force.

Forces from hail impacting an aluminum plate are the interest of this paper. In addition to calculating the force of a hail stone impacting an aluminum plate at a normal angle, normal forces will be calculated for hail stones at different impact angles and at different speeds. This includes the forces of hail stones that are traveling at supersonic speeds. The paper concludes with the examination of all of the calculated forces and their corresponding angles.

9.2 Force Reconstruction Theory

Force reconstruction is used to describe the inverse problem where the system's dynamic properties and response to a force is known. From the system's equations of motion,

$$\bar{X}(\omega) = \mathbf{H} \cdot \bar{F}(\omega), \tag{9.1}$$

where $\bar{F}$ is the input force, $\mathbf{H}$ is the system's dynamic properties, and $\bar{X}$ is the system's response from the input force, the system's dynamic properties can be inverted to calculate the input force on that system that caused the measured response as shown in

$$\mathbf{H}^{-1} \cdot \bar{X}(\omega) = \bar{F}(\omega). \tag{9.2}$$

Although the inversion process appears straightforward, Eq. (9.2) typically fails due to the $\mathbf{H}$ matrix being ill conditioned. It is ill conditioned because the solution for the force matrix is generally not unique. Also, the $\mathbf{H}$ matrix is generally rank deficient for some frequency lines, so it cannot be accurately inverted. The following subsections present two algorithms that solve this inverse problem. These algorithms are dubbed SWAT algorithms and they solve the non-uniqueness problem by reducing the force vector size down to six. The six vectors represent the sum of the forces and moments acting at the center of gravity which yields a unique solution.

9.2.1 Formulation of SWAT (Sum of Weighted Acceleration Technique)

The derivation of the SWAT algorithm that is used to calculate the sum of the external forces begins with the second order linear equations of motion,

$$\mathbf{M}\ddot{\bar{x}} + \mathbf{C}\dot{\bar{x}} + \mathbf{K}\bar{x} = \bar{F}, \tag{9.3}$$

where $\mathbf{M}$ is the mass matrix, $\mathbf{C}$ is the damping matrix and $\mathbf{K}$ is the stiffness matrix of the system. Modal substitution is then used to estimate the physical response with modal degrees of freedom shown by

$$\boldsymbol{\phi}\bar{q} \approx \bar{x}, \tag{9.4}$$

where $\boldsymbol{\phi}$ is the mode shapes of the system and $\bar{q}$ is the generalized modal coordinates. The modal approximation shown in Eq. (9.4) is substituted into Eq. (9.3) to get

$$\mathbf{M}\boldsymbol{\phi}\ddot{\bar{q}} + \mathbf{C}\boldsymbol{\phi}\dot{\bar{q}} + \mathbf{K}\boldsymbol{\phi}\bar{q} = \bar{F}. \tag{9.5}$$

At this point, Eq. (9.5) is premultiplied by the transpose of the rigid body modes to get

$$\boldsymbol{\phi}_r^T \mathbf{M} \boldsymbol{\phi} \ddot{\bar{q}} + \boldsymbol{\phi}_r^T \mathbf{C} \boldsymbol{\phi} \dot{\bar{q}} + \boldsymbol{\phi}_r^T \mathbf{K} \boldsymbol{\phi} \bar{q} = \boldsymbol{\phi}_r^T \bar{F}. \tag{9.6}$$

Because there is no internal damping or internal stiffness forces for the rigid body degrees of freedom, Eq. (9.6) simplifies to

$$\boldsymbol{\phi}_r^T \mathbf{M} \boldsymbol{\phi} \ddot{\bar{q}} = \boldsymbol{\phi}_r^T \bar{F} \tag{9.7}$$

due to

$$\boldsymbol{\phi}_r^T \mathbf{C} = 0 \ \& \ \boldsymbol{\phi}_r^T \mathbf{K} = 0. \tag{9.8}$$

The physical degrees of freedom are substituted for the modal degrees of freedom using the relationship in Eq. (9.4) into Eq. (9.7) to get

$$\boldsymbol{\phi}_r^T \mathbf{M} \ddot{\bar{x}} = \boldsymbol{\phi}_r^T \bar{F}. \tag{9.9}$$

Each row of Eq. (9.9) shows that the weighted sum of the measured accelerations can be equates to the sum of the external forces on the test object for a specific direction. The weights are represented by a linear combination of the mass matrix. The mass matrix is orthogonal with respect to the mode shapes as shown by

$$\boldsymbol{\phi}_r^T \mathbf{M} \boldsymbol{\phi} = [M_r \ 0]. \tag{9.10}$$

The mode shape matrix which includes both rigid body and elastic mode shapes is inverted and multiplied by both sides of Eq. (9.10) to get

$$\boldsymbol{\phi}_r^T \mathbf{M} = [M_r \ 0] \boldsymbol{\phi}^+ \tag{9.11}$$

where the superscript + denotes the pseudo-inverse of the matrix. Equation (9.11) is substituted back into Eq. (9.9) to get

$$[M_r \ 0] \boldsymbol{\phi}^+ \ddot{\bar{x}} = \boldsymbol{\phi}_r^T \bar{F}. \tag{9.12}$$

Equation (9.12) shows that the sum of the weighted accelerations can be used to calculate the sum of the external forces. These forces are the sum of the forces in the three principal translational directions that act at the center of gravity and the three moments around those three principal axes. The weights for the accelerations are the mass properties of the test object multiplied with the rigid body vector components of the pseudo-inverse of the mode shape matrix that includes rigid body and elastic mode shapes.

9.2.2 *Formulation of SWAT-TEEM (Sum of Weighted Acceleration Technique—Time Eliminated Elastic Modes)*

The derivation of the SWAT-TEEM algorithm utilizes the same beginning steps as SWAT until Eq. (9.9). At this point a weighting matrix, $\boldsymbol{w}$, is defined as

$$\boldsymbol{w}^T = \boldsymbol{\phi}_r^T \mathbf{M}. \tag{9.13}$$

and substituted into Eq. (9.9) to obtain

$$\boldsymbol{w}^T \ddot{\bar{x}} = \boldsymbol{\phi}_r^T \bar{F}. \tag{9.14}$$

To solve for the weighting vector, an assumption of the input force is made. In the case where the structure is impacted by an external force and then in a free state, there are no external forces after impact and the accelerations of the system would be in an exponential decay and Eq. (9.14) after the impact simplifies to

$$w^T \ddot{x} = 0. \tag{9.15}$$

To obtain a non-trivial solution, information about the rigid body modes need to be included because they were not present in the free decayed response. The rigid body constraint is formed by post-multiplying Eq. (9.13) by the rigid body shapes to get

$$w^T \boldsymbol{\phi}_r = \boldsymbol{\phi}_r^T \mathbf{M} \boldsymbol{\phi}_r, \tag{9.16}$$

which can be simplified to

$$w^T \boldsymbol{\phi}_r = M_r. \tag{9.17}$$

Equation (9.17) is added to Eq. (9.15) to get

$$w^T \left[\boldsymbol{\phi}_r \; \ddot{x}\right] = \left[M_r \; 0\right]. \tag{9.18}$$

Equation (9.18) is solved for the six weighting vectors in a constrained least squares problem with the rigid body term being the constraint. The pseudo-inverse of the rigid body shapes and the responses of the free decayed response are multiplied by both sides to solve for the weighting vector. These weight vectors are then substituted back into Eq. (9.14) to solve the for the external forces. Because the pseudo-inverse includes the time domain response of the system which are a linear combination of the mode shapes, the mode shapes of the system do not need to be separately calculated.

9.2.3 Condition of the Inverse in SWAT and SWAT-TEEM

The SWAT algorithm inverts the mode shape matrix as shown in Eq. (9.11). Since the response matrix must be inverted, one must be sure that it is not rank deficient, so there must be at least as many sensors as modes in the bandwidth of interest. The inversion of the mode shape matrix has to be well conditioned to avoid error in the computation. The condition number of the mode shape matrix is controlled through examination of the mode shapes and placing accelerometers in a way that the accelerometers span of the space of the mode shapes. This could be done by over instrumenting a modal test or using a finite element model to place accelerometers to minimize the condition of the shapes.

SWAT-TEEM inverts the rigid body mode shapes and the time response of a freely decaying signal after impact. Although the algorithm does not directly invert the mode shapes, the algorithm inverts the responses which contain a linear combination of all the modes excited by the forcing function. Therefore, SWAT-TEEM also needs to have the instrumentation span the space of all mode shapes that are active in the bandwidth of the forcing function.

9.3 Test Item Description

The test article was an 7075 aluminum plate. The dimensions of the plate were 12”×11”×2” and a photograph of the back side of the plate instrumented for the hail impact test can be found in Fig. 9.1. Twenty one 1/4” holes were drilled and tapped at the measured locations so that the instrumentation could be screwed at the desired measured locations. The coordinates of the nodes used in both the characterization and hail impact tests are given in Table 9.1. The node locations were chosen by checking the condition number of the mode shape matrix calculated by a FEM at candidate locations to ensure that the instrumentation produces a full rank matrix. The weight of this plate in this configuration without gauges, wires or tape was 26.497 lbs. Figure 9.1 also shows the direction of the X and Y axes. The Z axis is a resultant of the X and Y axes with respect to the right hand rule. The origin of the coordinate system is at the geometric center of the plate.

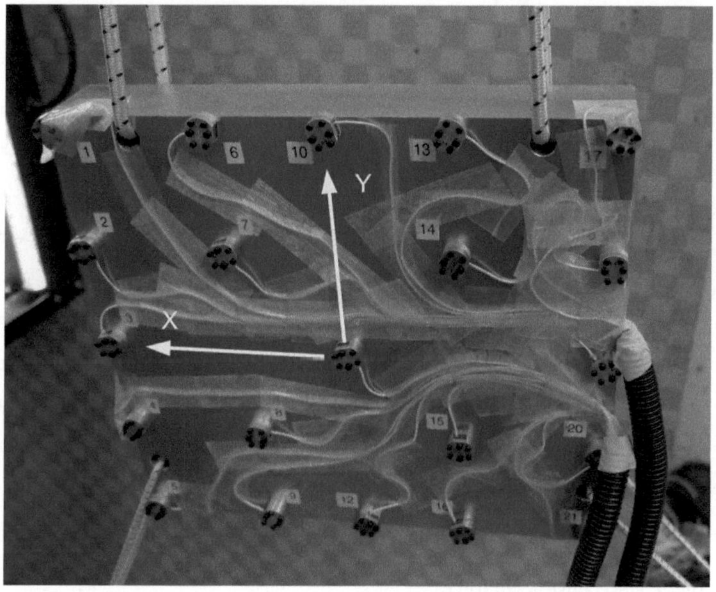

Fig. 9.1 Photo of the plate's nodes and coordinate system

Table 9.1 Coordinates for the plate nodes

Node	Coordinate system	Coordinates X (in.)	Y (in.)	Z (in.)	Node	Coordinate system	Coordinates X (in.)	Y (in.)	Z (in.)
1	Cartesian	5.75	5.25	−1.25	12	Cartesian	0.00	−5.25	−1.25
2	Cartesian	5.75	2.50	−1.25	13	Cartesian	−2.50	5.25	−1.25
3	Cartesian	5.75	0.00	−1.25	14	Cartesian	−2.50	2.50	−1.25
4	Cartesian	5.75	−2.50	−1.25	15	Cartesian	−2.50	−2.50	−1.25
5	Cartesian	5.75	−5.25	−1.25	16	Cartesian	−2.50	−5.25	−1.25
6	Cartesian	2.50	5.25	−1.25	17	Cartesian	−5.75	5.25	−1.25
7	Cartesian	2.50	2.50	−1.25	18	Cartesian	−5.75	2.50	−1.25
8	Cartesian	2.50	−2.50	−1.25	19	Cartesian	−5.75	0.00	−1.25
9	Cartesian	2.50	−5.25	−1.25	20	Cartesian	−5.75	−2.50	−1.25
10	Cartesian	0.00	5.25	−1.25	21	Cartesian	−5.75	−5.25	−1.25
11	Cartesian	0.00	0.00	−1.25					

9.4 Characterization Test

The purpose of the characterization test was to experimentally measure the elastic mode shapes of the aluminum plate for the SWAT analysis. Its other purpose was to compare the SWAT and SWAT-TEEM forces to a measured hammer force to determine the algorithms' accuracy. Three rigid body forces were measured and calculated from this analysis. The three rigid body forces that were calculated were the force in the Z direction, the moment about the Y axis and the moment about the X axis.

The analysis during the characterization portion of this test series consisted of fitting modal parameters to the data so that the mode shapes could be used for SWAT. Also, because a measured force was applied to the plate, a validation test of SWAT and SWAT-TEEM was run to validate the MATLAB code and determine the frequency band limitation of the algorithms by comparing the reconstructed forces to the measured force. All of the modal parameters fit from the data used the Synthesize Modes and Correlate (SMAC) algorithm developed by Hensley and Mayes [4].

Figure 9.2 shows the SWAT and SWAT-TEEM reconstructed forces in the Z direction against the hammer strike in the Z direction. These reconstructions can be examined in the frequency domain in Fig. 9.3. In the frequency domain, it was noted that SWAT-TEEM was an excellent representation of the force up to 9,000 Hz and performed better than SWAT which was an excellent representation of the force up to 4,000 Hz. This cutoff frequency was chosen because the quality of the reconstruction degraded past 10 kHz because the instrumentation did not span the space of the motion past 10 kHz. Therefore, SWAT-TEEM was the algorithm used in the hail impact test.

Another check on the validity of the reconstructed forces would be to calculate the location of the hammer hit by dividing the reconstructed moments by the reconstructed force to get the moment arm distance from the center of gravity of the plate.

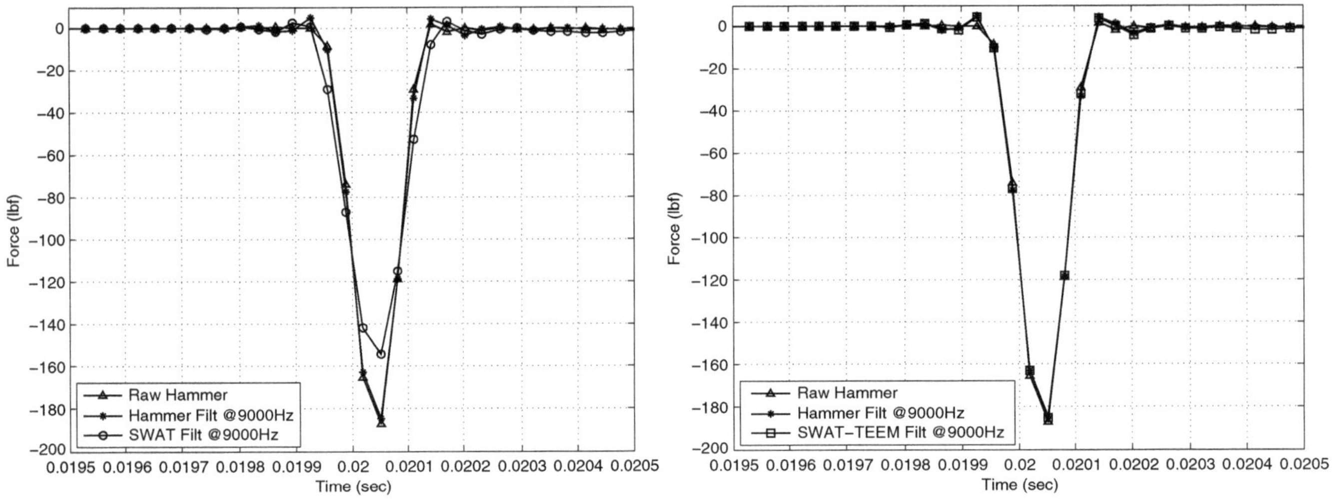

Fig. 9.2 Reconstructed SWAT (*left*) and SWAT-TEEM (*right*) force compared to the hammer hit

Fig. 9.3 Reconstructed forces compared to the hammer hit in frequency domain

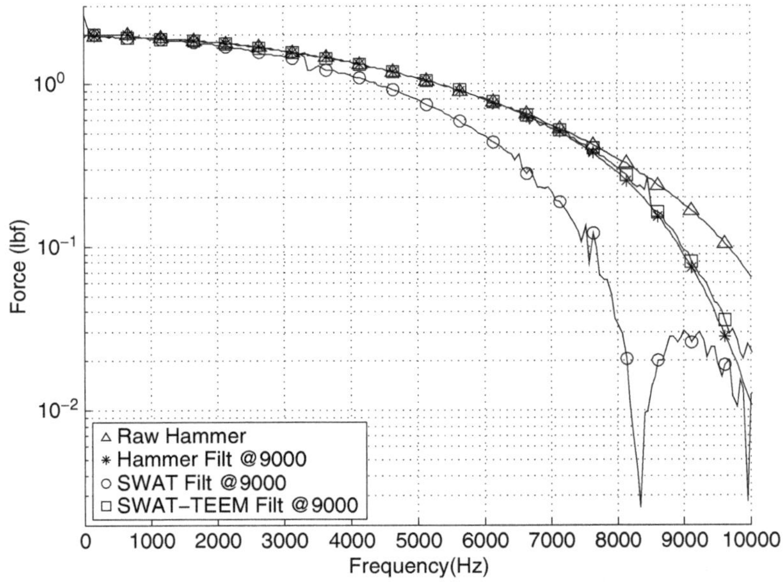

The moment arms were calculated using the force and moments at the time of the peak force. The calculated moment arms were 0.002 in. in the X direction and 0.016 in. in the Y direction. The hammer hit was located at node 11, which had coordinates of approximately (0,0).

Although the characterization test and hail impact test measured three rigid body forces, it was possible to measure all six rigid body forces. This would require more instrumentation to span the space of the in plane rigid body and elastic modes. This would provide input on the tangential forces that were caused by a hail stone. This was not done for this test series due to the lack of resources in instrumentation and data acquisition channels.

9.5 Hail Impact Test

The purpose of this test was to characterize the force of a hailstone striking 7075 aluminum. To characterize the impact, a test series was developed to calculate forces that had different impact speeds and attack angels. Attack angles were defined to be the angle between the travel of the hailstone and the surface of the plate as shown in Fig. 9.4.

The hailstones were made and shot per ASTM F320-10 [1]. The stones were cast into spheres with a nominal 0.8" diameter. At this size, the stones were to have a nominal weight of 4.38 g. They were cast by soaking a small cotton ball with water and placing it in the casting and filling up the rest of the cavity with water.

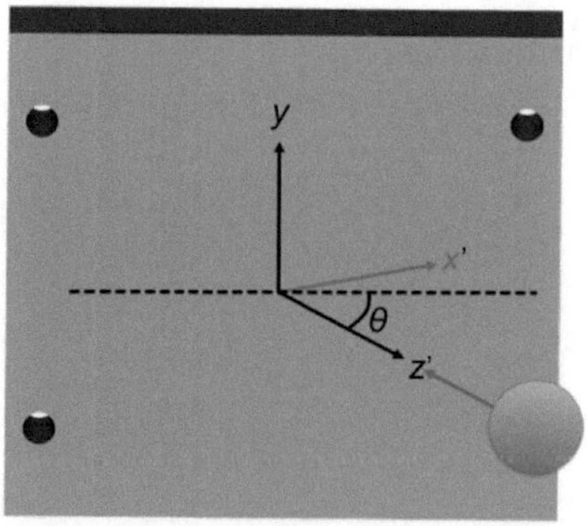

Fig. 9.4 Diagram of how the angle of attack was defined

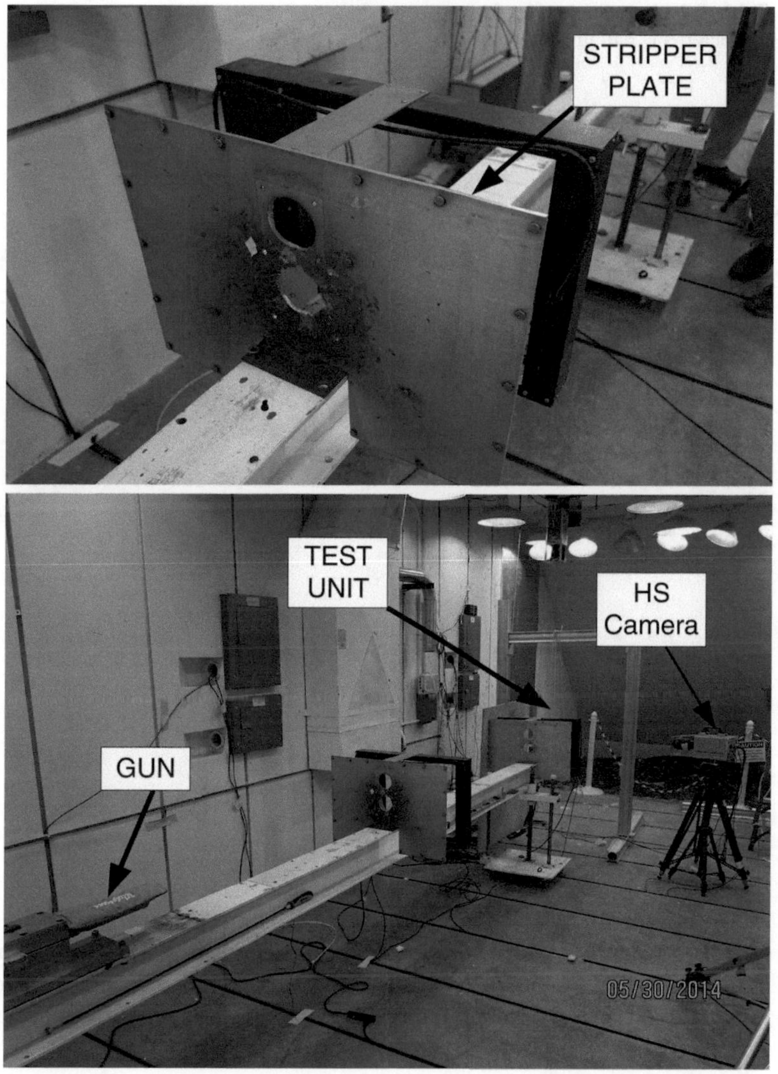

Fig. 9.5 Photo of the terminal ballistics facility setup with stripper plate (*top*) and overall setup (*bottom*)

The hailstone was propelled by shooting a hail stone/pusher assembly out of a gun with gunpowder. After the assembly was shot out of the gun, the pusher would separate from the hail stone and hit a stripper plate. The hail stone would continue flying and hit the instrumented aluminum plate that was suspended from bungees. A photo of this aluminum plate can be seen in Fig. 9.1. High Speed (HS) video was taken of the hail stones hitting the plate and the camera was set up next to the instrumented plate to capture the video. A photo of the stripper plate and the overall setup can be seen in Fig. 9.5. SWAT-TEEM was then used on the plate to calculate the force and moments that the hail stone imparted on the plate.

Fig. 9.6 Photo of a Endevco 7270A-M6 20k gauge

Table 9.2 Actual test series for characterizing hail impact forces

Run number	Angle	Speed $\left(\frac{\text{meter}}{\text{sec}}\right)$	Hail mass (g)	Kinetic energy (J)
6	10°	286	4.29	176
7	10°	272	4.22	156
9	10°	246	4.18	127
11	10°	231	4.31	115
17	10°	506	4.20	537
20	10°	627	4.34	853
22	10°	605	4.28	784
23	10°	639	4.20	858
24	10°	677	4.36	1000
25	10°	582	4.26	721
26	10°	482	4.18	486
27	10°	484	4.27	500
32	45°	252	4.19	133
33	45°	237	4.18	117
34	45°	271	4.41	162
35	45°	526	4.46	616
38	45°	534	4.24	605
40	90°	253	4.33	140
41	90°	263	4.31	149
42	90°	272	4.43	164
43	90°	162	4.39	57
44	90°	150	4.45	50
49	90°	200	4.27	86
50	90°	365	4.41	294

To measure the accelerations of the nodes of the plate during the hail impact environment, Endevco 7270A-M6 20k gauges were screwed into tapped holes on the back side of the plate. A photo of one of these gauges can be seen in Fig. 9.6. These gauges were chosen because they had a high dynamic range and were mechanically isolated which aided in suppressing the internal resonance. A photo of the plate suspended with all of the accelerometers installed can be seen in Fig. 9.1.

9.5.1 Results

The final test matrix can be seen in Table 9.2. Table 9.2 included information on the angle of attack as defined in Fig. 9.4, the measured speed of the hail stone per run, the measured mass of the hail stone per shot, and the magnitude of hail stone's kinetic energy as defined by

$$|KE| = \frac{(mass) \times (speed)^2}{2}. \tag{9.19}$$

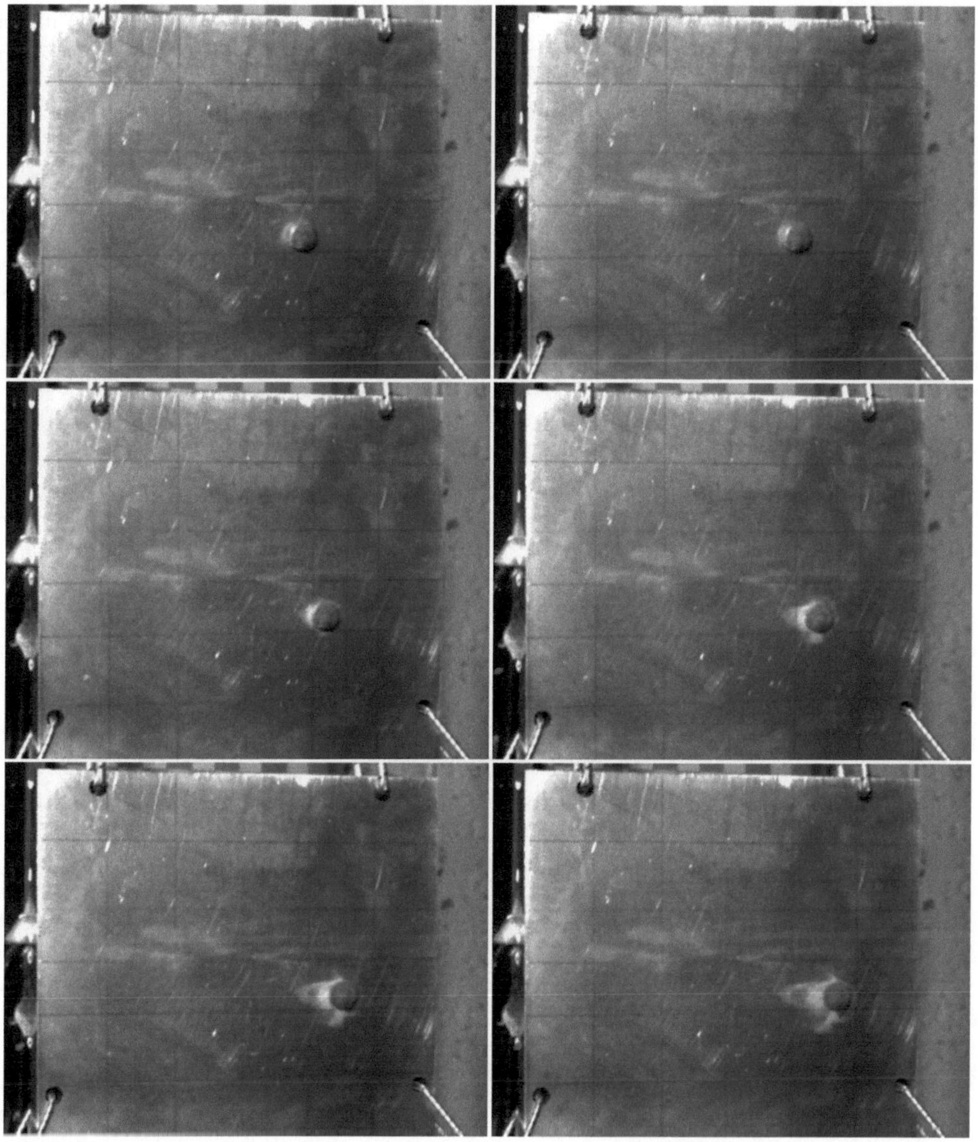

Fig. 9.7 Still frames of Run 11 (time step of 33.3 μs)

For each of the runs listed in Table 9.2, the force in the Z direction, moment about the X axis and moment about the Y axis were calculated with SWAT-TEEM. After examination of the forces in the frequency domain, it was determined that the algorithm was only accurate to 9,000 Hz and a lowpass filter was applied that had a cutoff frequency of 9,000 Hz. This was due to a resonance at 10 kHz not being filtered out by SWAT-TEEM. This filtered frequency was consistent with the results from the characterization test.

Because the translational force in the Z direction and moments about the X and Y axis were measured, the location of the hail impact was calculated as it was in the characterization test. This calculation was done at time of the peak force. This result was then checked against video taken of the hail strike. A grid was drawn on the plate to aid with locating the actual impact location. A sample of snapshots of a video can be found in Fig. 9.7.

The comparison between all of the calculated and measured locations of impact were documented and can be found in Table 9.3. Table 9.3 gives the X and Y coordinates of the calculated and measured impact locations for all of the successful impacts from Table 9.2.

The forces calculated with a 9 kHz low pass filter were concatenated into plots with respect to the impact angle. These results can be found in Figs. 9.8, 9.9, and 9.10. It was noted that there was significant filter ring for all the shots except the low energy 10° shots where most of the energy was contained under 9 kHz.

Table 9.3 Comparison between hail impact location determined by SWAT-TEEM and high speed video

Run number	Difference between Video and SWAT-TEEM	
	Δ X direction (in.)	Δ Y direction (in.)
6	0.0	0.1
7	0.0	0.3
9	0.0	0.1
11	0.1	0.4
17	0.1	0.9
20	0.5	0.1
22	0.2	0.2
23	0.3	0.3
24	0.5	0.5
25	0.4	0.8
26	0.3	0.2
27	0.4	0.2
32	0.0	0.2
33	0.3	0.7
34	0.3	0.3
35	0.0	0.3
38	0.9	0.9
40	0.1	0.3
41	0.2	0.1
42	0.1	0.4
43	0.2	0.4
44	0.0	0.3
49	0.3	0.4
50	0.2	0.1

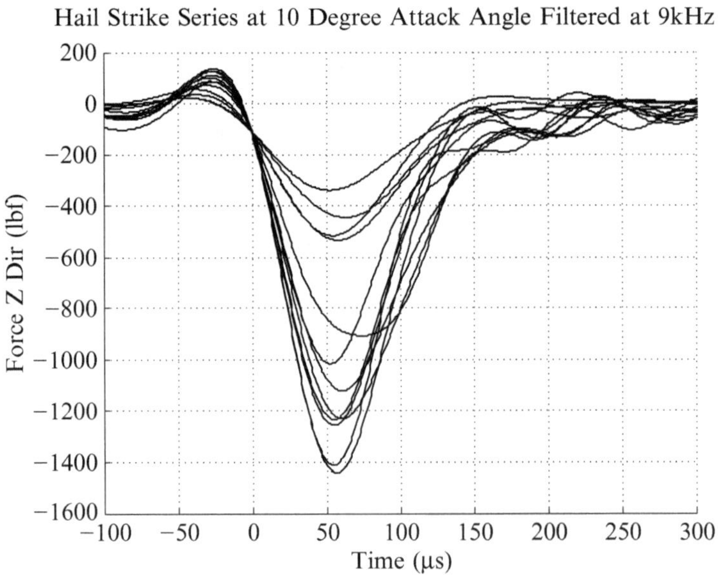

Fig. 9.8 Summary of the forces in the time domain with a 10° impact angle

The peak force of the impact was also analyzed with respect to the kinetic energy of the hail stone. Figure 9.11 displays all of the shots with the peak force as a function of the kinetic energy of the hail stone. The figure also contains linear regression lines that were fit to each of the impact angles. It can be seen in Fig. 9.11 that there was more variability shot to shot for the 10° impact angle shots, however, the trends for every angle of attack appeared to be linear.

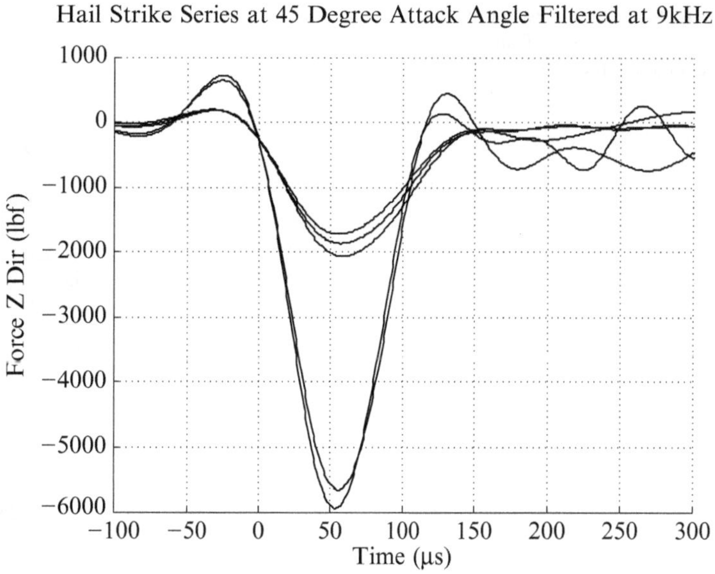

Fig. 9.9 Summary of the forces in the time domain with a 45° impact angle

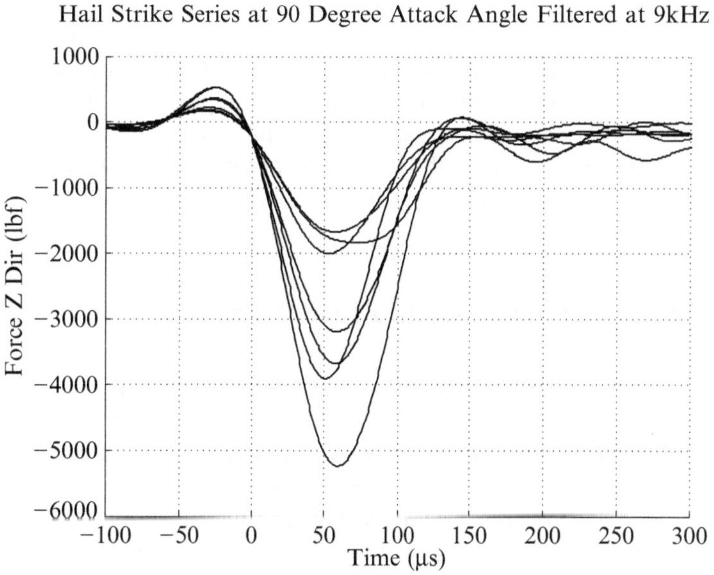

Fig. 9.10 Summary of the forces in the time domain with a 90° impact angle

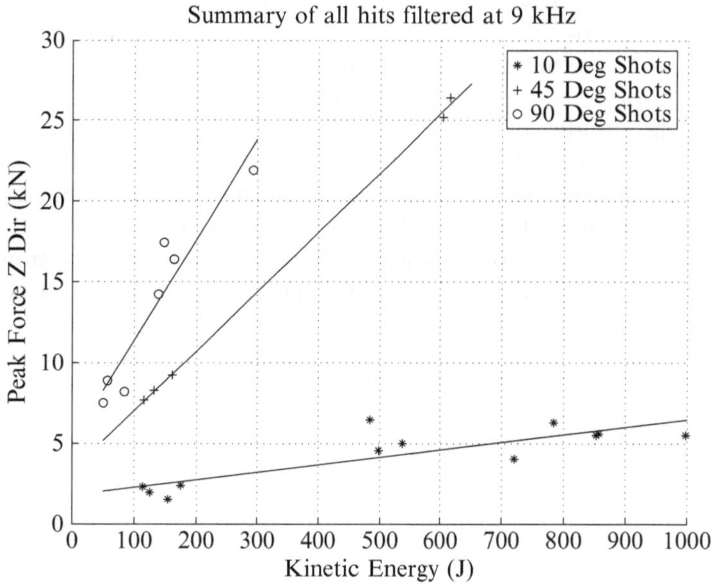

Fig. 9.11 Characterization of peak force with respect to kinetic energy of all runs

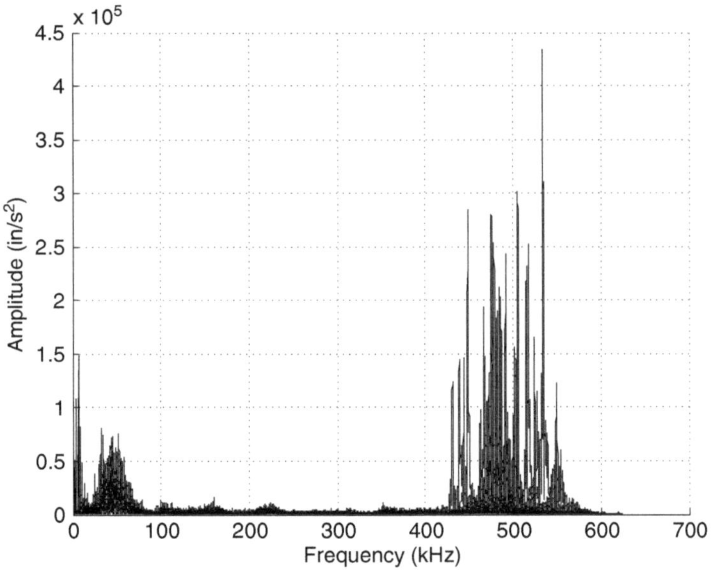

Fig. 9.12 Frequency domain of accelerometers for Run 50 displaying the internal resonances

Higher speed shots were planned for attack angles of 45° and 90°. However, the Endevco 7270A-M6 20k gauges began to overload due to their internal resonances being excited from the high frequency input from the hail stone traveling at high speeds and the impulse being shorter. Figure 9.12 shows the frequency responses of the accelerometers from Run 50 which shows significant energy at the internal resonance frequencies. The fact that the internal resonances were excited was significant because the M6 configuration of the 7270 was mechanically isolated so that the internal resonances weren't supposed to be excited as much as the original configuration.

9.6 Conclusions and Future Work

The characterization test showed the useful bandwidths of the SWAT and SWAT-TEEM algorithms for the accelerometer configuration was 9,000 Hz for SWAT-TEEM and 4,000 Hz for SWAT. The characterization test also revealed that the SWAT-TEEM algorithm had a superior frequency response for the hail impact test.

Man-made hail stones were created and shot at an aluminum plate. The forces caused by the hail impacts were calculated through the SWAT-TEEM algorithm. The forces were filtered at 9,000 Hz due to the limitation of the measured degrees of freedom spanning the space of the mode shapes outside that bandwidth.

Forces were calculated for hail impacts at different angles and speeds. The speeds ranged from around 152 m/s to 671 m/s. The angle of attacks were measured from the plane of the plate and were 10°, 45° and 90° (normal). The kinetic energy in the Z direction was calculated for each shot and was plotted with respect to the peak force that it caused. This relationship showed to be linear for each angle of attack.

It was determined after the test that a variety of speeds would have been useful so that more statistics could be performed on the linearity of the relationship between the kinetic energy of the hail stones versus the peak force applied.

The analysis throughout this paper calculated three of the six degree of freedom forces, translation in the Z direction, moment about the X axis and moment about the Y axis. With the proper instrumentation, all six rigid body forces could had been measured. This would had provided the friction component of the impact for the non-normal impacts.

Due to the internal resonances being excited in the accelerometers, data was not taken on the high speed large angle setups. Gauges that had more damping on its natural frequency should have been used so that data could have been acquired.

References

1. ASTM Standard F320-10 (2010) Standard test method for hail impact resistance of aerospace transparent enclosures. ASTM International, West Conshohocken. Doi: 10.1520/F0320-10. http://www.astm.org
2. Carne TG, Bateman VI, Dohrmann CR (1990) Force reconstruction using the inverse of the mode-shape matrix. Technical Report SAND90–2737, Sandia National Laboratories
3. Gregory DL, Priddy TG, Smallwood DO (1986) Experimental determination of the dynamic forces acting on non-rigid bodies. In: SAE Technical Paper Series, Paper No. 861791, Aerospace Technology conference and Exposition, Long Beach, CA, October 1986
4. Hensley DP, Mayes RL (2006) Extending smac to multiple references. In: Proceedings of the 24th International Modal Analysis Conference, pp. 220–230, January 2006
5. Kim H, Welch DA, Kedward KT (2003) Experimental investigation of high velocity ice impacts on woven carbon/epoxy composite panels. Compos A Appl Sci Manuf 34(6):25–41
6. Mayes RL (1994) Measurement of lateral launch lads on re-entry vehicles using swat. In: Proceedings of the 12th International Modal Analysis Conference, pp. 1063–1068, Feburary 1994
7. Pereira JM, Padula SA, Revilock DM, Melis ME (2006) Forces generated by high velocity impact of ice on a rigid structure. Technical Report NASA/TM-2006-214263, National Aeronautics and Space Administration
8. Tippmann JD, Kim H, Rhymer JD (2013) Experimentally validated strain rate dependent material model for spherical ice impact simulation. Int J Impact Eng 57:43–54

Chapter 10
Testing and Validation of the Dynamic Inertia Measurement Method

Alexander W. Chin, Claudia Y. Herrera, Natalie D. Spivey, William A. Fladung, and David Cloutier

Abstract The Dynamic Inertia Measurement (DIM) method uses a ground vibration test setup to determine the mass properties of an object using information from frequency response functions. Most conventional mass properties testing involves using spin tables or pendulum-based swing tests, which for large aerospace vehicles becomes increasingly difficult and time-consuming, and therefore expensive, to perform. The DIM method has been validated on small test articles but has not been successfully proven on large aerospace vehicles. In response, the National Aeronautics and Space Administration Armstrong Flight Research Center (Edwards, California) conducted mass properties testing on an "iron bird" test article that is comparable in mass and scale to a fighter-type aircraft. The simple two-I-beam design of the "iron bird" was selected to ensure accurate analytical mass properties. Traditional swing testing was also performed to compare the level of effort, amount of resources, and quality of data with the DIM method. The DIM test showed favorable results for the center of gravity and moments of inertia; however, the products of inertia showed disagreement with analytical predictions.

Keywords Dynamic inertia measurement • Frequency response function • Ground vibration test • Mass properties • Moment of inertia

Nomenclature

AFRC	Armstrong Flight Research Center
CAD	Computer-aided design
CG	Center of gravity
CRV	Crew return vehicle
DFRC	Dryden Flight Research Center
diff	Difference
DIM	Dynamic inertia measurement
DOF	Degree of freedom
ETA	Engineering test article
F	Force
FRF	Frequency response function
F_x	Force in x-axis
F_y	Force in y-axis
F_z	Force in z-axis
g	Gravitational acceleration
GVT	Ground vibration test
Hz	Hertz
I	Inertia matrix
IMAT	Interface between MATLAB® Analysis, and Test
I_{xx}	Moment of inertia about the x-axis

A.W. Chin (✉) • C.Y. Herrera • N.D. Spivey
NASA Armstrong Flight Research Center, P.O. Box 273, Edwards Air Force Base, CA 93523, USA
e-mail: alexander.w.chin@nasa.gov; claudia.herrera-1@nasa.gov; natalie.d.spivey@nasa.gov

W.A. Fladung • D. Cloutier
ATA Engineering, Inc., 13290 Evening Creek Dr. South, Ste. 250, San Diego, CA 92128, USA
e-mail: bfladung@ata-e.com; dave.cloutier@ata-e.com

A. Wicks (ed.), *Shock & Vibration, Aircraft/Aerospace, and Energy Harvesting, Volume 9*,
Conference Proceedings of the Society for Experimental Mechanics Series, DOI 10.1007/978-3-319-15233-2_10

I_{xy}	Product of inertia between x and y-axis
I_{xz}	Product of inertia between x and z-axis
I_{yy}	Moment of inertia about the y-axis
I_{yz}	Product of inertia between y and z-axis
I_{zz}	Moment of inertia about the z-axis
I_{xz}	Moment of inertia about the xz-axis
L	Length
m	Mass
M	Mass matrix
MOI	Moment of inertia
NACA	National Advisory Committee for Aeronautics
NASA	National Aeronautics and Space Administration
N_x	Moment about the x-axis
N_y	Moment about the y-axis
N_z	Moment about the z-axis
POI	Product of inertia
PSMIF	Power spectrum mode indicator function
T	Period
w	Weight
x	Time or frequency
$\ddot{x}$	Linear acceleration in x-axis
X_{CG}	Center of gravity location along the x-axis
Y_{CG}	Center of gravity location along the y-axis
Z_{CG}	Center of gravity location along the z-axis
$\ddot{y}$	Linear acceleration in y-axis
$\ddot{z}$	Linear acceleration in z-axis
$\ddot{\theta}$	Angular acceleration
$\ddot{\theta}_x$	Angular acceleration about x-axis
$\ddot{\theta}_y$	Angular acceleration about y-axis
$\ddot{\theta}_z$	Angular acceleration about z-axis
60K3S	60,000-lb Starr Soft Support

10.1 Introduction

The mass properties of a vehicle include the mass, center of gravity (CG), moments of inertia (MOIs), and products of inertia (POIs). This information is important to understanding and controlling the flight dynamics of the vehicle. The mass and CG can usually be determined through a weight and balance procedure, while MOI and POI require dynamic testing. Analytical models can also provide mass properties information, but must be sufficiently detailed as a realistic representation of the system to be accurate. As examples, vehicle modifications and changes are not always tracked in the computer-aided design (CAD) analytical model; and vehicles can be acquired without their corresponding analytical models. If changes to the vehicle are not explicitly modeled, the analytical mass properties will not be accurate. Remodeling an entire vehicle, however, can be cumbersome and costly to complete. Thus, it becomes necessary to experimentally test for the mass properties of a vehicle.

Spin-balance tables can provide accurate approximations of the CG and MOI, but these become increasingly difficult to use as the size of the object being tested increases. When spin-balance tables are not available or practical, pendulum-based methods are often used; however, pendulum-based methods require significant amounts of labor, materials, and time, leading to high cost and risk.

Frequency response function (FRF) testing has gained interest as an alternative methodology for determining mass properties using a ground vibration test (GVT) setup. Frequency response function testing analyzes the dynamic response of a test article and is often used to identify mode shapes and natural frequencies of objects. The Dynamic Inertia Measurement (DIM) method utilizes FRF information to determine mass properties [1].

The DIM method has been in development at the University of Cincinnati and has shown success on a variety of small-scale test articles such as automobile brake rotors, steel blocks, and other custom fixtures from the university [2, 3]. Attempts to apply the DIM method to larger test articles, however, have met with limited success [4]. This paper documents efforts to mature the DIM technology for application toward full-scale aerospace vehicles in conjunction with GVTs.

10.2 DIM Background Theory

The mass properties of an object are determined by measuring all forces and moments acting on a body and the rigid body motion caused by these forces and moments. The DIM method measures the inertia properties of an object by analyzing the FRFs measured during a GVT. The FRFs are measurements that normalize the response (acceleration and reaction force) to the excitation force in the frequency domain. A simulated free-free boundary condition GVT provides the appropriate test environment such that all the reaction forces can be measured. The advent of six degree-of-freedom (DOF) force sensors has enabled the measurement of all the reaction forces and moments on the test article [5–7].

The DIM method uses the rigid body forces, moments, and linear and angular accelerations to calculate the inertia matrix. Equation 10.1 shows Newton's second law simplified for constant mass, which defines the relationship between forces, mass, and linear accelerations.

$$\{F\} = [M]\{\ddot{x}\} \tag{10.1}$$

Equation 10.2 shows Euler's second law for defining the relationship between moments, moments and products of inertia, and angular accelerations. For this solution, the cross terms were ignored because the test articles are assumed to be rigid to an extent that the vehicle rotation rate terms were small. Note that this assumption would not hold for large, flexible structures.

$$\{N\} = [I]\left\{\ddot{\theta}\right\} \tag{10.2}$$

Applying the small angle assumption to the moment arms and combining the force and moment equations for six degrees of freedom yields the 6×6 mass matrix for full rigid body motion as shown in Equation 10.3. All forces, moments, and accelerations are measured quantities. The forces and moments are measured from DIM-related sensors. The accelerations are measured from GVT sensors. The ten unknown terms in the mass matrix (M) are the mass (m), CG location (X_{CG}, Y_{CG}, Z_{CG}) with respect to some point P, moments of inertia (I_{xx}, I_{yy}, I_{zz}) calculated about P, and products of inertia (I_{xy}, I_{xz}, I_{yz}) calculated about P.

$$\left\{\begin{matrix} F_x \\ F_y \\ F_z \\ N_x \\ N_y \\ N_z \end{matrix}\right\} = \begin{bmatrix} m & 0 & 0 & 0 & mZ_{CG} & -mY_{CG} \\ 0 & m & 0 & -mZ_{CG} & 0 & mX_{CG} \\ 0 & 0 & m & mY_{CG} & -mX_{CG} & 0 \\ 0 & -mZ_{CG} & mY_{CG} & I_{xx} & -I_{xy} & -I_{xz} \\ mZ_{CG} & 0 & -mX_{CG} & -I_{xy} & I_{yy} & -I_{yz} \\ -mY_{CG} & mX_{CG} & 0 & -I_{xz} & -I_{yz} & I_{zz} \end{bmatrix} \left\{\begin{matrix} \ddot{x} \\ \ddot{y} \\ \ddot{z} \\ \ddot{\theta}_x \\ \ddot{\theta}_y \\ \ddot{\theta}_z \end{matrix}\right\} \tag{10.3}$$

10.3 Previous DIM Validation Efforts on Large-Scale Test Articles

The National Aeronautics and Space Administration (NASA) Dryden Flight Research Center (DFRC) which was recently renamed to the NASA Armstrong Flight Research Center (AFRC) attempted to apply the DIM method on two large-scale test articles. The first test article was the X-38 Crew Return Vehicle (CRV); the second test article was an assembly of steel I-beams and plates referred to as the "initial iron bird."

Fig. 10.1 The X-38 CRV undergoing DIM testing

10.3.1 X-38 Crew Return Vehicle

In the late 1990s NASA DFRC researchers applied the DIM two separate times on the X-38 CRV. Challenges arose due to flexible modes of internal vehicle components. Although the DIM technology clearly required further maturation, the results compared favorably with those obtained through experimental testing. Figure 10.1 shows the X-38 CRV undergoing DIM testing.

10.3.2 Initial Iron Bird Test Article

During the summer of 2010, an experiment the sole purpose of which was to validate the DIM method for large-scale vehicles was conducted at the NASA DFRC. The test article was an iron bird assembled from an assortment of welded steel beams and plates and referred to as the initial iron bird.

This testing effort supported the procurement of the specialized instrumentation needed to perform the DIM test and the development of the DIM processing algorithms.

A laser tracker was used for precise measurements of the locations and orientations of the sensors. Several sets of GVT accelerometers were installed and multiple test conditions were performed. Some problems were experienced during the test and analysis. In order to interface the initial iron bird to the soft supports, steel pedestals were inserted between the soft supports and the initial iron bird, because of soft-support extension limits. The pedestals and soft supports introduced local modes that limited the frequency range at which the DIM method could be applied and analyzed. In addition, the initial iron bird was not ideal for highly accurate analytical mass property predictions because it consisted of an assortment of welded plates having various holes.

Fig. 10.2 The initial iron bird test article undergoing DIM testing

Fig. 10.3 The iron bird test article

The results of the effort were mixed, with several lessons learned to improve the validation effort. The biggest problems to correct included avoiding the appearance of local modes and creating an experimental test article to yield high confidence in its analytical mass properties. Figure 10.2 shows the DIM test setup from the initial iron bird test article.

10.4 Iron Bird Test Article

Two 8500-lb, 20-ft long, W14X426 steel I-beams were bolted together off-center to model the approximate mass of a fighter-type aircraft. The test article which was dubbed the "iron bird;" is shown in Fig. 10.3. The iron bird was intentionally simple in design to ensure high reliability of its analytical mass properties. Holes were drilled into the iron bird at designated

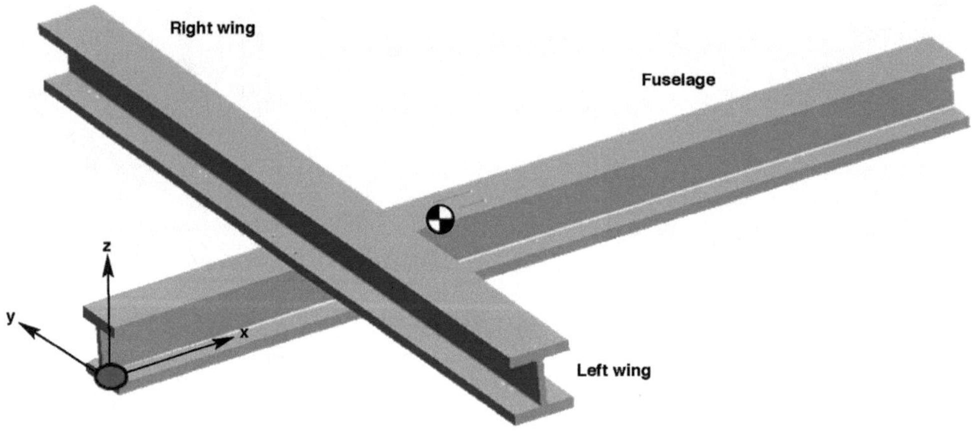

Fig. 10.4 The CAD model showing the coordinate system for the iron bird

Table 10.1 The CAD model analytical mass properties of the iron bird

Mass, kg	X_{CG}, m	Y_{CG}, m	Z_{CG}, m	I_{xx}, kg*m^2	I_{yy}, kg*m^2	I_{zz}, kg*m^2	I_{xy}, kg*m^2	I_{xz}, (kg*m^2)	I_{yz}, kg*m^2
7,716.5	2.3	0.0	0.5	12,700.6	17,207.2	28,503.1	0.0	−1,398.8	0.0

locations to attach interface ball plates that were to rest on aircraft jacks and the soft-support system. Holes were also drilled to attach any required fixture (s) for MOI pendulum testing. Steel was chosen for its rigidity and lower cost.

The reference coordinate system used for the iron bird places the origin at the forward lower surface of the test article. Positive X points aft from the nose, positive Y points in the direction of the right wing, and positive Z points upward on the test article. The coordinate system is defined in the CAD model shown in Fig. 10.4.

10.5 Test Plan

Three methods for determining the mass properties of the 17,000-lb iron bird test article were used for comparison. First, an analytical model was created using the solid modeling CAD program PTC Creo (Pro/ENGINEER®) (PTC Inc., Needham, Massachusetts). Second, pendulum swing tests were performed. Third, the DIM method was implemented. These three separate and independent approaches were used to document the level of effort involved for each method. For expediency, however, the pendulum swing tests relied on the analytical vertical Z_{CG}. The DIM test obtained data completely independently of any analytical and pendulum mass properties data.

10.6 Analytic Model

Pro/ENGINEER® was used to analytically model the iron bird test article and obtain the mass properties. Care was taken to apply as many realistic details to the CAD model as possible, including all holes and added interface attachments. The simplicity of the iron bird test article design was to ensure the analytical CAD model could be treated as the *truth model*. Table 10.1 shows the CAD model analytic mass properties of the iron bird.

10.7 Pendulum Swing Test

Weight and balance measurements in conjunction with pendulum swing tests have been the conventional experimental method for obtaining mass properties of aerospace vehicles for almost a century. This analytically straightforward method focuses on measuring weight, periods, and pendulum lengths. Many problems can develop, however, when trying to meet the requirements for a successful test. The stiffness and integrity of the lifting hardware must be certified for NASA or any other lifting operations. In addition, the pendulum test hardware must be nearly frictionless to ensure minimal damping for accurate pendulum period measurements.

Table 10.2 Experimental weight, X_{CG}, and Y_{CG}

	Measured	Analytic	Diff analytic, %	Uncertainty
Mass, kg	7, 698.8	7,716.5	−0.23	+/−47.36
X_{CG}, m	2.286	2.286	0.00	$+/-2.09\times10^{-3}$
Y_{CG}, m	−0.000762	0	n/a	$+/-2.07\times10^{-3}$

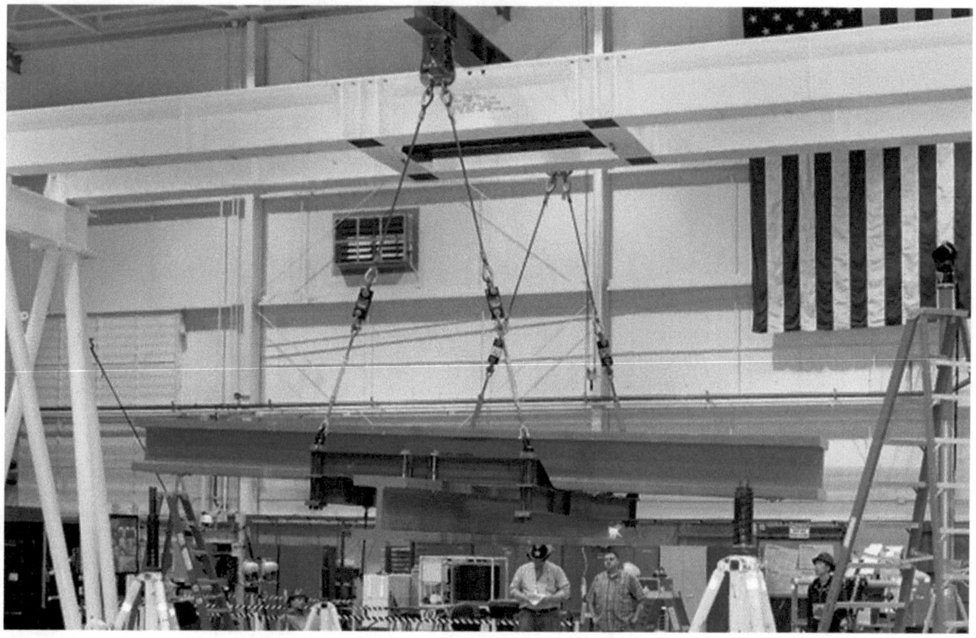

Fig. 10.5 The iron bird undergoing I_{xx} pendulum testing

The X_{CG} and Y_{CG} were measured using a balance of forces calculation. The Z_{CG} was not experimentally verified due to the level of effort required to tilt the iron bird at multiple angles for Z_{CG} calculation. The experimentally-measured weight, X_{CG}, and Y_{CG} are shown in Table 10.2. Uncertainties were calculated using the law of propagation of uncertainty methodology.

Classical pendulum equations were used to determine the moments of inertia. The methodology and equations referenced multiple National Advisory Committee for Aeronautics (NACA) papers [8–13]. In order to obtain the moments of inertia of the iron bird, all tests also required swinging the fixture by itself in order to subtract out the fixture mass properties from the total combined iron bird and fixture assembly. The moments of inertia about the x-axis and y-axis used a compound pendulum setup and is calculated from Equation 10.4, where terms with a subscript of “1” represent the combined test article and fixture; terms with a subscript of “2” represent the fixture only; and “L” is the length from the pivot point to the CG of the respective subscripts.

$$I_{xx\ ETA} = \left[\frac{w_1 T_1{}^2 L_1}{4\pi^2} - \frac{w_1 L_1{}^2}{g}\right] - \left[\frac{w_2 T_2{}^2 L_2}{4\pi^2} - \frac{w_2 L_2{}^2}{g}\right] \tag{10.4}$$

The z-axis MOI uses a bifilar torsional pendulum setup and is calculated from Equation 10.5. The “L” term is the suspended length from the pivot point.

$$I_{zz\ ETA} = \frac{T_1{}^2 w_1 A^2}{16\pi^2 L} - \frac{T_2{}^2 w_2 A^2}{16\pi^2 L} \tag{10.5}$$

Figure 10.5 shows the iron bird test article undergoing I_{xx} pendulum testing.

Figure 10.6 shows the iron bird test article undergoing I_{zz} pendulum testing.

Obtaining the product of inertia, I_{xz}, required the iron bird to rotate in the bifilar pendulum in a tilted pitch configuration. Due to limitations of the test setup, only a tilt of 5 deg was achievable, instead of the minimum 15 deg needed. This condition led to a large I_{xz} error compared to the analytical predictions. The results of the iron bird MOI and POI are summarized in Table 10.3. The large I_{xz} uncertainty comes from compounding uncertainties from I_{xx} and I_{zz} measurements. A small tilt angle exacerbates the uncertainty as well.

Fig. 10.6 The iron bird undergoing I_{zz} pendulum testing

Table 10.3 Summary of iron bird pendulum MOI results

	Measured MOI, kg*m^2	Analytic MOI, kg*m^2	Diff analytic, %	Uncertainty, %
I_{xx} iron bird	12,265	12,708	−3	8
I_{yy} iron bird	17,932	17,196	4	10
I_{zz} iron bird	27,848	28,507	−2	3.40
I_{xz} iron bird	−2,767	−1,398	98	242.70

Several difficulties with the test setup and the data were encountered using the pendulum method. Weight exceeding safety *close calls* was an ever-present concern. Oscillations damped out more quickly than was expected and the frequency oscillation was not constant. Post-test analysis revealed that the main culprit for the observed frequency changes was the non-ideal design of the pendulum test hardware.

10.8 Dynamic Inertia Measurement Test

The iron bird DIM testing was conducted at the NASA AFRC Flight Loads Laboratory (FLL) from September 16, 2013 through September 24, 2013. ATA Engineering, Inc. (San Diego, California) was contracted to assist with the iron bird DIM testing and to perform analysis of the data, in order to utilize their previous experience with the DIM method [14, 15]. The equipment required and test description required for the DIM method test is described below.

10.8.1 Equipment

The DIM method requires many of the same sensors (accelerometers, force transducers, and soft-support load cells) and equipment (shakers, soft-support system) that are needed to perform a GVT. In addition to the GVT equipment, a few specialized sensors, such as 6-DOF force sensors, 3-DOF force transducers, laser tracker, and seismic accelerometers, are required for DIM. The 6-DOF force sensors ensure all reaction forces and moments needed for the DIM calculation are

Fig. 10.7 Left, the 6-DOF reaction force sensor; and right, the 3-DOF soft support load cell

measured. The seismic accelerometers have a higher sensitivity and lower noise floor which may be needed to accurately record low-amplitude responses. The DIM processing will evaluate and compare the need for the seismic accelerometers versus typical GVT accelerometers. A data acquisition system was used to acquire and record data from all the sensors installed during DIM testing. Each sensor location and orientation was measured with a laser tracker.

10.8.1.1 Soft-Support System

The 60,000-lb Starr Soft Support (60K3S) system is a specialized piece of equipment that acts as both an aircraft jack and a nitrogen bladder soft-support system. The soft suspension system was used to support the iron bird in order to simulate free-free boundary conditions as accurately as possible. The soft-support isolation system consists of an aircraft-jacking device with three jacking points, each point having an individual motor and accommodating up to 20,000 lb, for a total capacity of 60,000 lb. The 60K3S system can be transported to an aircraft by forklift and placed at the jacking points using a pallet jack. The motors power the electric actuators, raising the aircraft off the ground until the landing gear can retract [16].

10.8.1.2 3-DOF Load Cells: Soft Supports

On each of the soft supports, a 3-DOF load cell (Interface 5200 series) (Interface Inc., Scottsdale, Arizona) was used to monitor the 60K3S side loads and ensure that load limits were not exceeded. This information was not used in the DIM calculation. These load cells are part of the standard instrumentation for using the soft supports for GVTs. The 3-DOF 60K3S load cells are shown in Fig. 10.7.

10.8.1.3 6-DOF Force Sensors: Reaction Forces

Three 6-DOF force sensors were custom-made for the NASA AFRC researchers by PCB Piezotronics, Inc. (Depew, New York). These unique sensors are an assembly of three 3-DOF piezoelectric dynamic force sensors. The load cells were placed between the iron bird and the soft-support system. The 6-DOF force sensors are shown in Fig. 10.7.

10.8.1.4 Shakers: Excitation

The DIM testing required up to two shakers for excitation during random and sine-sweep test runs. For all configurations, either one or two 110-lb MB Dynamics (Cleveland, Ohio) electrodynamic shakers were used to excite the structure. The shakers used plastic stingers to transmit a direct force with minimal bending moment. The stingers are rigid enough to transmit unattenuated axial force within the frequency range of interest. The stingers completed the connections between the force transducer and shakers. For tests runs using two shakers, the shakers were attached to the structure at a skewed angle. Shaker blocks with a 45-deg angle were attached to the structure, and then the force transducer was attached to the angled side of the block. A picture of a shaker is shown in Fig. 10.8.

Fig. 10.8 An MB Dynamics (Cleveland, Ohio) shaker

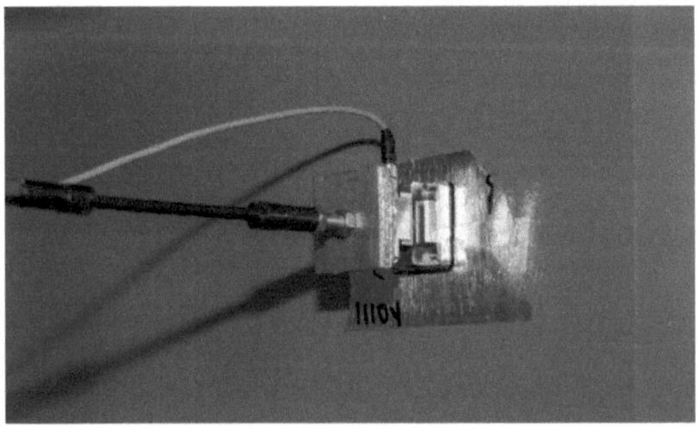

Fig. 10.9 A 3-DOF excitation force transducer

10.8.1.5 3-DOF and 1-DOF Force Sensors: Excitation

A 3-DOF excitation force sensor (PCB M/N 261A01) (PCB Piezotronics, Inc.) was used to measure the input force from the shaker. Only the force in-line with the shaker was used as the reference for the FRF measurements. The forces in the other two axes were treated as reaction forces. The transverse forces were typically only a few percent of the in-line force. A 3-DOF excitation force sensor is shown in Fig. 10.9.

10.8.1.6 Accelerometers

Three sets of accelerometers were installed for the DIM test: seismic accelerometers, single-axis accelerometers, and triaxial accelerometers. The primary set of DIM accelerometers consisted of single-axis seismic accelerometers (PCB M/N 393B04) (PCB Piezotrinics, Inc.) mounted in a triaxial configuration. These transducers were selected for their high sensitivity, low noise floor, and low-frequency characteristics to evaluate the necessity to capture acceleration for more massive objects. The second set consisted of single-axis 100 mV/g accelerometers (PCB M/N T333B) (PCB Piezotronics, Inc.) typically used for modal tests, which were also mounted in a triaxial configuration collocated with the seismic accelerometers. The purpose of this redundancy was to compare the DIM processing results between these two types of sensors and evaluate the adequacy of typical modal accelerometers versus the expensive seismic accelerometers. The third set consisted of 100 mV/g triaxial accelerometers (PCB M/N 356A16) (PCB Piezotronics, Inc.) mounted on the wings, fuselage, and soft supports for modal information. The first two sets of accelerometers installed for the DIM test were defined using a local Cartesian coordinate system aligned with the global Cartesian coordinate system. The third set was installed in the global Cartesian coordinate system for modal situational awareness and not used in the DIM calculation. A laser tracker system was used to determine the accelerometer locations. Figure 10.10 shows the three seismic and three T333B (PCB Piezotronics, Inc.) single-axis accelerometers, mounted in a triaxial configuration, and the triaxial accelerometers.

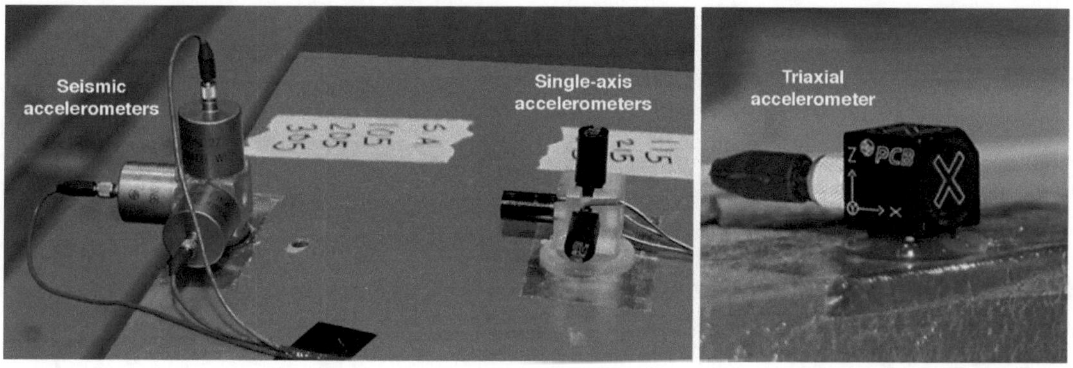

Fig. 10.10 Left, the three seismic and three single-axis accelerometers; and right the triaxial accelerometer

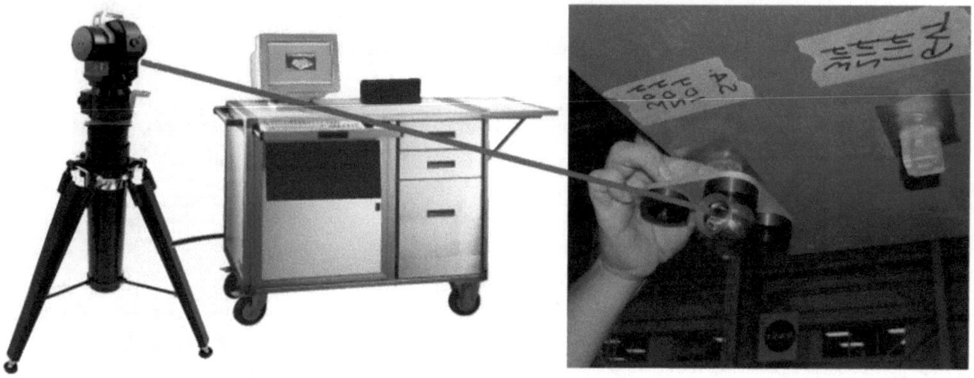

Fig. 10.11 The laser tracker system for sensor locations

10.8.1.7 Laser Tracker System

The DIM method involves resolving the measured DOFs to rigid body accelerations and forces at a single point. Performing this resolution requires the accurate position and orientation of the measured accelerations, excitation forces, and reaction forces. In the most general sense, a procedure was developed to create a local coordinate system at each measurement node with respect to the global coordinate system for all of the DIM sensors of interest. The NASA AFRC has a laser tracker system that can accurately and precisely measure the coordinates of a point in space. This system was used to measure the accelerometers, 6-DOF force sensors, and force transducers using a custom tool designed to capture three points relative to a corner of the sensor to be measured, as shown in Fig. 10.11. The output text file from the laser tracker software was then used to automatically generate an MSC Nastran (MSC. Software Corporation, Newport Beach, California) bulk data file, and the output of the Nastran run provided the transformation matrices from the measurement DOFs to the rigid body node. This method decreased the possibility of errors due to misalignment and facilitated the bookkeeping of the sensor local coordinate systems.

10.8.2 Test Model

The final set of accelerometers totaled 144 DOFs at 51 different node locations. Even though all test DOFs were nominally coincident with the global Cartesian coordinate system, a laser tracker was used to precisely define a local displacement coordinate system for each sensor location used for the DIM calculations. The laser tracker was used to define local displacement coordinate systems for the collocated DIM accelerometers (seismic and T333B) (PCB Piezotronics, Inc.), the 3-DOF force transducer and the 6-DOF reaction force sensors. The DOFs for the test article and soft-support system sensor locations which were not used in the DIM calculations were measured by the laser tracker but were defined in the global coordinate system. Figure 10.12 shows the test display model and accelerometer locations. Each arrow in the figure represents the DOFs at an accelerometer location. The blue elements represent the three instrumented soft supports.

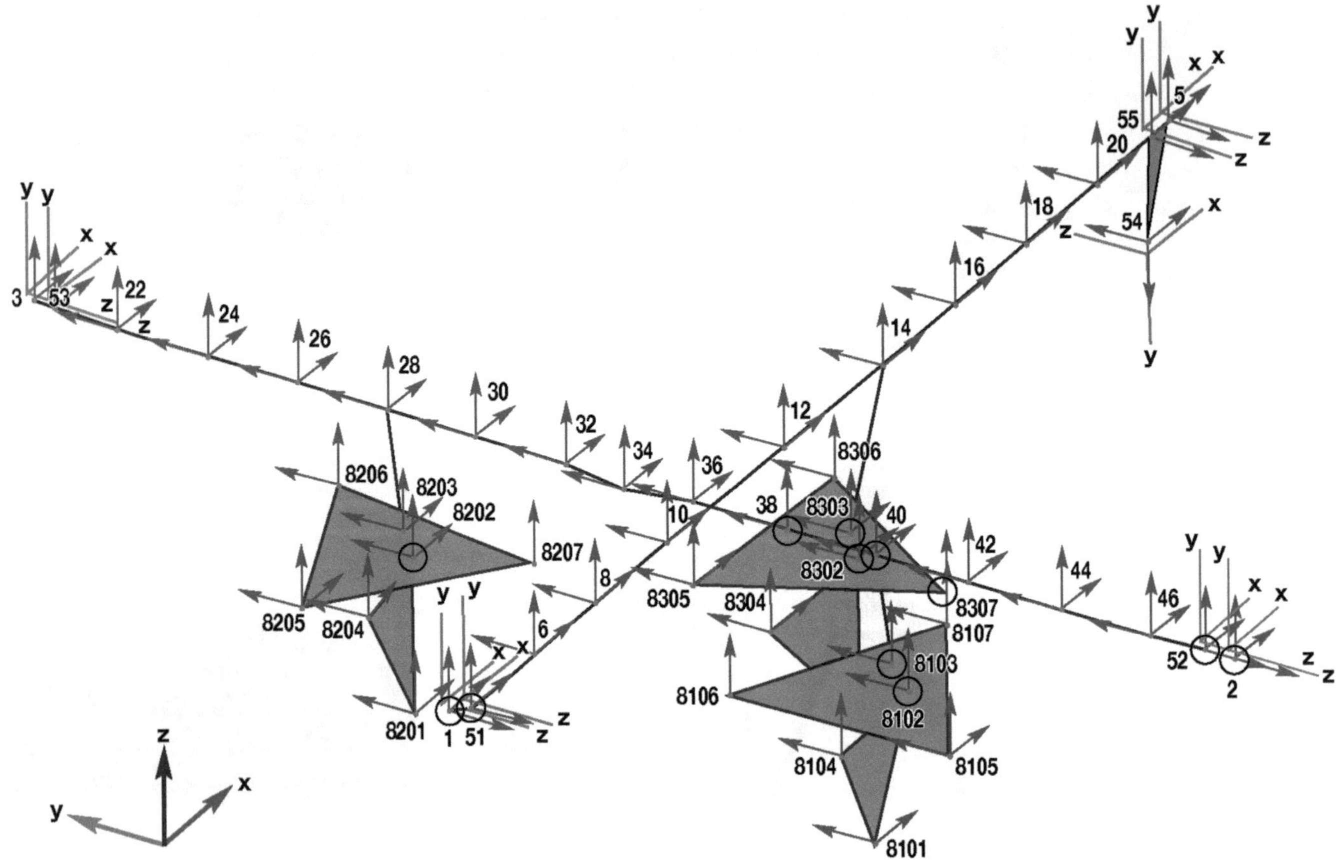

Fig. 10.12 The test display model and accelerometer locations

10.8.3 *Test Description*

A total of 12 different DIM analysis cases were conducted through the course of 54 test runs. These runs included check-out, single-shaker, multi-shaker, and quiescent runs.

During preliminary test runs, unexpected soft-support modes in the frequency range of 614 Hz were observed. Additional accelerometers were installed on the 60K3S to characterize these modes. Based on the observed mode shapes, the lower portion of the soft support was constrained to the supporting frame before testing was resumed.

The added constraints to the 60K3S increased the stiffness, and the 60K3S modes moved from the frequency range of 6–14 Hz to the frequency range of 10–18 Hz to aid in the frequency separation necessary for the DIM calculation. The power spectrum mode indicator functions (PSMIFs), a summation of the squared-magnitude of the FRF, plotted in Fig. 10.13 show the effect of the added constraints with the original PSMIF (solid blue) compared to the modified PSMIF (dashed blue). The other PSMIF lines were taken at the start of each testing day to verify that the 60K3S modes were still beyond the 10 Hz range.

For reference, modes extracted from the multiple-shaker random run (run 43) are provided in Table 10.4. All six iron bird rigid body modes were extracted and were at less than 2 Hz. The first set of 60K3S modes were in the frequency range of 10–18 Hz. The first flexible mode of the iron bird was at 18.7 Hz.

Several parameters during the DIM testing were changed to evaluate and compare the results of the DIM calculations to determine which parameters worked well and which did not. Each DIM analysis case consisted of two or more test runs. Several types of shaker excitation were used for the DIM testing: single-shaker and multiple-shaker burst random, true random, and sine-sweep excitation. The shakers were also placed at various locations (nose, right wing, left wing, and tail) around the test article to determine the ideal excitation locations for the DIM calculation. The list of cases is shown in Table 10.5.

Figure 10.14 shows the iron bird test article undergoing DIM testing on soft supports with two shakers for excitation.

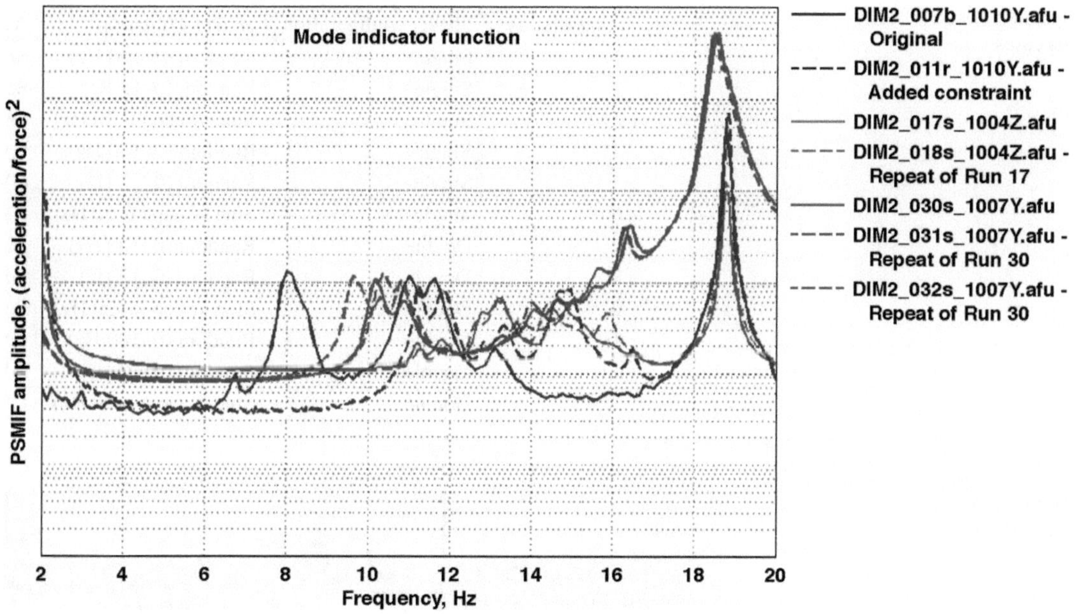

Fig. 10.13 Power spectrum mode indicator functions showing the effect of added constraints on the 60K3S

Table 10.4 Modes extracted from the multiple-shaker random run of the iron bird

Mode	Frequency, Hz	Damping, % critical	Description
1	0.72	8.52	Rigid body pitch
2	0.98	6.32	Rigid body roll
3	1.25	3.81	Rigid body yaw
4	1.56	6.21	Rigid body vertical
5	1.80	2.99	Rigid body fore-aft
6	1.99	2.91	Rigid body lateral
7	10.08	1.99	Starboard 60K3S canister X+/Y−
8	11.13	1.27	Port 60K3S canister X+/Y−
9	11.47	1.46	Starboard 60K3S canister X+/Y+
10	12.41	0.98	Aft 60K3S canister RZ
11	12.96	1.77	Aft 60K3S canister X+/Y−
12	13.78	1.27	Aft 60K3S canister X+/Y+
13	13.79	1.52	Starboard 60K3S X+/Y−
14	14.88	1.00	Port 60K3S X+/Y+
15	15.61	1.07	Starboard 60K3S X+/Y+
16	16.37	0.72	Port 60K3S lateral
17	18.23	2.03	Aft 60K3S lateral
18	18.73	0.24	Iron bird first flexible mode

10.9 Results

Post-processing tools were used to extract the FRFs from the test data. The FRFs were analyzed using the ATA Engineering IMAT + Signal™ analysis software where "IMAT" stands for "Interface between MATLAB®, Analysis, and Test."

10.9.1 Dynamic Inertia Measurement Analysis

The PSMIF for the two-shaker random 0–100-Hz excitation analysis case is shown in Fig. 10.15. Two PSMIFs are shown: the blue function is the PSMIF for only the DOFs on the iron bird; the green function is the PSMIF for all DOFs, which also includes the DOFs on the 60K3S. While the soft-suspension modes in the frequency range of 10–17 Hz are clearly seen in

Table 10.5 Dynamic inertia method analysis cases

Analysis case	Shaker configuration	Shaker excitation	Accelerometers
1	Single	Random (0–100 Hz)	Seismic
2	Single	Random (0–12 Hz)	Seismic
3	Single	Sine sweep (1–20 Hz)	Seismic
4	Single	Random (0–100 Hz)	GVT
5	Single	Random (0–12 Hz)	GVT
6	Single	Sine sweep (1–20 Hz)	GVT
7	Double	Random (0–100 Hz)	Seismic
8	Double	Random (0–12 Hz)	Seismic
9	Double	Sine sweep (1–20 Hz)	Seismic
10	Double	Random (0–100 Hz)	GVT
11	Double	Random (0–12 Hz)	GVT
12	Double	Sine sweep (1–20 Hz)	GVT

Fig. 10.14 The iron bird undergoing DIM testing on soft supports

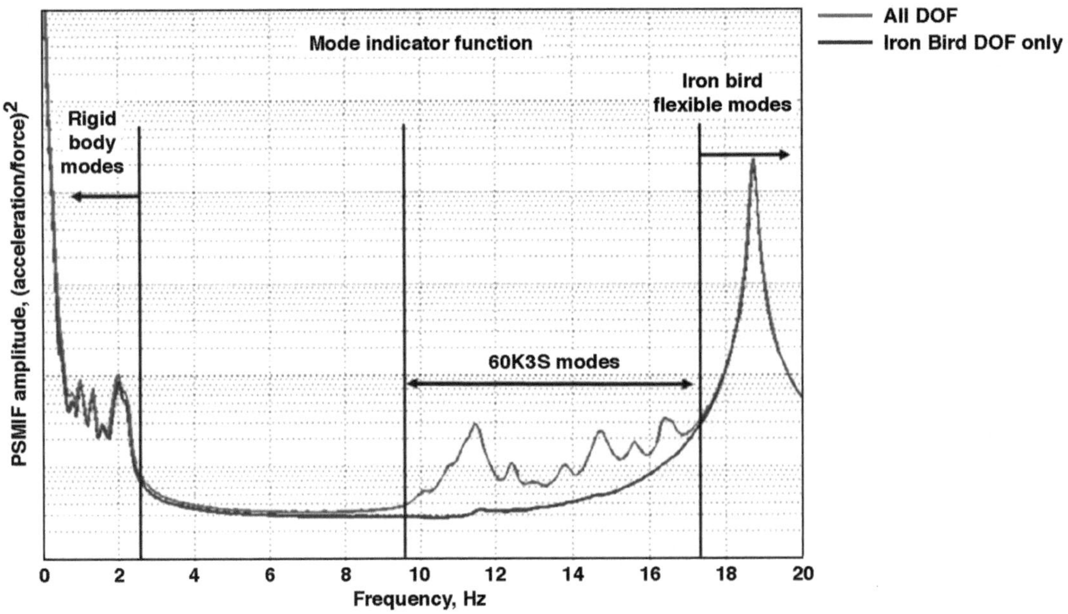

Fig. 10.15 Power spectral mode indicator function for the two-shaker random 0–100 Hz excitation analysis case

the PSMIF of all DOFs, they are localized modes of the soft supports and have little response in the iron bird DOFs. The rigid body modes of the iron bird on the soft supports are at 2 Hz and below, and the flexible modes are above 18 Hz, which is sufficient separation for estimating the inertia properties in this frequency range. The existence of the soft-suspension modes starting at 10 Hz and becoming prominent at 12 Hz, however, limited the frequency range over which the DIM calculations were performed. Two frequency ranges were analyzed and evaluated to determine the effects of frequency range on DIM analysis results.

10.9.2 Comparison of Seismic and GVT T333B (PCB Piezotronics, Inc.) Accelerometers

One objective of the DIM test was to determine whether typical GVT accelerometers would be sufficient for accurate DIM analysis or if higher sensitivity seismic accelerometers would be required. The DIM analysis cases were repeated and compared using seismic accelerometers (DIM analysis case 1) and typical modal test T333B (PCB Piezotronics, Inc.) accelerometers (DIM analysis case 4). A method of evaluating the error in DIM calculations is to perform a normalized error function calculation [7]. Figure 10.16 shows the normalized error functions calculated using the seismic accelerometers for DIM analysis case 1.

No DOFs were consistently and significantly large enough for all test cases to warrant removing them from the mass properties calculations. Figure 10.17 shows the normalized error functions for DIM analysis case 4 using T333B (PCB Piezotronics, Inc.) GVT accelerometers. As expected, the overall mass properties results from the seismic accelerometers were consistently more in agreement with analytical predictions due to the lower noise floor than that of the T333B (PCB Piezotronics, Inc.) GVT accelerometers. The seismic accelerometers were thought to be required because of their higher sensitivity. The higher sensitivity creates a lower noise floor, which produces acquisition of cleaner data. The cleaner data lessen the variation in the results when compared to the T333B (PCB Piezotronics, Inc.) accelerometers.

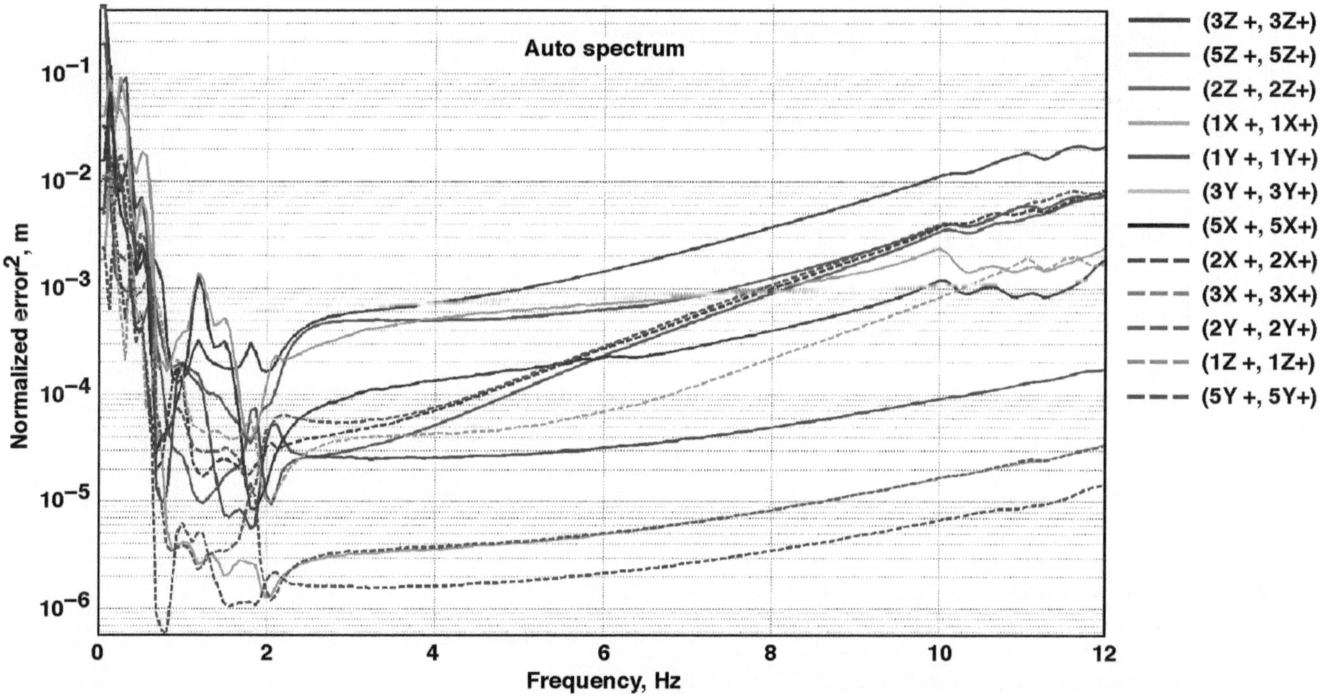

Fig. 10.16 Normalized error functions calculated using the seismic accelerometers for DIM analysis case 1

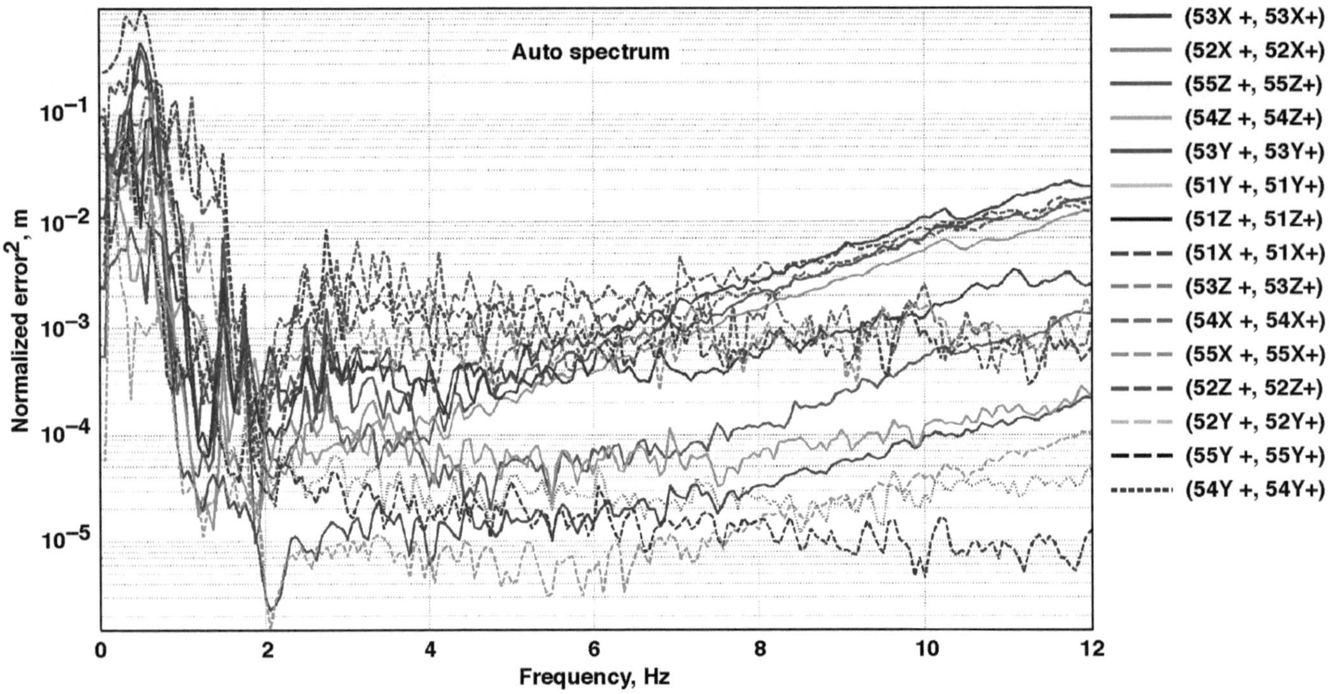

Fig. 10.17 Normalized error functions using the GVT accelerometers for DIM analysis case 4

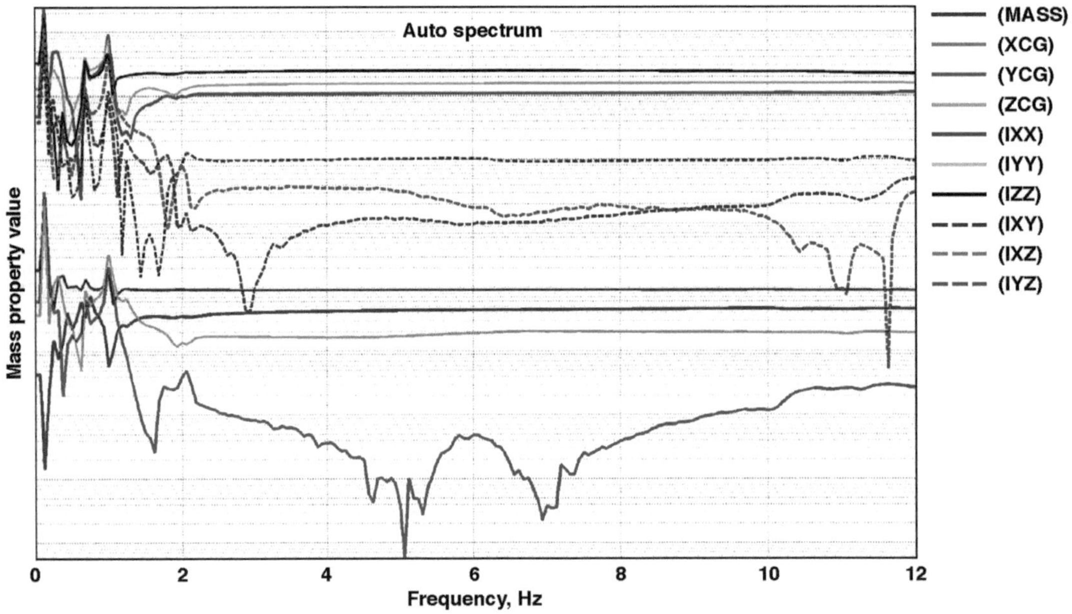

Fig. 10.18 Computed mass properties for DIM analysis case 1

10.9.3 Dynamic Inertia Measurement Results

The computed mass, MOI, POI, and CG values are plotted as a function of frequency for DIM analysis case 1 in Fig. 10.18 for a 2–12 Hz DIM analysis. The mass, X_{CG} and Z_{CG}, three MOIs, and I_{xz} functions are relatively flat from 2 Hz to 12 Hz. The Y_{CG}, I_{xy}, and I_{yz} functions exhibit greater fluctuations, but since these values are nominally zero and the estimated values are very small compared to the other CG and POI values, these fluctuations are to be expected.

The mass properties were compared for all DIM analysis cases to determine the ideal frequency band within which to compute average values. From examination of the mass property functions, the 2–12 Hz frequency band was fairly flat for all non-zero properties. A second smaller frequency band of 8–10 Hz was also used for DIM processing, as this seemed to be the most consistent region over all of the analysis cases. In general, this smaller 8–10 Hz frequency range produced slightly larger mass properties results that were slightly larger than the results produced by the DIM analysis of the 2–12 Hz band.

Several parameters of the DIM analyses were varied during testing to investigate which produce the more accurate results. The 3-DOF force transducer provided a measurement for the excitation force in all three translational directions, which included the off-axis components of the input force. In the results presented above for the DIM analysis cases, the off-axis forces measured by the 3-DOF excitation force transducer were treated as additional reaction forces in the inertia calculations. For the two-shaker configurations, however, one of the inputs was measured with a single-axis force transducer because only one 3-DOF force transducer was available. To evaluate their effect on the computed mass properties, two DIM analysis cases were processed excluding the off-axis reaction forces. The two DIM analysis cases studied were the 0–100 Hz random excitation with seismic accelerometers for a single shaker (analysis case 1) and for two shakers (analysis case 7). The mass properties calculations for both cases are across the smaller frequency range of 8–10 Hz. The results for the two DIM analysis cases with and without the off-axis forces are listed in Table 10.6. For both DIM analysis cases, the computed mass was less when not using the off-axis forces. For DIM analysis case 7, however, in which only one shaker had measured off-axis force, the MOIs increased slightly and were closer to DIM analysis case 1. Further study is recommended of the effects of the off-axis forces.

Table 10.7 summarizes the mass properties calculation across the 2–12 Hz frequency range for all DIM analysis cases. These cases are calculated using the off-axis forces measured by the 3-DOF force transducer.

Table 10.8 summarizes the mass properties calculation across the smaller 8–10 Hz frequency range for all DIM analysis cases. These cases are also calculated using the off-axis forces measured by the 3-DOF force transducer.

10.9.4 Comparison of Results from Different Methods

Table 10.9 summarizes the results from analytical, pendulum swings, and DIM methods to find mass properties. Case 1 of the DIM method was used for comparison. The symmetrical geometry of the iron bird made the I_{xy} and I_{yz} POI negligible; they were not tested.

The DIM method yielded results that matched within approximately 5 % of the analytical iron bird mass, CG, and MOI. The I_{xz} POI did not match as well, having errors exceeding 20 %, however, the DIM I_{xz} results were still better than the 98-percent error from the pendulum-based testing results due to test setup limitations (that is, shallow tilt angle).

The results for the DIM method were in good agreement with the analytical predictions, but the DIM method also demonstrated other benefits. The DIM method enabled more of the mass properties to be measured using less hardware, fewer test configurations, and less testing time. The time, resources, and labor required to execute both the pendulum oscillation and the DIM methods were noted to compare these two methods in these aspects.

Conventional pendulum testing methods require orienting the test article in several different configurations. The measurement of weight and CG is required for MOI testing, which calls for a different setup than that used for MOI. The MOIs each require a different orientation (that is, test setup) for each axis: I_{xx}, I_{yy}, I_{zz}, and I_{xz}. Additionally, any fixtures attached to the test article as part of the setup must also be tested to measure their contribution to the total measured inertia. Without considering delays, and including test setup and teardown, the measurement of the mass properties using conventional methods required approximately 30 days.

The DIM test requires setup and teardown times similar to that required for GVT, but with the additional DIM sensors the DIM and GVT data can be acquired simultaneously. As well, the DIM method can be executed using a single vehicle configuration and without extra fixtures. No additional data are needed for this method, and the DIM method has been shown to be able to provide the full set of mass properties. Without considering delays and including test setup and teardown, the measurement of the mass properties using the DIM method required approximately 15 days.

A similar comparison can be made for the cost associated with each method. An appreciable amount of the equipment that is required to execute the DIM method consists of equipment that is already used for execution of GVTs. A GVT requires accelerometers, soft-support systems, shakers, and force transducers. A GVT may be expanded to encompass a DIM test by adding 6-DOF force sensors, a 3-DOF force transducer, and seismic accelerometers, which are currently thought to be required for successful and more accurate DIM calculations. All aircraft and flight test experiments call for a GVT as part of the airworthiness process at some point in the project life cycle. The additional time necessary to set up and tear down for a DIM test is minimal, since most of the involved activities are already required when performing a GVT. A GVT can, therefore, be expanded to a DIM test without causing significant inconvenience or schedule addition and can provide more of

Table 10.6 Comparison of mass properties using DIM with and without off-axis forces

	8–10 Hz frequency range															
Analysis case	MASS, kg	Mass, % error	X_{CG}, m	X_{CG}, % error	Y_{CG}, m	Z_{CG}, m	Z_{CG}, % error	I_{XX}, kg*m^2	I_{XX}, % error	I_{YY}, kg*m^2	I_{YY}, % error	I_{ZZ}, kg*m^2	I_{ZZ}, % error	I_{XY}, kg*m^2	I_{XZ}, kg*m^2	I_{XZ}, % error	I_{YZ}, kg*m^2
1	7,981	3.40 %	2.273	−0.60 %	−0.0178	0.485	2.00 %	12,349.4	−2.70 %	17,616.9	2.30 %	26,952.1	−5.50 %	−172.7	−1,106.2	−21.00 %	155.1
1 (no off-axis force)	7,624	−1.20 %	2.278	−0.30 %	−0.0051	0.462	−2.70 %	12,290.9	−3.30 %	17,470.6	1.50 %	26,981.4	−5.30 %	−117.1	−1,123.7	−19.80 %	149.2
7	7,568	−1.90 %	2.271	−0.70 %	0.0737	0.259	−45.30 %	8,896.3	−30.00 %	17,060.9	−0.80 %	22,240.6	−22.00 %	−1,144.2	2,838.6	−302.80 %	−1,267.1
7 (no off-axis force)	7,214	−6.50 %	2.283	−0.10 %	0.0432	0.320	−32.50 %	9,949.8	−21.70 %	17,441.3	1.40 %	24,815.9	−12.90 %	−468.2	1,387.1	−199.20 %	−5.9

Table 10.7 Summary of mass properties using DIM for the 2–12 Hz frequency range

	2–12 Hz frequency range																
Analysis case	MASS, kg	Mass, % error	X_{CG}, m	X_{CG}, % error	Y_{CG}, m	Z_{CG}, m	Z_{CG}, % error	I_{XX}, kg*m^2	I_{XX}, % error	I_{YY}, kg*m^2	I_{YY}, % error	I_{ZZ}, kg*m^2	I_{ZZ}, % error	I_{XY}, kg*m^2	I_{XZ}, kg*m^2	I_{XZ}, % error	I_{YZ}, kg*m^2
1	7,623	−1.20 %	2.278	−0.30 %	−0.005	0.462	−2.60 %	12,290.9	−3.20 %	17,470.6	1.50 %	27,010.7	−5.20 %	−120.0	−1,120.8	−19.90 %	187.3
2	7,631	−1.10 %	2.278	−0.30 %	−0.010	0.472	−0.70 %	12,290.9	−3.20 %	17,382.8	1.00 %	27,010.7	−5.20 %	−155.1	−1,114.9	−20.30 %	228.3
3	7,600	−1.50 %	2.278	−0.40 %	−0.013	0.470	−1.00 %	12,290.9	−3.30 %	17,382.8	1.00 %	27,010.7	−5.20 %	−155.1	−1,094.5	−21.90 %	275.1
4	7,474	−3.10 %	2.299	0.60 %	0.028	0.445	−6.20 %	12,203.1	−3.90 %	16,680.5	−3.10 %	27,537.4	−3.40 %	−90.7	−951.1	−32.10 %	−275.1
5	7,523	−2.50 %	2.299	0.50 %	0.025	0.460	−3.40 %	12,203.1	−3.90 %	16,622.0	−3.30 %	27,537.4	−3.40 %	−114.1	−942.3	−32.70 %	−245.8
6	7,509	−2.70 %	2.299	0.50 %	0.023	0.457	−3.70 %	12,203.1	−3.90 %	16,651.2	−3.20 %	27,508.2	−3.50 %	−111.2	−924.7	−33.80 %	−207.8
7	7,209	−6.60 %	2.278	−0.30 %	0.064	0.297	−37.70 %	9,627.9	−24.30 %	17,207.2	−0.10 %	24,289.1	−14.80 %	−640.9	1,630.0	−216.60 %	−298.5
8	7,316	−5.20 %	2.271	−0.70 %	0.066	0.335	−29.40 %	10,213.1	−19.60 %	17,265.8	0.30 %	23,703.8	−16.80 %	−693.6	1,586.1	−213.40 %	−418.5
9	7,119	−7.70 %	2.291	0.20 %	0.046	0.328	−30.90 %	9,598.6	−24.40 %	17,265.8	0.40 %	23,879.4	−16.20 %	−772.6	1,808.5	−229.20 %	−550.2
10	7,263	−5.90 %	2.299	0.60 %	0.056	0.353	−25.60 %	10,096.1	−20.50 %	16,534.2	−3.80 %	26,191.3	−8.10 %	−421.4	719.9	−151.40 %	−485.8
11	6,562	−15.00 %	2.296	0.50 %	0.038	0.462	−2.40 %	11,003.3	−13.30 %	16,651.2	−3.20 %	26,337.6	−7.60 %	−301.4	248.7	−117.80 %	−310.2
12	7,146	−7.40 %	2.309	1.00 %	0.043	0.386	−18.60 %	10,096.1	−20.50 %	16,387.8	−4.80 %	25,547.5	−10.40 %	−561.9	860.4	−161.50 %	−834.0

Table 10.8 Summary of mass properties using DIM for the 8–10 Hz frequency range

	8–10 Hz frequency range																
Analysis case	MASS, kg	Mass, % error	X_{CG}, m	X_{CG}, % error	Y_{CG}, m	Z_{CG}, m	Z_{CG}, % error	I_{XX}, kg*m^2	I_{XX}, % error	I_{YY}, kg*m^2	I_{YY}, % error	I_{ZZ}, kg*m^2	I_{ZZ}, % error	I_{XY}, kg*m^2	I_{XZ}, kg*m^2	I_{XZ}, % error	I_{YZ}, kg*m^2
1	7,981	3.40 %	2.273	−0.60 %	−0.018	0.485	2.00 %	12,349.4	−2.70 %	17,616.9	2.30 %	26,952.1	−5.50 %	−172.7	−1,106.2	−21.00 %	155.2
2	7,979	3.40 %	2.273	−0.50 %	−0.023	0.500	5.20 %	12,349.4	−2.70 %	17,558.4	2.00 %	26,981.4	−5.40 %	−187.3	−1,082.8	−22.70 %	178.5
3	7,985	3.50 %	2.276	−0.40 %	−0.023	0.500	5.30 %	12,349.4	−2.70 %	17,558.4	2.10 %	27,010.6	−5.20 %	−172.7	−1,091.6	−21.90 %	298.5
4	7,796	1.00 %	2.299	0.60 %	0.033	0.457	−3.80 %	12,290.9	−3.30 %	16,856.1	−2.00 %	27,888.6	−2.10 %	−143.4	−916.0	−34.50 %	−351.2
5	7,811	1.20 %	2.299	0.60 %	0.028	0.480	1.30 %	12,261.6	−3.50 %	16,826.8	−2.20 %	27,888.6	−2.10 %	−143.4	−904.3	−35.30 %	−310.2
6	7,815	1.30 %	2.301	0.60 %	0.025	0.483	1.60 %	12,261.6	−3.50 %	16,856.1	−2.10 %	27,917.9	−2.00 %	−125.8	−910.1	−35.00 %	−219.5
7	7,568	−1.90 %	2.271	−0.70 %	0.074	0.259	−45.30 %	8,896.3	−30.00 %	17,060.9	−0.80 %	22,240.6	−22.00 %	−1,144.2	2,838.6	−302.80 %	−1,267.1
8	7,616	−1.30 %	2.268	−0.80 %	0.061	0.315	−33.70 %	10,330.2	−18.50 %	18,231.5	6.00 %	24,757.3	−13.10 %	−263.4	1,267.1	−190.60 %	377.5
9	7,554	−2.10 %	2.278	−0.30 %	−0.003	0.368	−22.20 %	10,008.3	−21.20 %	17,792.5	3.40 %	23,089.3	−19.00 %	−737.5	2,112.9	−251.00 %	−860.4
10	7,527	−2.50 %	2.306	0.90 %	0.061	0.345	−27.20 %	9,569.3	−24.60 %	16,709.7	−2.90 %	25,108.5	−11.90 %	−784.3	1,451.5	−203.70 %	−1,182.3
11	7,058	−8.50 %	2.311	1.10 %	0.038	0.439	−7.60 %	11,647.1	−8.30 %	17,909.6	4.10 %	29,820.0	4.60 %	529.7	−1,372.5	−1.80 %	1,293.5
12	7,337	−4.90 %	2.316	1.30 %	0.000	0.457	−3.40 %	10,944.7	−13.70 %	17,178.0	−0.20 %	26,688.8	−6.30 %	−105.4	58.5	−104.20 %	−439.0

Table 10.9 Summary of mass properties calculations derived by three different methods

Mass, kg	Mass, error	X_{CG}, m	X_{CG}, % error	Y_{CG}, m	Y_{CG}, % error	Z_{CG}, m	Z_{CG}, % error	I_{XX}, kg*m^2	I_{XX}, % error	I_{YY}, kg*m^2	I_{YY}, % error	I_{ZZ}, kg*m^2	I_{ZZ}, % error	I_{XZ}, kg*m^2	I_{XZ}, % error
Analytical															
7,716.513	–	2.286	–	0	–	0.475	–	12,700.6	–	17,207.2	–	28,503.1	–	−1,398.8	–
Pendulum swing															
7,698.823	−0.2 %	2.287	0 %	−0.03	–	0.475	0 %	12,261.6	−3.5 %	17,938.8	4.3 %	27,859.3	−2.3 %	−2,768.4	−97.9 %
Dynamic inertia measurement															
7,622.620	−1.2 %	2.278	−0.3 %	−0.01	–	0.462	−2.6 %	12,290.9	−3.2 %	17,470.6	1.5 %	27,010.7	−5.2 %	−1,120.8	19.9 %

the mass properties than could be acquired using the pendulum methods. As well, conventional pendulum methods require massive fixtures, extensive lift procedures and reviews, and more testing time, but only provide a limited number of the mass properties.

It is recognized that state-of-the-art DIM test execution requires equipment that is more expensive than the equipment required for the pendulum tests. The 6-DOF force sensors are custom-designed, one-of-a-kind equipment, and as a result necessitate high recurring costs for calibration. The seismic accelerometers are currently thought to be required because of the low noise floor they exhibit, and the 3-DOF excitation force transducer is considered required because the off-axis forces are used as additional reaction forces in the DIM calculations. These are more specialized types of equipment than the typical GVT accelerometers, and as such are more expensive and also necessitate more costly calibration procedures. The laser tracker is a high-precision measurement instrument and is also considered to be expensive. Most of these pieces of equipment, however, are considered to be one-time costs, and with further development the DIM method may show promise in becoming more feasible.

10.10 Recommendations and Considerations

This iron bird testing series was a successful exercise; however, the DIM method is not yet a fully mature technology for large aerospace vehicles. ATA Engineering provided the following list of recommendations and considerations for future DIM method tests, DIM processing, and DIM theory to further advance the technology readiness level (TRL) of the DIM method on large aerospace vehicles.

10.10.1 Testing Recommendations and Considerations

1. Retest with the iron bird supported on bladder-type air springs instead of the 60K3S system for this test. Using this other type of soft support should eliminate the cluster of modes in the 5–8 Hz frequency range and give a broader frequency range over which to estimate the inertia properties.
2. Predict MOI perturbation by adding a mass with known inertia properties to the structure.

10.10.2 Processing Recommendations and Considerations

1. Process the measurements assuming that the accelerometers were mounted in the global coordinate system.
2. Investigate how many accelerometers are sufficient to produce accurate inertia results.
3. Evaluate whether 3-DOF reaction forces are sufficient or 6-DOF reaction forces are necessary.
4. Reformulate the inertia equations for known terms (for example, mass, 2 of 3 CG coordinates).
5. Develop a non-subjective procedure to select the "flat spot" in the inertia functions.
6. Develop and explore uncertainty analysis methods for DIM calculations such as a Monte Carlo or Cramer-Rao bounds.
7. Implement spatial filtering with the ATA Engineering IMAT + Signal™ spVIEW™ user tools.

10.10.3 Basic Theory Recommendations and Considerations

1. Perform analysis or modeling to better understand the effects of force path and to explain more completely the differences of exciting on the structure and exciting through the force sensors.
2. Investigate the influence of gravity at low frequency. Is the calibration of the 6-DOF force sensors dependent on the orientation of the calibration mass in the gravitation field? It is known that the output of sensors depends upon orientation, but the current implementation does not compensate for gravity.

10.10.4 Additional Recommendations

1. Evaluate the effects of the 3-DOF force transducer off-axis reaction forces on the accuracy of the computed mass properties.
2. Investigate the required redundancy for the excitation of the six rigid body modes in relation to the typical multiple-shaker configurations used during GVT.

10.11 Conclusions

The Dynamic Inertia Measurement (DIM) method shows promise for mass properties testing applications involving large aerospace vehicles. There were sources of error that required mitigation; for example, the soft-support system introduced modes into the test data. As well, the DIM method was found to be sensitive to different shaker configurations and test setups. Several recommendations were therefore made with regard to the method of DIM testing.

Performing the DIM method on the "iron bird" test article advanced the maturity level of the method toward future use on full-scale aerospace vehicles. The next step in the maturation of the DIM method would be to apply the technique to a full-scale aerospace vehicle.

Conventional pendulum-based mass properties testing was performed to compare results and level of effort with the DIM method. The pendulum methods created a variety of operational challenges. Several lift procedures and reviews were required to move the iron bird test article into the pendulum-testing configuration due to the safety-critical nature of the setup, and the design of the pendulum test setup itself had many shortcomings. For example, design flaws in the test hardware introduced friction into the test setup that created a frequency shift and deteriorated the quality of the data. In spite of these challenges, usable data were collected and used to estimate mass properties.

The iron bird test article was supported on three soft supports that simulated free-free boundary conditions. The rigid body modes were below 2 Hz, and the first flexible mode was above 18 Hz, which was sufficient separation for the DIM method; however, there was a cluster of soft-support modes in the 10–17 Hz frequency range. These modes were localized suspension modes having little effect on the calculated inertia functions below 12 Hz. This configuration allowed a maximum 2–12 Hz frequency band from which to estimate inertia and center of gravity values. The results were fairly consistent across the six single-shaker test cases, and the relative error compared to analytical values was typically only a few percent, except for the one non-zero product of inertia. The results for the six skewed, double-shaker test cases, however, exhibited greater relative error and larger variance among the test runs. Therefore, in this iron bird case, single-shaker configurations provided the best results.

The DIM mass properties testing method requires expensive sensors and equipment. Fortunately, much of the necessary equipment is already available if ground vibration testing has been performed. Additionally, performing the DIM test can simultaneously provide the same modal characteristics data used for ground vibration testing analysis. After comparing the labor and time needed to perform each test, the DIM test was determined to be capable of experimentally determining mass properties twice as fast as the conventional pendulum method. In addition, the conventional pendulum method contains much higher schedule and vehicle risk, requires more procedural reviews and multiple pieces of specialized hardware and interface frames, and multiple testing configurations. The DIM method, with further development, may prove to be a more efficient approach to estimating the mass properties of a large aerospace vehicle.

Acknowledgements The authors thank the NASA Aeronautics Research Mission Directorate Aerosciences Project and Aeronautics Test Program for funding the Dynamic Inertia Measurement research. The authors also thank the testing support personnel at the Flight Loads Laboratory at the Armstrong Flight Research Center for conducting the Dynamic Inertia Measurement and pendulum tests. Special thanks also go to aerospace engineers Bob Clarke and Adam Harding of the Armstrong Flight Research Center for their support and expertise in performing the conventional pendulum mass properties tests.

References

1. Toivola J, Nuutila O (1993) Comparison of three methods for determining rigid body inertia properties from frequency response functions. In: Proceedings of the 11th international modal analysis conference, Kissimmee, pp 112–1132
2. Witter MC,(2000) Rigid body inertia property estimation using the dynamic inertia method. Master of Science thesis, Department of Mechanical Engineering, University of Cincinnati

3. Lazor DR (2004) Considerations for using the dynamic inertia method in estimating rigid body inertia property. Master's of Science thesis, Department of Mechanical Engineering, University of Cincinnati
4. Peterson WL (2004) Mass properties measurement in the X-38 project. SAWE paper no. 3325, category 6.
5. Stebbins MA, Brown DL (1998) Rigid body inertia property estimation using a six-axis load cell. In: Proceedings of the 16th international modal analysis conference, Santa Barbara, pp 900–906
6. Gatzwiller KB, WitterMC, Brown DL (2000) A new method for measuring inertial properties. In: Proceedings of the 18th international modal analysis conference, San Antonio, pp 1056–1062
7. Witter MC, Brown DL, Dillon M (1999) A new method for RBP estimation – the dynamic inertia method. SAWE paper no. 2461, category no. 6
8. Green MW (1927) Measurement of the moments of inertia of full scale airplanes. NACA Technical Note No. 265
9. Miller MP (1930) An accurate method of measuring the moments of inertia of airplanes. NACA Technical Note No. 351
10. Miller MP, Soule HA (1931) Moments of inertia of several airplanes. NACA Technical Note No. 375
11. Soule HA, Miller MP (1933) The experimental determination of the moments of inertia of airplanes. NACA Report No. 467
12. Turner HL (1950) Measurement of the moments of inertia of an airplane by a simplified method. NACA Technical Note 2201
13. Gracey W (1941) The additional-mass effect of plates as determined by experiments. NACA Report No. 707
14. Fladung WA, Napolitano KL, Brillhart RD (2010) Final report on dynamic inertia measurement method testing, unpublished internal document. ATA Engineering, San Diego
15. Cloutier D, Fladung WA (2013) Final report on dynamic inertia measurement method testing on iron bird 2, unpublished internal document. ATA Engineering, San Diego
16. Holland JA (2014) A safe, advanced, adaptable isolation system that eliminates the need for critical lifts. NP-2009-08-02-DFRC, http://www.nasa.gov/pdf/484129main_Soft-Support-TOP.pdf. Accessed 30 Oct 2014

Chapter 11
Estimation of Amplitude-Dependent Resonance and Damping in MEMS Shock Accelerometers

Jason R. Foley, Thomas J. Lagoski, Jontia Brown, and Jonathan Hong

Abstract Understanding the dynamic response of accelerometers is critical to interpreting data obtained in impulsive loading conditions. While this information is commonly provided by manufacturers, the estimated properties are typically obtained at levels below the full range of the sensor. Using high bandwidth operating data of varying amplitudes, the damping ratio and principle resonant frequency of shock accelerometers are estimated using a simple dynamic model. While the resonance is shown to be essentially amplitude-invariant, the damping ratio is shown to vary with the excitation amplitude. Possible causes related to the frequency content of the excitation and the presence of repeated roots are discussed.

Keywords Shock accelerometers • Microelectromechanical systems (MEMS) • Sensors • Damping • Operating modal analysis

Nomenclature

τ	Pulse width/duration
ω	Angular/radian frequency
ϕ	Phase response (frequency domain)
m	Mass
c	Damping coefficient
i	Imaginary number
k	Spring constant
s	Complex frequency (Laplace variable)
t	Time
x	Input function (time domain)
u	Base displacement coordinate
z	Relative displacement coordinate
A	Amplitude
H	Frequency response function

11.1 Introduction

Accelerometers are ubiquitous for measurements of structural dynamics and other dynamic events. Several types of transduction mechanisms can be used for sensing acceleration; piezoelectric, capacitive, and servo (or force balance) [1] are all common classes of accelerometers. Piezoresistive (PR) sensors have long been used in high shock environments [2] and are desirable in these applications for two primary reasons. First, piezoelectric elements have been shown to be susceptible to domain depoling in extreme shock events [3]. PR microelectromechanical systems (MEMS) are also readily manufactured from silicon dies using conventional electronic machining processes.

J.R. Foley (✉) • T.J. Lagoski • J. Brown
Air Force Research Laboratory, AFRL/RWMF, 306 W. Eglin Blvd., Bldg. 432, Eglin AFB, FL 32542-5430, USA
e-mail: jason.foley.1@us.af.mil

J. Hong
Applied Research Associates, Inc., 956 W. John Sims Pkwy., Niceville, FL 32578, USA

A. Wicks (ed.), *Shock & Vibration, Aircraft/Aerospace, and Energy Harvesting, Volume 9*,
Conference Proceedings of the Society for Experimental Mechanics Series, DOI 10.1007/978-3-319-15233-2_11

While commonly used for a variety of test applications, accelerometers designed for high-shock environments generally utilize stiffer elements with correspondingly high resonant frequencies (>50 kHz) and extremely low damping ratios ($\xi \ll 0.01$) that are effectively undamped [4]. The latest generation of shock accelerometers, developed and produced in the past decade, have increased the damping ratio significantly even though they are still relatively lightly damped ($\xi \sim 0.01$). Representative commercially-available sensors include the Meggitt Model 727 and PCB Piezotronics Model 3991 [5].

The details of the damping mechanisms of these devices depend on the specific design. However, they are generally specified and modeled as possessing a constant damping ratio. This paper tests this hypothesis by estimating the damping ratios for sensors at varying levels of excitation. A nonlinear model is proposed and preliminary results are calculated and discussed using experimental data.

11.2 Analysis

The accelerometer is modeled as a linear, damped, single degree of freedom (SDOF) system as shown in Fig. 11.1. The mathematical description is a one-dimensional simple harmonic oscillator where x is the motion of the spring within the sensor, u is the base motion, and z is the relative coordinate, i.e., $z = x - u$. The corresponding equation of motion for the system is then

$$m\ddot{z} + c\dot{z} + kz = -m\ddot{u}, \tag{11.1}$$

where m is the mass, c is the damping, and k is the stiffness of the dynamic element [6].

A state-space model is appropriate to reduce the problem into a simple linear system of equations. We begin with the following two substitutions:

$$z_1 = z \text{ and } z_2 = \dot{z}\,. \tag{11.2}$$

It immediately follows that

$$\dot{z}_1 = z_2. \tag{11.3}$$

Likewise, the governing equation of motion in terms of z_1 and z_2, solved for $\ddot{z}_2$, is

$$\dot{z}_2 = -\frac{k}{m}z_1 - \frac{c}{m}z_2 - \ddot{u}. \tag{11.4}$$

The linear system of equations is then

$$\begin{bmatrix} \dot{z}_1 \\ \dot{z}_2 \end{bmatrix} = \begin{bmatrix} 0 & 1 \\ -\frac{k}{m} & -\frac{c}{m} \end{bmatrix} \begin{bmatrix} z_1 \\ z_2 \end{bmatrix} + \begin{bmatrix} 0 \\ -\ddot{u} \end{bmatrix}. \tag{11.5}$$

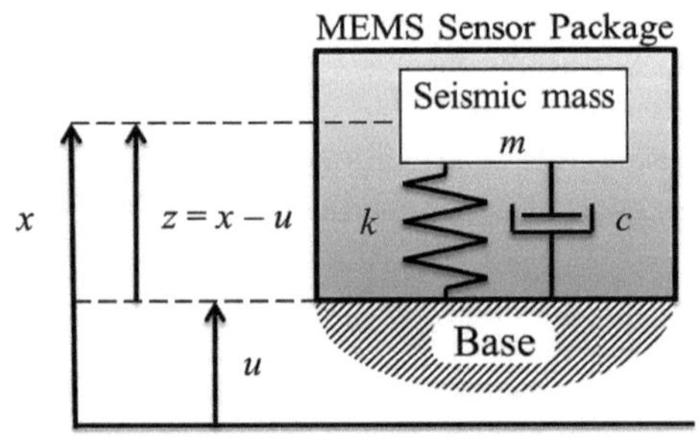

Fig. 11.1 Schematic of the coordinate system and components of a simplified MEMS accelerometer with a dynamic seismic mass

Table 11.1 Common analytic base excitation waveforms and the resulting response of a system with a natural frequency (f_n) of 80 kHz and damping ratio (ζ) of 0.05 (i.e., $Q = 10$)

Base excitation function	Duration τ [μs]	Amplitude A [g]	Base excitation and seismic response waveform
Haversine	50	1,000	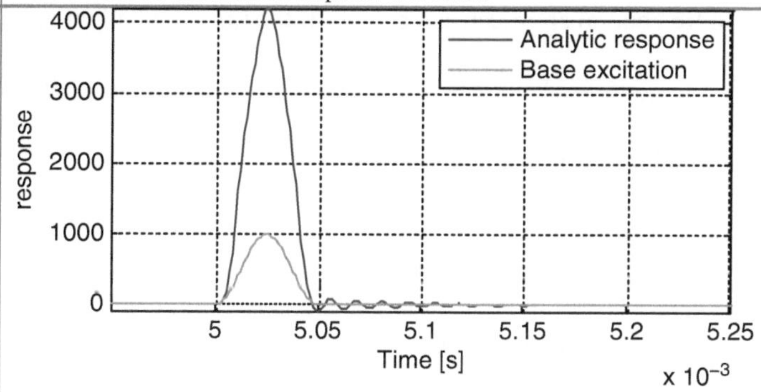
Rectangular pulse	50	1,000	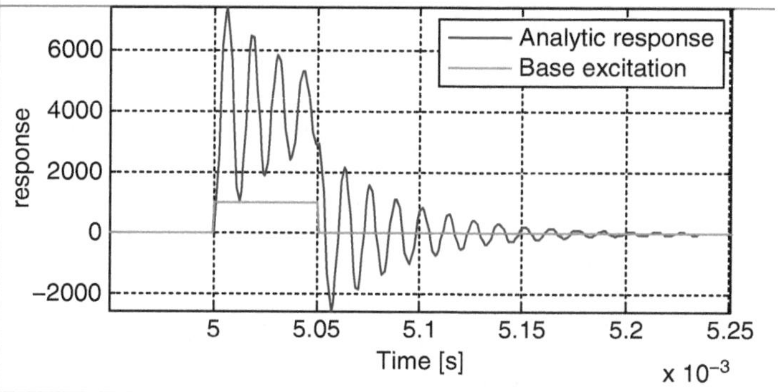
Impulse	1	1,000	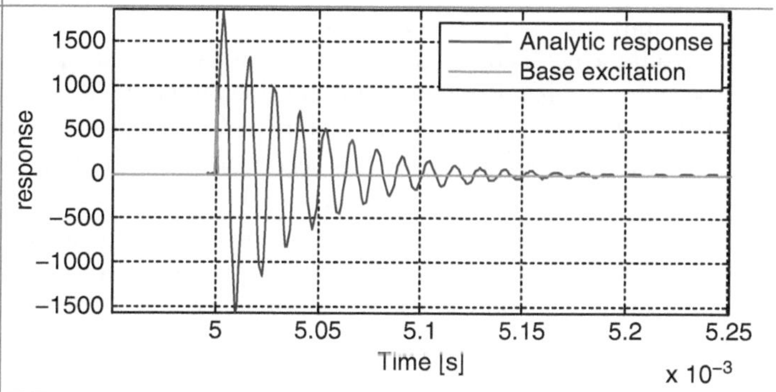

The time dependent response of the sensor can be found if the initial conditions are defined. For a sensor that begins at rest, the set of initial conditions becomes

$$\begin{aligned} z_1(0) &= 0, \\ z_2(0) &= 0, \\ \dot{z}_1(0) &= 0, \\ \dot{z}_2(0) &= \ddot{u}(0). \end{aligned} \tag{11.6}$$

The differential system of equations (Eq. 11.5) with initial conditions (Eq. 11.6) is solved in Matlab using the ODE solver *ode23t*. This particular solver is chosen due to the lack of numerical damping. Each solution takes approximately one minute on a desktop machine.

Table 11.1 shows several base excitations and the corresponding acceleration response of the seismic mass. Common analytic functions that describe base excitations include rectangular, Gaussian, and haversine pulses of finite duration. The base excitation is given as a haversine function,

$$\ddot{u}(t) = A \text{ haversin}(t, \tau) = A\sin^2\left(\frac{\pi t}{\tau}\right), \tag{11.7}$$

where τ is the pulse width. The haversine is advantageous because a piecewise function remains smooth and continuously differentiable function even if the function is identically zero outside of the pulse interval.

It is assumed that the base excitation is measured independently from the sensor being analyzed.

A more general approach can be used if the time-dependent applied force $f(t)$ is known. In this case, the governing ODE is

$$m\ddot{z} + c\dot{z} + kz = f(t). \tag{11.8}$$

Taking the Fourier transform of Eq. 11.8 yields

$$-m\omega^2\tilde{z} + i\omega c\tilde{z} + k\tilde{z} = \tilde{f}(\omega), \tag{11.9}$$

where $\tilde{f}(\omega)$ is the frequency-domain representation of the applied force. The frequency response function (H_{zf}) of the response z due to the input force can be quickly obtained from the governing equation through simple arithmetic:

$$H_{zf}(\omega) \equiv \frac{\tilde{z}(\omega)}{\tilde{f}(\omega)} = \frac{1}{-m\omega^2 + i\omega c + k}. \tag{11.10}$$

It should be noted that this result only holds if the mass, stiffness, and damping are all independent of frequency.

11.3 Experiment

A traditional 1.81 m (6 ft) long, 2.54 cm (1 in.) diameter titanium Hopkinson bar [7] is modified to include a "flyaway" disk at the end of the bar as shown in Fig. 11.2. The flyaway is designed to remain coupled to the bar (via a vacuum collar) until it the initial stress pulse is applied to the system [8], after which it decouples. This creates an effectively single-event loading profile which is optimal for transient dynamic analysis of accelerometers. The unit under test (UUT) and reference (REF) accelerometers are mounted side-by-side on the flyaway. The UUT is a prototype Kulite Semiconductors GMD-280-100KG-P MEMS accelerometer [9], and the reference accelerometer is a Meggitt Sensing Systems 7270A-60KG [4]. The 7270A-60KG is an undamped MEMS accelerometer and is chosen as the reference due to its higher resonant frequency (800 kHz) and corresponding high linear bandwidth.

The accelerometers are powered, filtered, and amplified using a Precision Filter 28000 chassis with 28144A Quad-Channel Wideband Transducer Conditioner cards [10] in constant voltage mode. The signal conditioner is run in bypass mode, which provides an analog antialiasing filter at approximately 800 kHz. A National Instruments chassis with PXI-6133 multifunction input/output cards is used to digitize the analog data at 2.5 MSa/s with 16 bits of vertical resolution [11].

The Hopkinson bar is impacted with a striker at velocities ranging from 5 to 20 m/s measured by a laser extensometer. Pulse shaping is accomplished using index cards (typically 1, 2, or 3) between the striker and the bar. Experiments are run several times to provide adequate statistics for the subsequent analysis; typical acceleration profiles for such a run are shown

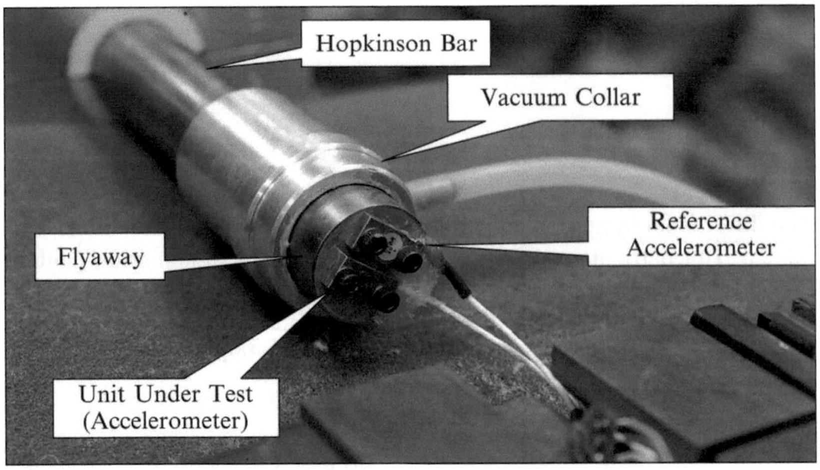

Fig. 11.2 Annotated picture of the experiment showing the side-by-side arrangement of the unit under test and the reference accelerometer. The Hopkinson bar, flyaway, and vacuum collar are also indicated

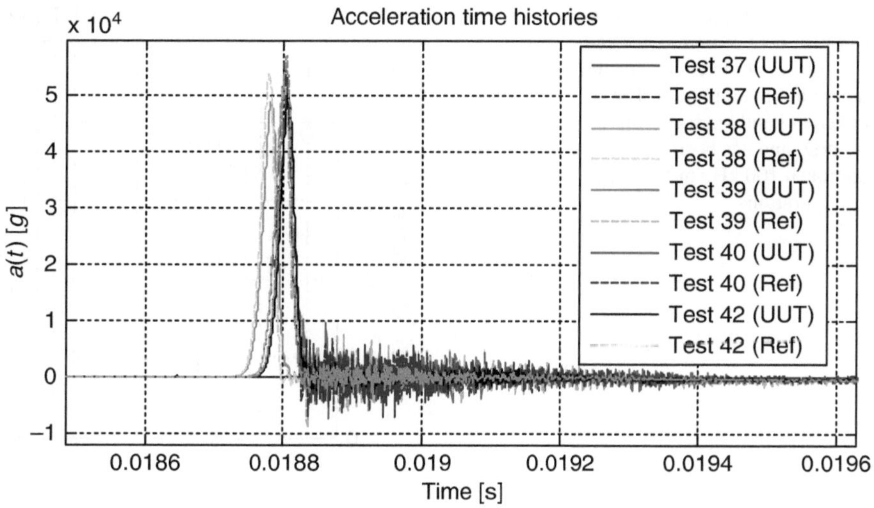

Fig. 11.3 Typical acceleration profiles for a range of tests with nominally identical excitation

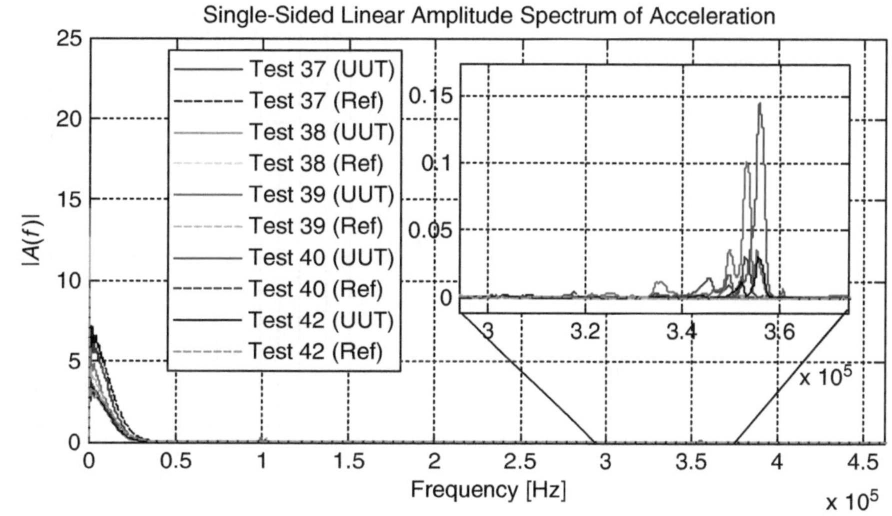

Fig. 11.4 Autopower profiles from tests 37–42; the inset shows the response at the expected resonant frequency for the UUT

in Fig. 11.3. The linear spectra from the same series of tests are shown in Fig. 11.4. In general, the impact energy excites a broad range of frequencies up to the expected resonance of each device.

The coherence of the system is also estimated using the auto- and cross-power spectra of the measurements [6]. As shown in Fig. 11.5, a coherent bandwidth of at least 50 kHz, and in some cases exceeding 100 kHz, is present for all measurements. The local modes of the flyaway are expected in the 50 kHz range based on similar analysis in Ref. [12]. The coherence beyond these modes is not surprising due to the nearly ideal side-by-side mounting configuration and the use of Hopkinson bar to ensure a nearly one-dimensional (i.e., planar) loading profile.

11.4 Analysis

We now further examine the dynamics of the system. The UUT is assumed to be a "black box" SDOF with a frequency response function described by Eq. (11.10). The averaged autopower spectra (from the five tests shown in Fig. 11.4) are given in Fig. 11.6, this time displayed on a semi-logarithmic scale. Two high frequency (>100 kHz) peaks are immediately evident. The UUT has a strong peak in the neighborhood of 350 kHz, the predicted fundamental mode of the sensor based on the manufacturer's preliminary datasheet [9], that is not present in the reference. Likewise, the reference accelerometer has a peak of 816 kHz, again in the manufacturer's predicted range of resonant frequencies between 750 and 850 kHz.

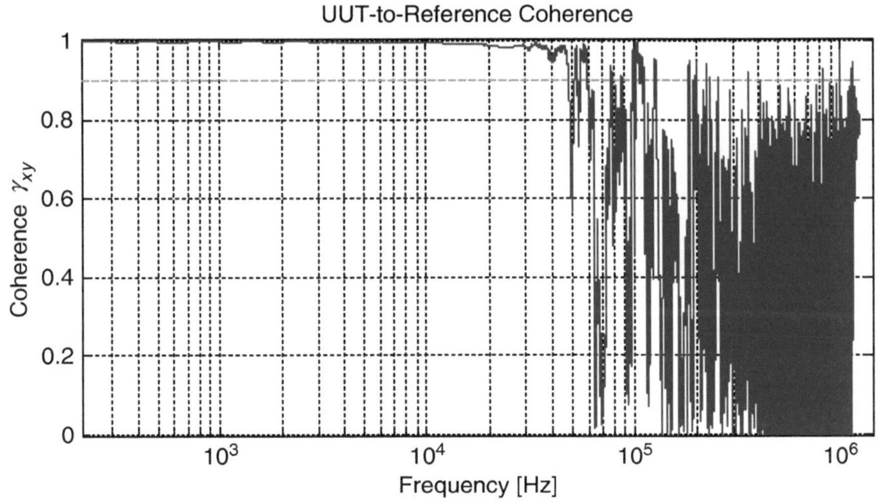

Fig. 11.5 A typical coherence estimate between reference and UUT accelerometers for five tests at maximum amplitude. The coherent bandwidth exceeds 50 kHz, and has been shown to approach 100 kHz in low-noise configurations

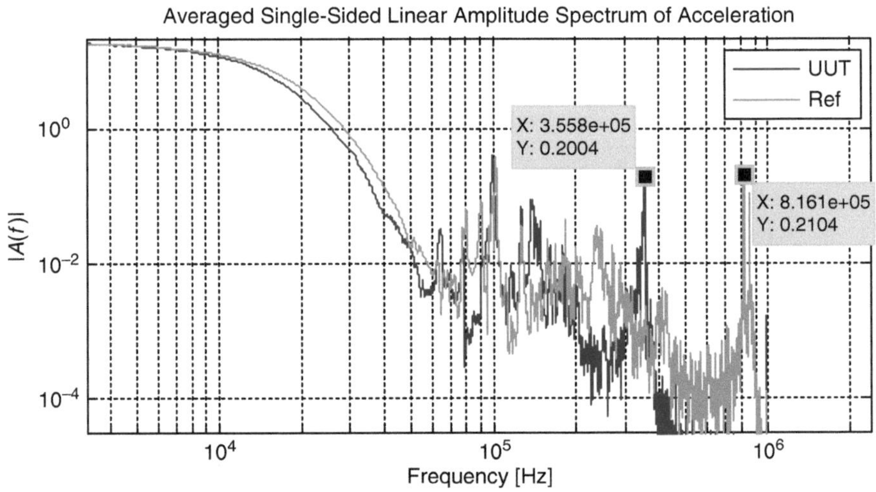

Fig. 11.6 Autopower profile from high amplitude tests; the peaks corresponding to the resonant frequency of the UUT and reference sensors are indicated

The frequency response function of the UUT with respect to the reference is examined next. The frequency response function is given as the ratio of the output $Y(\omega)$ to the input,

$$H_{yx}(\omega) \equiv \frac{Y(\omega)}{X(\omega)}, \tag{11.11}$$

in the frequency domain. Assuming the reference as the input, an amplified signal present in the UUT due to sensor resonance will create a corresponding peak in the frequency response function. An FRF from a high-amplitude test (peak acceleration of approximately 100 kg) is shown in Fig. 11.7. A SDOF frequency response function (Eq. 11.10) is fit to the data using a Matlab curve-fitting by varying the fundamental frequency and damping ratio. Even with coarse tolerance, a fit of the FRF amplitude yields best-fit values of $f_n = 343.9$ kHz and $\zeta = 0.00185$ ($Q \sim 27000$).

The results change markedly if low amplitude data is examined. Figure 11.8 gives the average FRF of a lower amplitude test (peak acceleration of approximately 20 kg). Due to the increased amount of programming material and low impact velocity, the input stress pulse has minimal high frequency content to excite the resonance of the UUT. As seen in Fig. 11.8, the resulting FRF has a barely discernible peak and considerable noise. The results of the fitting are more telling: the fit of the lower amplitude data yields best-fit values of $f_n = 344.8$ kHz and $\zeta = 0.0415$ ($Q \sim 1200$).

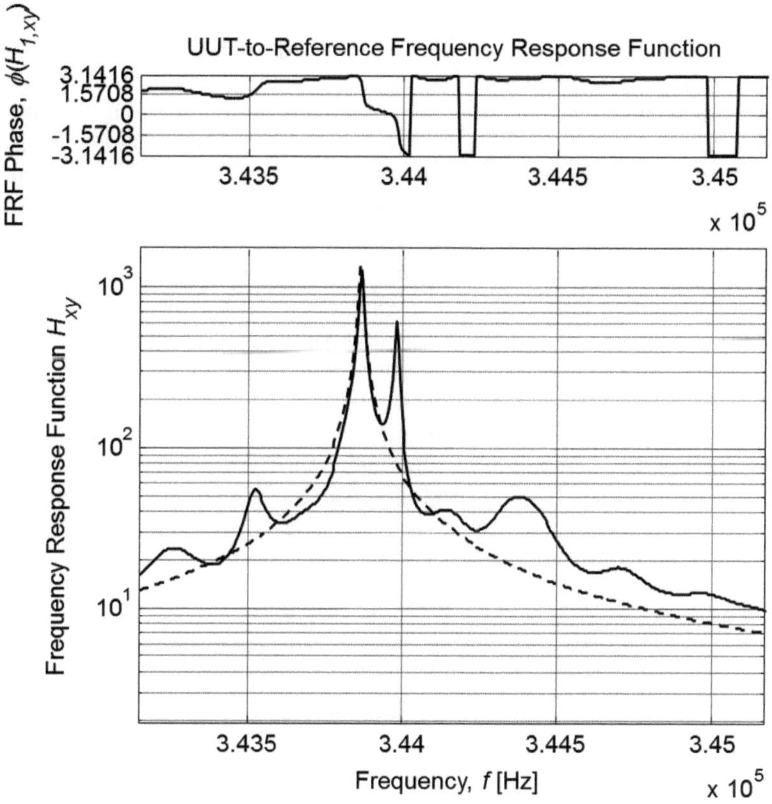

Fig. 11.7 The FRF from a high amplitude experiment. The experimental data (*solid line*) is overlaid with the FRF of a SDOF oscillator (*dashed line*) with $f_n = 343.9$ kHz, $\zeta = 0.00185$ ($Q \sim 27000$)

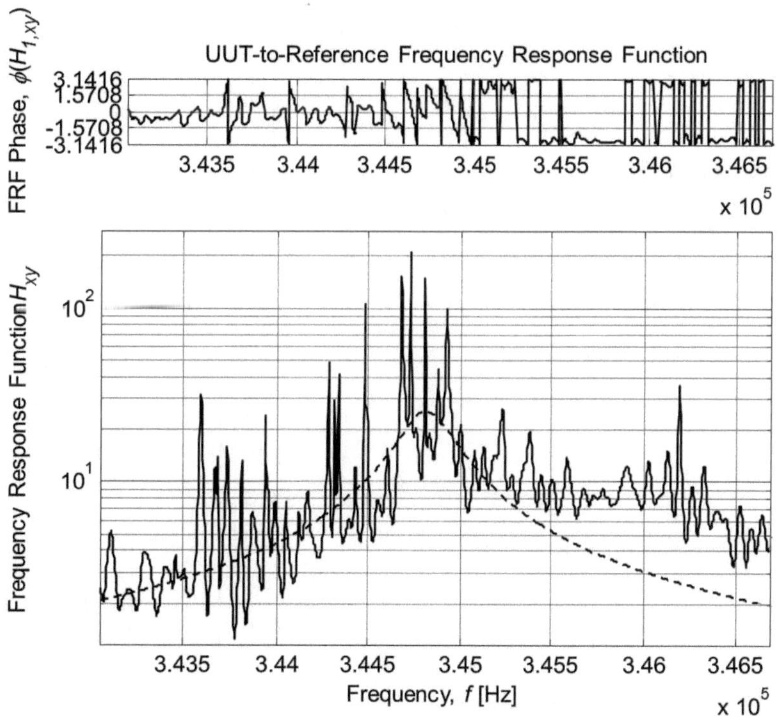

Fig. 11.8 The FRF from a low amplitude experiment. The experimental data (*solid line*) is overlaid with the FRF of a SDOF oscillator (*dashed line*) with $f_n = 344.8$ kHz, $\zeta = 0.0415$ ($Q \sim 1200$)

11.5 Discussion

Several issues arise when examining and attempting to draw conclusions on the data presented above. The inconsistency of the estimates of the resonant frequency – and even more so the damping – raises questions that bear further investigation. The resonant frequency varied little between low and high amplitude data: 344.8 and 343.9 kHz, respectively. However, the damping ratio differs by a factor of over 20 times, i.e., $\zeta = 0.0415$ at low amplitude vs. $\zeta = 0.00185$ at high amplitude. When examining operating data with low frequency content, it is reasonable that high frequency resonances (such as those in MEMS devices) will not be readily excited. The input (or reference) spectrum must be compared with the noise floor near the expected resonance; even with a relatively high Q, the sensor data presented here did not achieve enough excitation (or equivalently signal-to-noise) to provide conclusive low amplitude results.

Repeated roots and/or closely spaced modes are another challenge in utilizing operating data to infer the dynamic parameters of a sensor. Figures 11.9 and 11.10 show two additional sets of data (again from the same sensor) at varying levels of input amplitude and frequency content. In the FRF in Fig. 11.9, two strong peaks are evident as was the case in Fig. 11.7. However, the higher frequency mode now has greater participation. Additionally, the values have also shifted significantly relative to either the low or high amplitude data. Likewise, Fig. 11.10 indicates the presence of additional modes or repeated roots. By considering the system a "black box" without insight into the design features, interpreting the dynamic response becomes exceedingly difficult. Estimating the various contributing modes, quantifying their participation in various dynamic events, and ultimately linking the response to detailed models of the device structure is the focus of ongoing research.

Future studies will also focus on the details of the device-level resonance, particularly any damping mechanisms such as squeeze film damping [13] commonly employed in MEMS devices. Due to expected nonlinearities in the response, both time-frequency analysis and traditional nonlinear analytic tools such as backbone curves will be applied to operating data to confirm and quantify the underlying mechanisms.

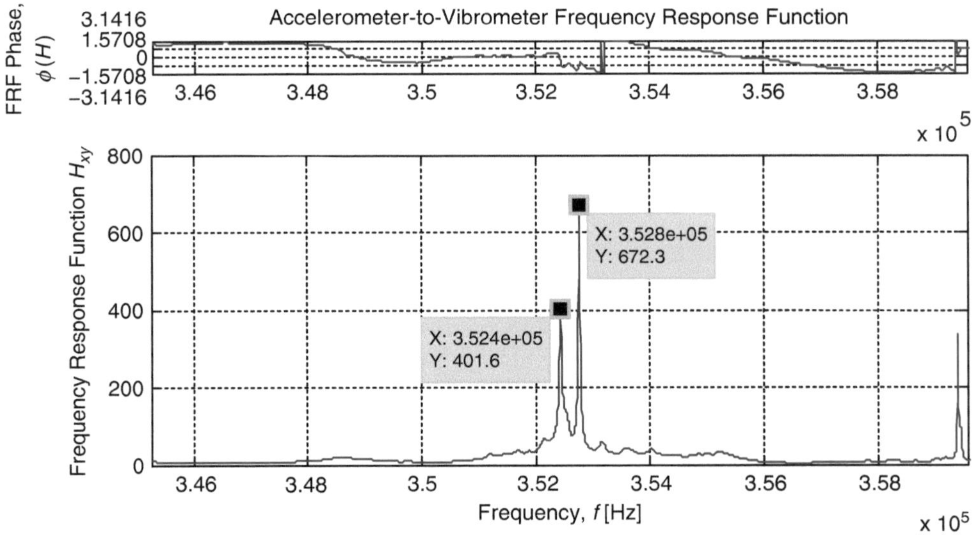

Fig. 11.9 Averaged FRF from intermediate range (40 kg) input; the presence of two nearby modes is implied by the closely spaced peaks and corresponding phase shifts

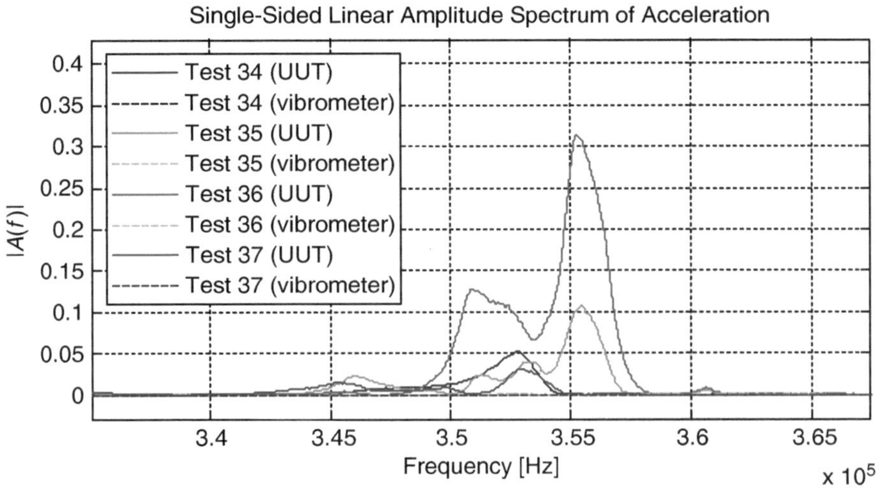

Fig. 11.10 Detailed autopower spectra in the neighborhood of the UUT resonance (350 kHz). Multiple, tightly spaced modes are potentially indicated in this distribution

11.6 Summary

Data on the expected dynamic properties of sensors are commonly provided by manufacturers. However, estimated properties are typically obtained at levels far below the full range. This paper analyzed high bandwidth operating data of varying amplitudes to identify the fundamental modes and dynamic characteristics of an undamped MEMS accelerometer. The damping ratio and principle resonant frequency of the shock accelerometers were estimated using a simple dynamic model for varying levels of excitation. The resonance was shown to be essentially independent of amplitude. In contrast, the damping ratio was shown to be strongly dependent on the excitation amplitude. The cause of the damping ratio change with respect to excitation was proposed to be a combination of signal-to-noise ratio and the lack of dynamic excitation in the input. Future studies are proposed to examine this issue and define the damping behavior in damped MEMS structures.

Acknowledgements The authors would like to acknowledge support of the project through the Air Force Research Laboratory and the Thurgood Marshall College Fund. The assistance of Kulite Semiconductors, specifically Steve Carter and Andy Bemis, is also gratefully acknowledged. Opinions, interpretations, conclusions, sensor and instrumentation selections, procedures, and recommendations are those of the authors and are not necessarily endorsed by the United States Air Force.

References

1. Wilson JS (2005) Sensor technology handbook. Elsevier, Amsterdam
2. Bateman VI, Brown FA, Davie NT, Nusser MA (1995) High shock, high frequency characteristics of a mechanical isolator for a piezoresistive accelerometer. In: Proceedings of the eighteenth transducer workshop, Colorado Springs, 20–22 June 1995
3. Walter PL (2001) The handbook of dynamic force, pressure and acceleration measurement. Endevco Corporation, San Juan Capistrano
4. Endevco Corporation (2006) Model 7270A Accelerometer Data Sheet, San Juan Capistrano
5. Sill RD (2007) Development of a damped piezoresistive MEMS high shock sensor. In: Proceedings of the 78th shock and vibration symposium, SAVIAC, Philadelphia
6. Scavuzzo RJ, Pusey HC (1995) Principles and techniques of shock data analysis. SAVIAC, Arlington
7. Bell W, Davie N (1987) Reverse Hopkinson bar testing. Sandia National Labs, Albuquerque
8. Togami TC, Bateman VI, Brown FA (1997) Evaluation of a Hopkinson bar fly-away technique for high amplitude shock accelerometer calibration. In: Proceedings of the 68th shock and vibration symposium, SAVIAC, Hunt Valley, p 11
9. Kulite Semiconductor Products, Inc. (2014) GMD-280-100KG-P Accelerometer specifications (preliminary)
10. Precision Filters, Inc. (2009) Precision 28144 quad-channel wideband transducer conditioner with voltage and current excitation (datasheet). Precision Filters, Inc., Ithaca
11. National Instruments (2003) NI PXI-6133 specifications. National Instruments, Austin
12. Foley JR, Dodson JC, Schmidt M, Gillespie P, Dick AJ, Idesman A, Inman, DJ (2009) Wideband characterization of the shock and vibration response of impact-loaded structures. In: Proceedings of the SEM IMAC XXVII, SEM, Orlando
13. Huang S, Borca-Tasciuc D-A, Tichy J (2014) Limits of linearity in squeeze film behavior of a single degree of freedom microsystem. Microfluid Nanofluid 16(6):1155–1163

Chapter 12
Development of a Mapping Function for a Low- to High-Amplitude Input

Joshua H. Campbell, Janet C. Wolfson, Jacob C. Dodson, Alain L. Beliveau, Jonathan Hong, and Greg Falbo

Abstract The Air Force Research Laboratory is continuing research to design a controllable pyroshock test to excite a test item over the entire frequency spectrum from low (10 Hz) to high (10 kHz) at different forcing levels. When a linear Frequency Response Function exists we would expect that we can design a multiple input test to achieve a desired output response based on superposition of the single input modal response of the test structure. Experimental modal analysis is performed to evaluate the differences in FRF's for the different input force levels generated by a modal hammer and a detonator. The final outcome of this research endeavor is to be able to perform a modal analysis on the test platform and use that information to define the input force location, magnitude, and time delay to achieve a desired output response. Initial studies at the output location will focus on utilizing the Shock Response Spectrum as the Figure of Merit, however, energy methods and other criteria may be considered. This paper will present progress in the development of the Design-a-Shock method and the difficulties that were encountered in trying to capture the input response of a detonator, the lessons learned from the testing, and a discussion on the characterization of the system response.

Keywords Bookshelf • Mapping • Simultaneous • Modal • Multi-axial

12.1 Introduction

The idea behind the Design-a-Shock is to determine if a relationship exists between an input from a force hammer and an input from a pyroshock event (detonator, multiple detonators, and pellet sized explosives) in the elastic range. The desire is to develop a scientific, repeatable, field experiment that can reproduce the desired forces in order to determine the failure mechanisms in the systems under test. A new test article was proposed and an initial computational study was performed and reported at the Society of Experimental Mechanic's IMAC XXVIII [1]. The analytical results were then compared to a near-field pyroshock test. This analysis led to the development of the Multi-Axial Pyroshock Plate (MAPP) test set-up. Initial pyroshock tests were performed on the MAPP test set-up where acceleration time histories were captured and Shock Response Spectra (SRS) were calculated. The success of the tests was determined by comparing the SRS from the tests with a desired SRS band. These initial tests show that the technology could simulate aspects of the SRS, however, the application of the pyroshock needed further refinement. In order to determine the optimum placement of pyroshock inputs an initial test series was developed to compare simple modal impacts to more complex near-field pyroshock inputs. A series of laboratory tests were performed on a sub-scale aluminum plate and the more complex full-scale MAPP test article. These tests involved using an impact hammer (input force) at different locations and capturing the frequency domain data at a variety of locations along a grid. The initial sub-scale laboratory tests focused on a specific input at 32 different locations and four output locations. Acceleration time histories were captured at each location. From that data the Frequency Response Functions

J.H. Campbell (✉) • J.C. Wolfson • J.C. Dodson
Air Force Research Laboratory, AFRL/RWMF; 306 W. Eglin Blvd., Bldg. 432, 32542-5430 Eglin AFB, FL, USA
e-mail: joshua.campbell.28@eglin.af.mil

A.L. Beliveau • J. Hong
Applied Research Associates, Inc., 956 W John Sims Pkwy, Niceville, FL 32578, USA
e-mail: alain.beliveau.ctr@us.af.mil

G. Falbo
LMS Americas, 5755 New King Drive Troy, MI 48098, USA
e-mail: greg.falbo@siemens.com

A. Wicks (ed.), *Shock & Vibration, Aircraft/Aerospace, and Energy Harvesting, Volume 9*,
Conference Proceedings of the Society for Experimental Mechanics Series, DOI 10.1007/978-3-319-15233-2_12

(FRF) and Shock Response Spectra (SRS) are calculated. From these previous test results, a new test series was developed with multiple sensor and detonator locations. The data from these initial test results has been gathered and analyzed to compare with the impact hammer and sub-scale laboratory data.

This paper will provide an overview of the MAPP field tests and discuss the results from that analysis. It will then provide a detailed discussion of the relationship between single and multiple pyroshock inputs, along with FRF and SRS analysis of these responses. This will demonstrate the feasibility of utilizing the multiple input aspects of this test methodology in achieving a higher magnitude of response depicted in the time and frequency domains. The final outcome of this research endeavor is to determine if one can perform a modal analysis on the test platform and use the information to predict an output response of an item under test utilizing input stimulus with much larger magnitudes. This will allow the engineer the opportunity to design the input force location, magnitude, and time delay based on a desired output.

12.2 Multi-axial Pyroshock Plate (MAPP) Set-Up

The test article consists of a $4' \times 8' \times 1''$ thick aluminum plate that is hung from an aluminum tube by turn buckles. They suspend the plate in simulated Free-Free boundary conditions. The aluminum tube that supports the MAPP test system spans $22'$ and is connected to concrete blast walls through the use of $3/8''$ expansion anchors. A "bookshelf" has been welded to the front of the plate; it is $6''$ long and is constructed out of aluminum tube stock. The nominal dimensions on the tube stock are $6'' \times 6'' \times ½''$. Figure 12.1a depicts the entire plate with the bookshelf. Figure 12.1b shows a closer view of the bookshelf and a steel plate that is attached to it. In order to allow for the connection of the steel plate, or any other test article, to the bookshelf a series of 8–$0.32''$ diameter thru holes were drilled into the bookshelf.

A standard x-, y-, z, coordinate system is assumed for this test article. The x-axis runs along the long side of the plate [along the bottom edge of the plate shown in Fig. 12.1a] while the y-axis is located parallel to the short side of the plate [along the left hand side of the plate in Fig. 12.1a]. The origin of the coordinate system is located at the lower left hand corner of the plate in the photo below. The z-axis comes out of the plate and runs along the $6''$ length of the bookshelf. This coordinate system will be used in the discussion of the results for the field tests.

12.2.1 Interface

Based on the observed damage that occurred in previous testing it became apparent that an interface must be used between the detonator and the plate. In order to not affect the application of the energy to the plate the interface must be made of the same material as the plate and be attached along the edges to ensure that there is not a material impedance problem between the two similar metals. This test series utilized two aluminum interfaces constructed out of 6,061 Aluminum. The interface was a minimum of $1½''$ in diameter and was attached to the MAPP test set-up by applying glue along the outside of the disc and between the disks. There was also a detonator holder attached to the outside disk for easy placement of the RP-87 detonator. Finally, a piece of Kapton tape was laid in-between the aluminum interface disks and the MAPP, as seen in Fig. 12.2.

Fig. 12.1 MAPP test article. (**a**) Overall MAPP test set-up; (**b**) view of "bookshelf" and steel plate

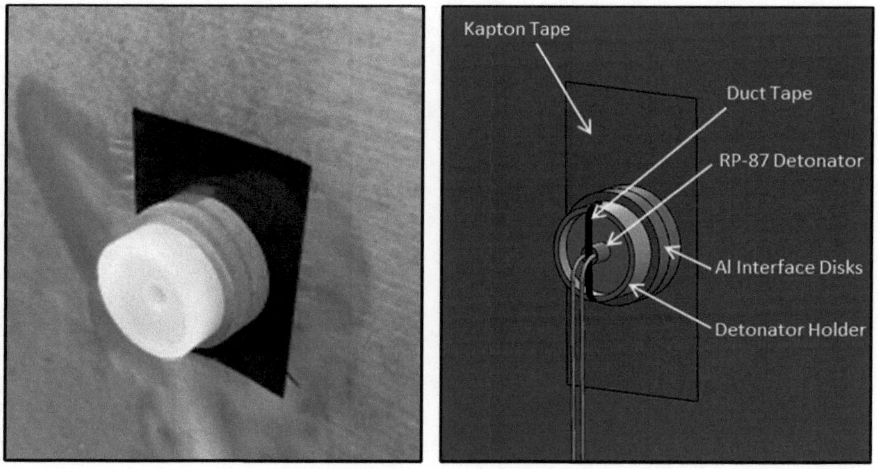

Fig. 12.2 Detonator interface pre and post test

12.3 Background

The initial test of the MAPP set-up occurred in April 2010 utilizing a pellet sized pyroshock input. These tests demonstrated that it was possible to match portions of a desired figure of merit. However, it led to desiring a better understanding of how the input can be utilized to predict a response. In other words, is there a mapping function between a modal and pyroshock input? To develop this relationship a complete modal study was performed on the MAPP test set-up in December 2011. Tri-axial accelerometers (consisting of 3 PCB 353B17 accelerometers on blocks) were glued to the plate on a $1' \times 1'$ grid as shown in Fig. 12.3. Figure 12.4 depicts a schematic of the test set-up and the layout of accelerometer locations. The AFRL Tri-axial accelerometer that was placed on the bookshelf is show in Fig. 12.5a. A tri-axial accelerometer created by placing 3 PCB 353B17 accelerometers on a titanium block is shown in Fig. 12.5b, while Fig. 12.5c shows 2 rows of accelerometers used in this test.

To perform this analysis a series of five impacts were captured at each accelerometer location. The Frequency Response Functions (FRF) at every location (input and output) was averaged together and the Auto Power Spectra (input location/ output location) and Cross Power Spectra (output location/ output location) were calculated using LMS Test.Lab. This data was saved and will be utilized to develop a mapping function between pyroshock and modal inputs.

12.3.1 Sub-scale Comparison Laboratory Study

The results from prior sub-scale laboratory tests show that multiple inputs can significantly affect the desired outputs as described by Wolfson [2]. Figure 12.6 shows the acceleration time histories for the z-axis output (orthogonal with the plate) for two points on the plate [Fig. 12.6a, b], and the simultaneous impacts at both points [Fig. 12.6c]. It can be seen that the individual impacts have a maximum force of less than 40 g's, however; if they are combined their overall force magnitude is higher at over 60 g's.

This point is further illustrated in the Frequency Response Functions (FRF's) shown in Fig. 12.7. In this plot, the red line represents the FRF at the output location from an input at one point. The green line represents an input at the second point, and the blue line is the combined input at both points. Figure 12.7 shows that there are some different principal modes in the two individual inputs, but there are many frequencies where they do have similar responses. When the inputs are combined the magnitude of the response at specific frequencies are generally increased, however, in some instances the overall behavior is reduced at that frequency.

The final comparison chart, and the one that shows the most significance, is in Fig. 12.7b. It shows the comparison of SRS from the individual and combined inputs. The SRS plots shown in Fig. 12.7b were calculated using LMS Test.Lab. This comparison shows a significant increase in the damage potential at the output point due to the simultaneous impacts. Above 1,000 Hz, there appears to be an order of magnitude increase in the SRS. This initial study has shown the effect that simultaneous impacts can have on the FRF's and the SRS's for the aluminum plate.

Fig. 12.3 Layout of accelerometers from initial modal tests

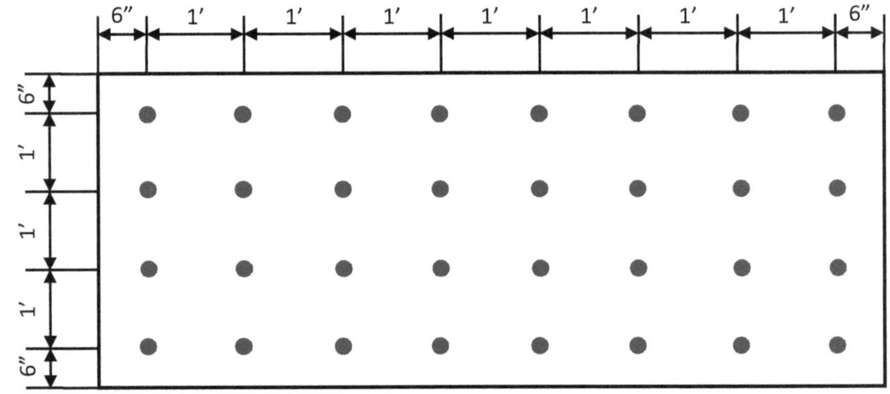

Fig. 12.4 Schematic of accelerometer layout for modal tests

12.4 Full-Scale Pyroshock Testing

The specific test configurations for each test series were chosen to determine the effect that certain distances had on both the input and the output. Understanding how the wave propagates through the MAPP and how the intensity died off with time was important in post-processing. In all of the configurations there was a control accelerometer that remained constant throughout the test series. The final configuration was a combined response that incorporated all of the detonator locations with common sensor locations.

12.4.1 Test Layout

Five different test layouts were utilized in this test series. Four layouts utilized 1 – RP83 detonator, 7270A-60 k accelerometers, 7270A-20 k accelerometers, and 7270A-6 k accelerometers. The location of the accelerometers with respect to the detonator was used to determine what variant to use. Three of the 6 k accelerometers were placed in the AFRL tri-axial mount and attached at the center of the bookshelf. Configuration 1, 2, and 3 show the layouts when 1 detonator is used. Configuration 4 shows the details of the multi-detonator setup (Fig. 12.8).

12.5 Analysis

After concluding with the field tests, the sensor data was used to create shock profiles in the time and frequency domains. Acceleration data was compared in the time domain to observed relationships between the single and multiple input configurations. FRF analysis and Cross Power Spectra analysis was the majority of interest in the frequency domain.

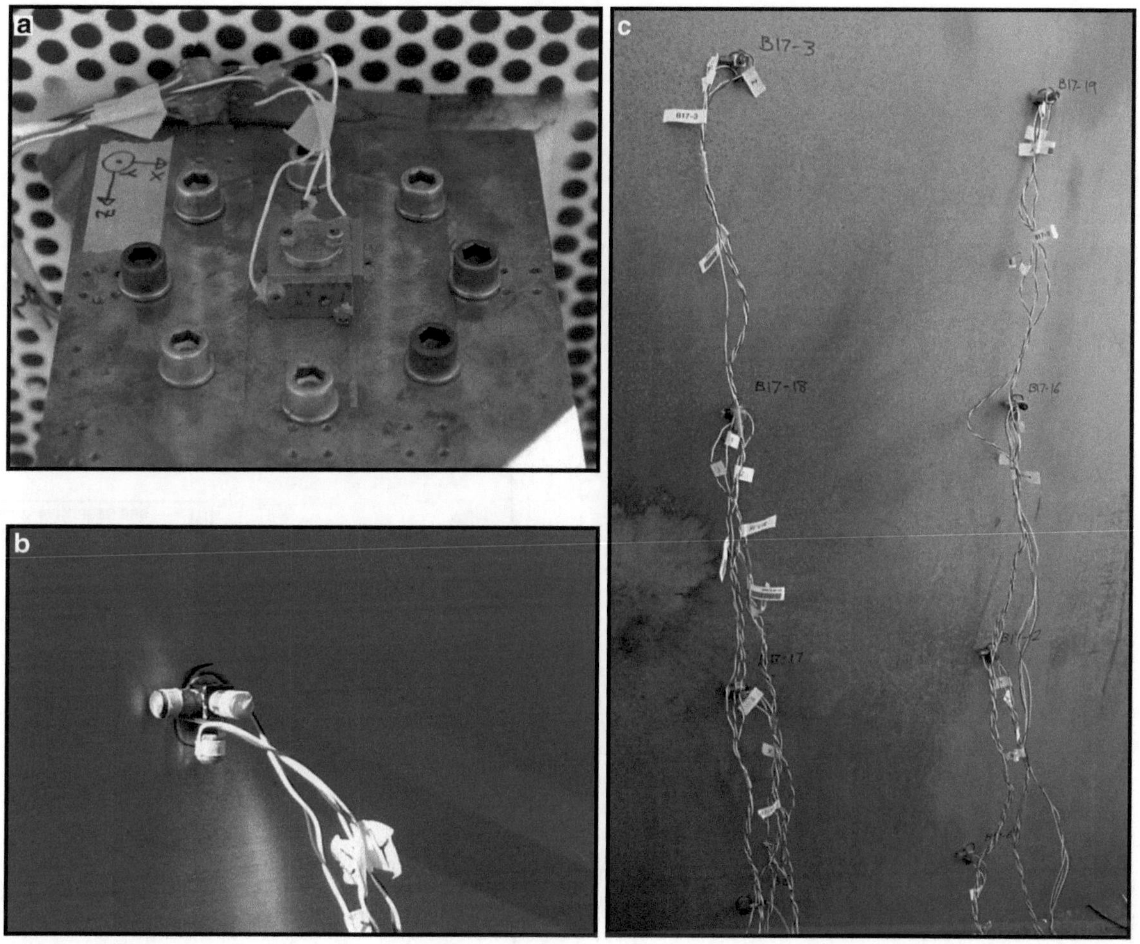

Fig. 12.5 Photos of accelerometers from initial modal tests. (**a**) AFRL tri-axial shock accelerometer mounting block; (**b**) typical tri-axial modal accelerometer mounting block; (**c**) two rows of tri-axial accelerometers

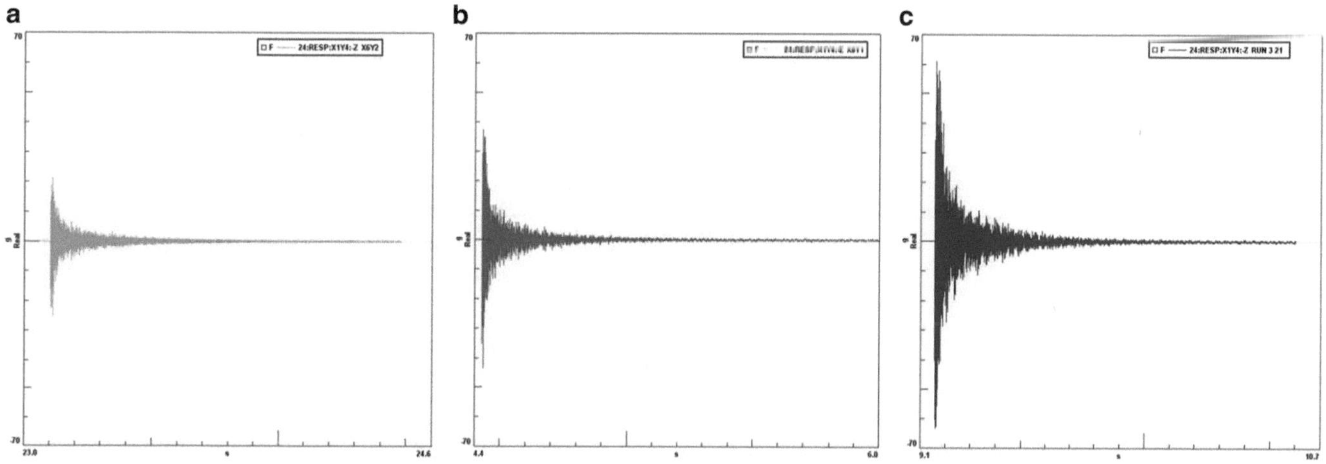

Fig. 12.6 Time history of individual and combined inputs. (**a**) Individual input; (**b**) individual input (**c**) combined input

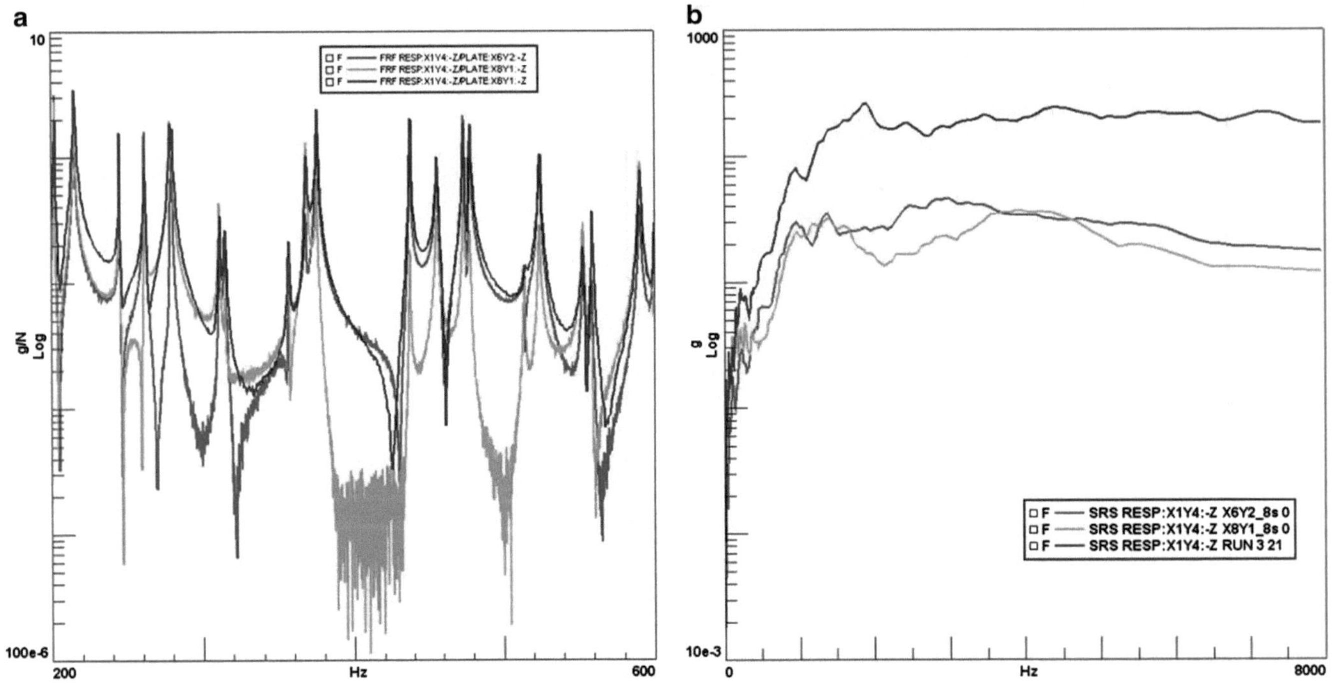

Fig. 12.7 Frequency response function and shock response spectra of points 1 and 2 and combined. (**a**) FRF of individual points and combined; (**b**) SRS of individual points and combined

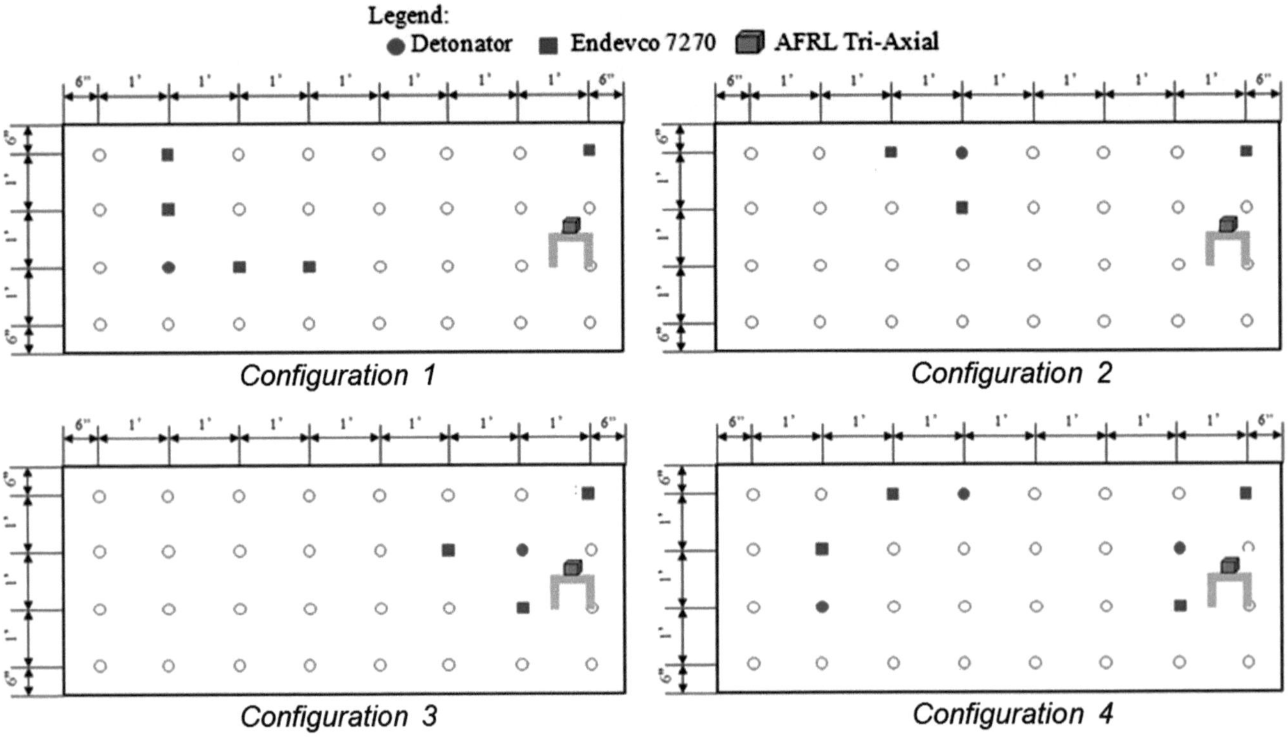

Fig. 12.8 Schematics of accelerometer configurations

This allowed us to superimpose frequency data from each configuration in order to determine the relationship between single and multiple inputs. Finally, the SRS was calculated and analyzed for maximum amplitudes at specific frequencies. This showed whether our tests reached the maximum SRS bands desired from previous full-scale testing.

12.5.1 Time Domain

Pyroshock is the desired method to excite the MAPP test apparatus because it differs from other types of mechanical shock. Compared to mechanical shock, there is very little rigid-body motion of a structure in response to pyroshock. The acceleration time-history of a pyroshock, measured on the structure, is oscillatory and approximates a combination of decayed sinusoidal accelerations with very short duration as seen in Fig. 12.9. The data has been trimmed to the start of the event and filtered at 800 kHz. In these cases the pyroshock acceleration time-history consists of a high-frequency, high amplitude shock that may

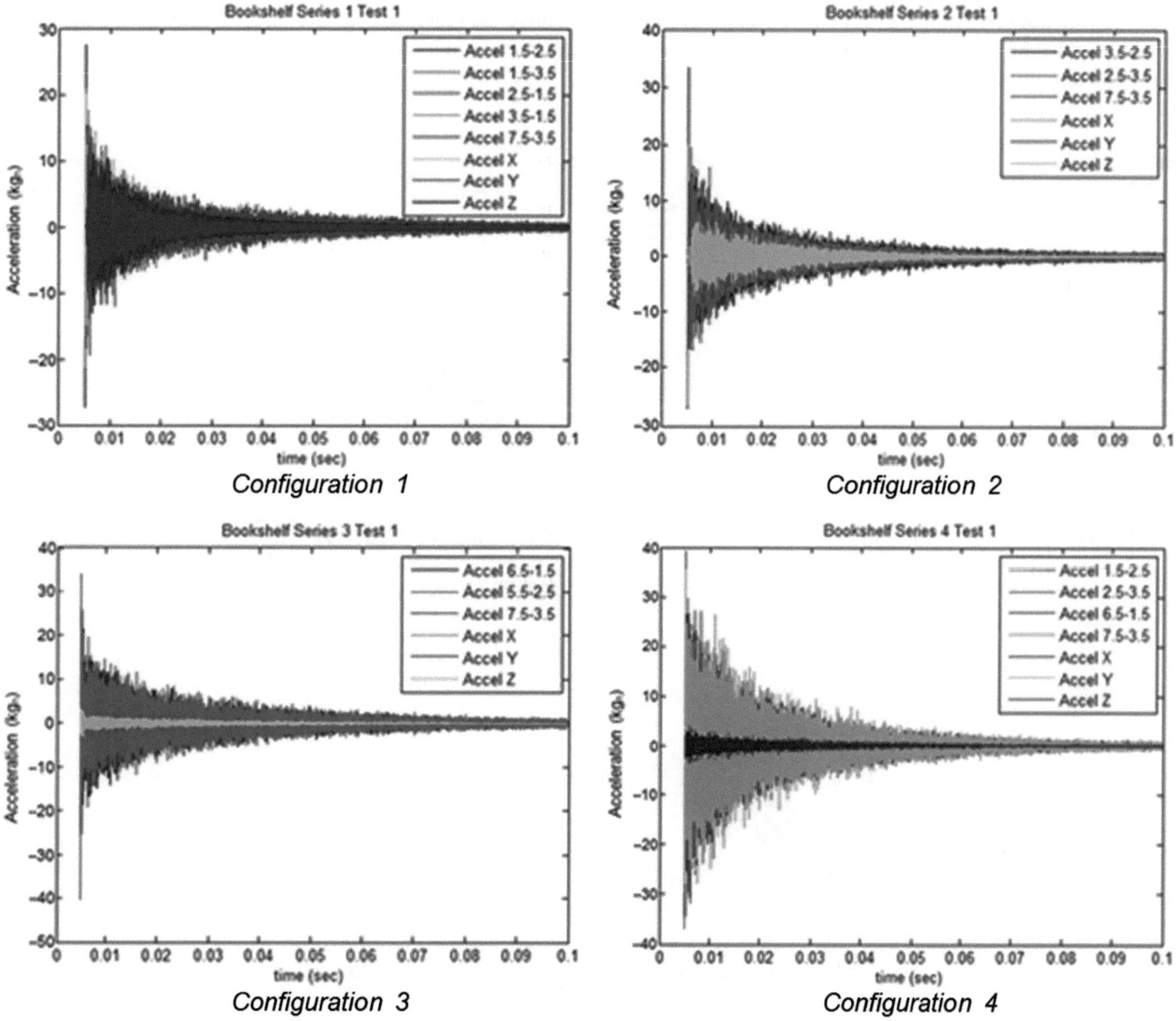

Fig. 12.9 Multi-axial pyroshock acceleration time histories

have transients of microseconds or less. This near-field energy is distributed over a wide range of frequencies and is typically not dominated by a few selected frequencies. The energy deposition time for a pyrotechnic event is very small and does not strongly excite the rigid body modes of the structure. The resulting stress waves, from the explosives, propagate through the test article and high-frequency energy is gradually attenuated due to various material and structural damping mechanisms. That high-frequency energy is then transferred or coupled into the lower frequency modes of the structure.

Unlike the results from the sub-scale laboratory tests, multiple inputs had some effect on the desired output depending on where the sensors were located. For the tri-axial block, there was no relationship with individual and simultaneous inputs in the time domain. For every configuration, the X, Y, and Z simultaneous input test had the second lowest magnitude of the 4 tests. Configuration 4 should be larger than the other 3 configurations if there was a direct mapping function between an impact hammer and pyroshock input (Fig. 12.10).

Similar to the sub-scale laboratory test, multiple inputs had an impact on many of the sensors located on the plate. The top-right corner sensor that is common among all of the configurations maintained the highest magnitude with the simultaneous input test. The combination of the single inputs may equal the Configuration 4 input but further analysis will be required. This was the same for two other sensors in Configuration 1 and 2. For the most part, there are some locations where the superposition of single inputs does tend to lead the correlating multiple input magnitude. However, this is not evident as a trend for the test series. Figure 12.11 shows some circumstances where the first 3 test series contain acceleration magnitudes lower than the 4th configuration. The possibility exists that, for these configurations, the superposition of single impacts may equate to a multiple impact.

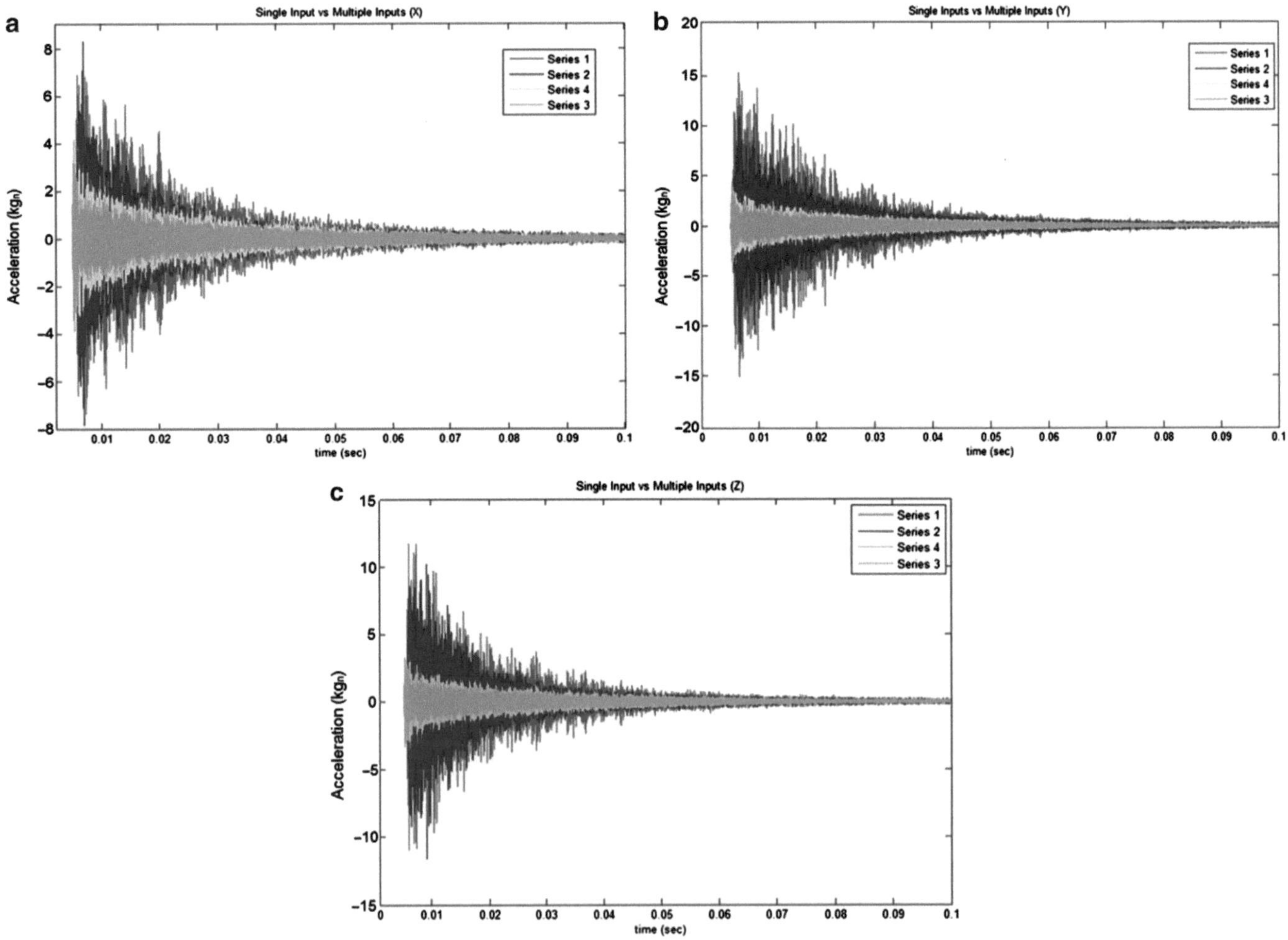

Fig. 12.10 Single and multiple inputs for tri-axial block. (**a**) Single and multiple inputs (X direction); (**b**) single and multiple inputs (Y direction); (**c**) single and multiple inputs for tri-axial block

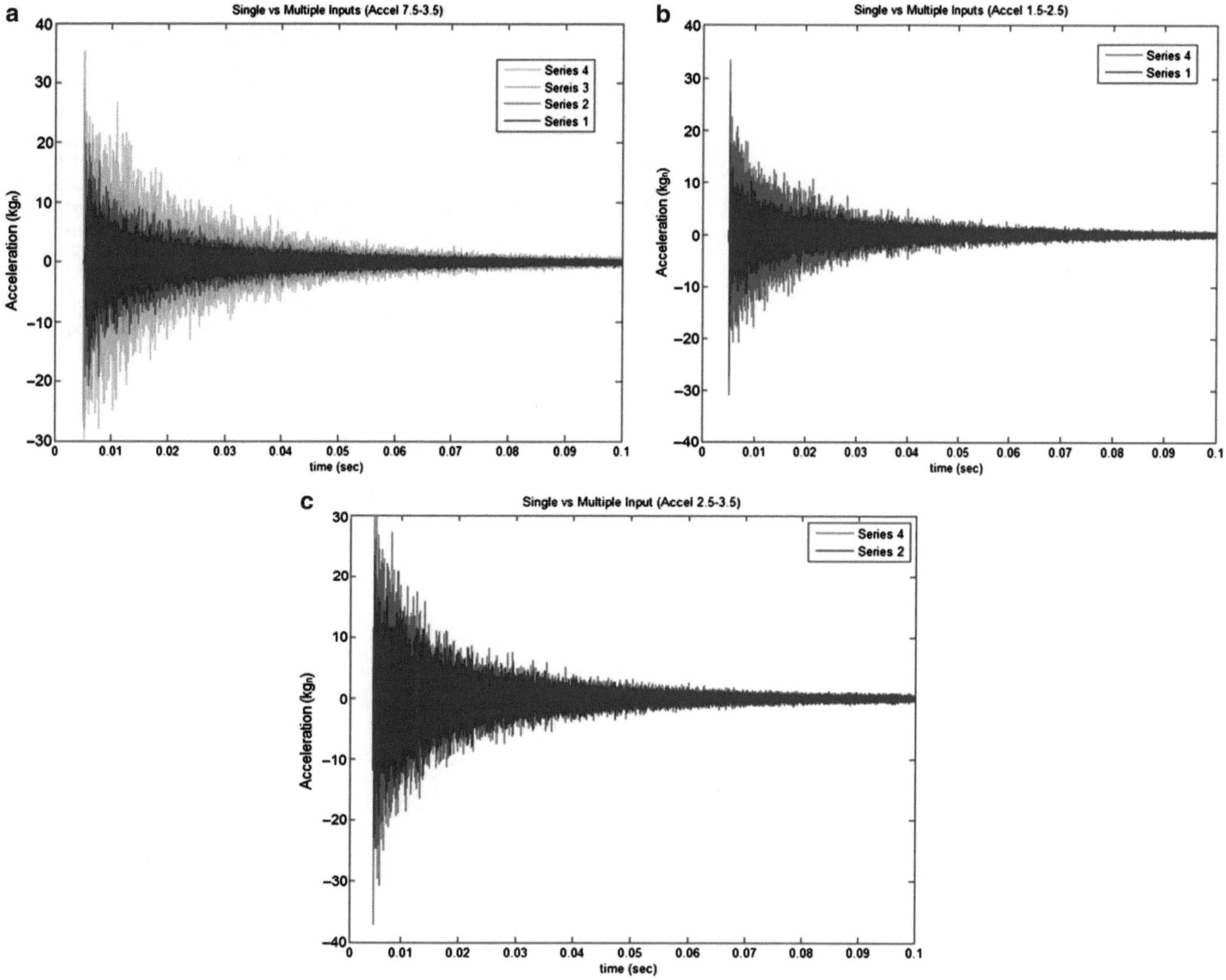

Fig. 12.11 Single and multiple inputs for plate sensors. (**a**) single and multiple inputs (7.5–3.5); (**b**) single and multiple inputs (1.5–3.5); (**c**) single and multiple inputs (2.5–3.5)

12.5.2 Frequency Response Function Analysis

The frequency response is the quantitative measure of the output spectrum of a system in response to a stimulus, and is used to characterize the dynamics of a system. It is a measure of magnitude and phase of the output as a function of frequency, in comparison to the input. For a system to be linear, the output response of an input will be at the same frequency with a certain magnitude and phase angle relative to the input. When you double the amplitude of the input, the amplitude of the output should also double. The FRF is significant for mechanical properties because it shows attenuation or amplification at a certain frequency for a structure.

When looking at the FRF, we are also looking at the coherence of the system to check for confidence in our measurement. As seen in Fig. 12.12, the FRF's don't change drastically between the series for the X-, Y- or Z- direction of the tri-axial block. This would imply a linear system but a closer check of the coherence would be required for a definitive statement. Certain frequency bands do have drastic magnitude changes between the single inputs and the multiple inputs. More analysis at these certain frequency bands along with a coherence check would lead to a possible conclusion of what frequency bands maintain the hypothesis that single input acceleration magnitudes sum up to the multiple input magnitude at given locations. In general, there is no relationship in the FRF from single to multiple inputs.

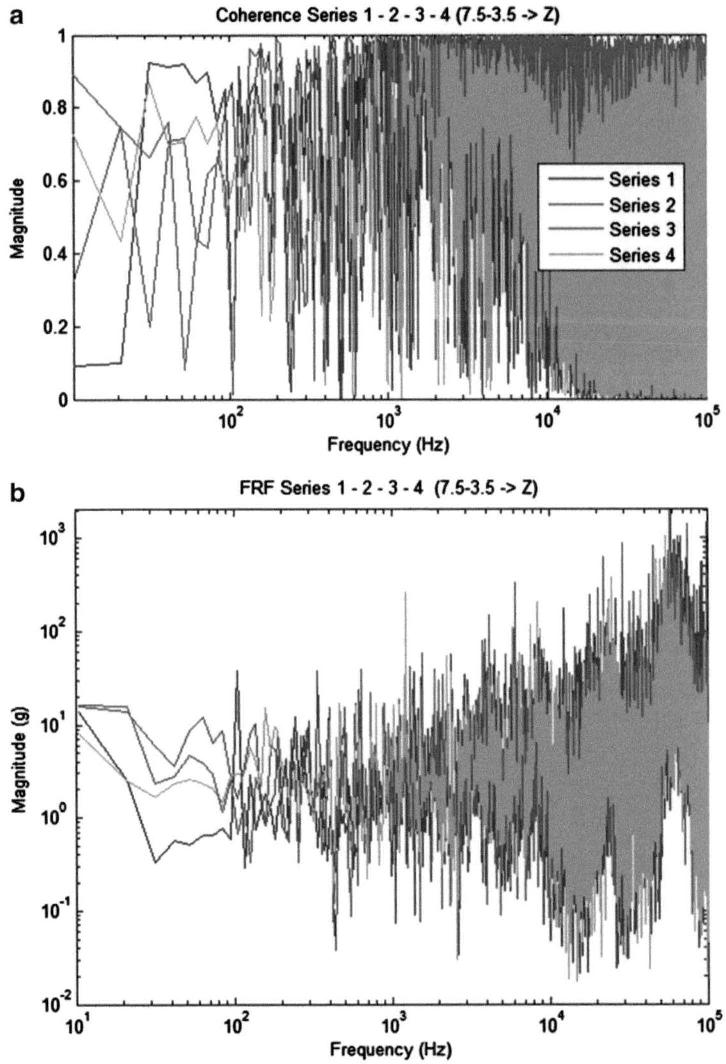

Fig. 12.12 FRF analysis with coherence. (**a**) Coherence; (**b**) FRF

12.5.3 Output Auto Power Spectra Analysis

When looking at the Auto Power Spectra Analysis, we take into account all of the outputs from the various locations and compare them. This is a frequency domain analysis of the cross-correlation between certain time series and their particular amplitudes. The Auto Power Spectra method is preferred because this is the output, which includes the systems response, instead of the input where vital information could be left out. Also, the phase is eliminated for Auto Spectra Analysis. We can use this to determine if the single input magnitudes (units of g^2) combine to the simultaneous input magnitudes. It is usually not helpful to look at the coherence of the system when doing auto spectra analysis because only the output is considered. Coherence should only be utilized when the input and output of a system are analyzed, such as with an FRF. As seen in Fig. 12.13a, in general there is no relationship between single and simultaneous inputs in regards to the output. A more detailed analysis would be required to find correlations in certain frequency bands for this particular phenomenon. However, there are certain bands that maintain the theory that Series 1, Series 2, and Series 3 all sums up to equal the Series 4 magnitude. This can be seen in Fig. 12.13b along with a magnitude calculation at 42.13 kHz. In general, there is no relationship between the single and simultaneous inputs across the entire frequency spectrum based on an Auto Power Spectra analysis. This also implies that the system tends to be non-linear.

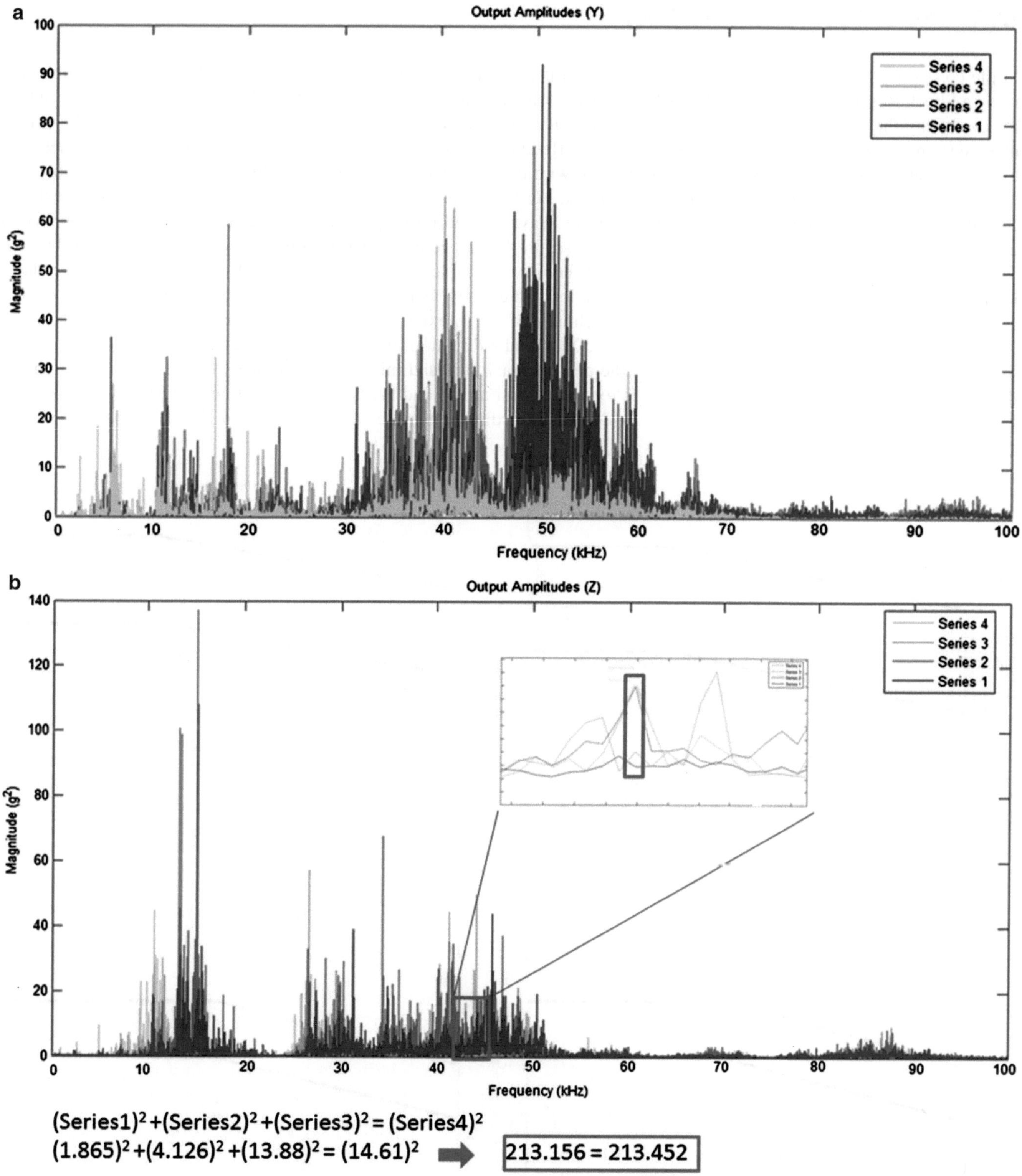

Fig. 12.13 Cross spectra analysis of tri-axial block. (**a**) Auto spectra analysis of output amplitudes (X direction); (**b**) Auto spectra analysis of output amplitudes (Z direction)

12.5.4 Shock Response Spectra Analysis

In the complex environment that AFRL is interested in there is an accepted methodology of developing test requirements using a Shock Response Spectrum (SRS), which was used as our initial figure of merit. The SRS has been proposed as a tool for evaluating the damage potential in a given acceleration time history. The SRS is defined using an array of 1-D spring-mass systems, each with a spring constant tuned to a different resonant frequency ($\omega = \sqrt{k/m}$). The maximum acceleration by an oscillator when coupled to a rigid base moving with the specified acceleration time history defines the "positive" or "negative" SRS depending on the direction of the shock. Further details on the SRS can be found in comprehensive reviews, e.g., Irvine [3]; other spectral analyses can be found in Scavuzzo and Pusey [4]. The positive and negative maximum SRS gives the maximum acceleration of the 1-D spring mass in the respective directions due to the acceleration time history. The SRS provides a measure of the effect of the pyroshock on a simple mechanical model with a single degree of freedom. Generally, a measured acceleration time-history is applied to the model and the maximum acceleration response is calculated.

The maximum positive SRS for each test and each axis of the tri-axial accelerometer were calculated and are shown in Fig. 12.14. Figure 12.14 compares the response of the X, Y, and Z directions for the accelerometer and tri-axial accelerometer in the component. The test data is plotted with the desired maximum and minimum SRS bands for this test series. One of the goals for the design-a-shock methodology is to create a test article and the necessary input needed to have the SRS fall within the maximum and minimum bands in the X, Y, and Z directions. The charts depicted in Fig. 12.14 show a variety

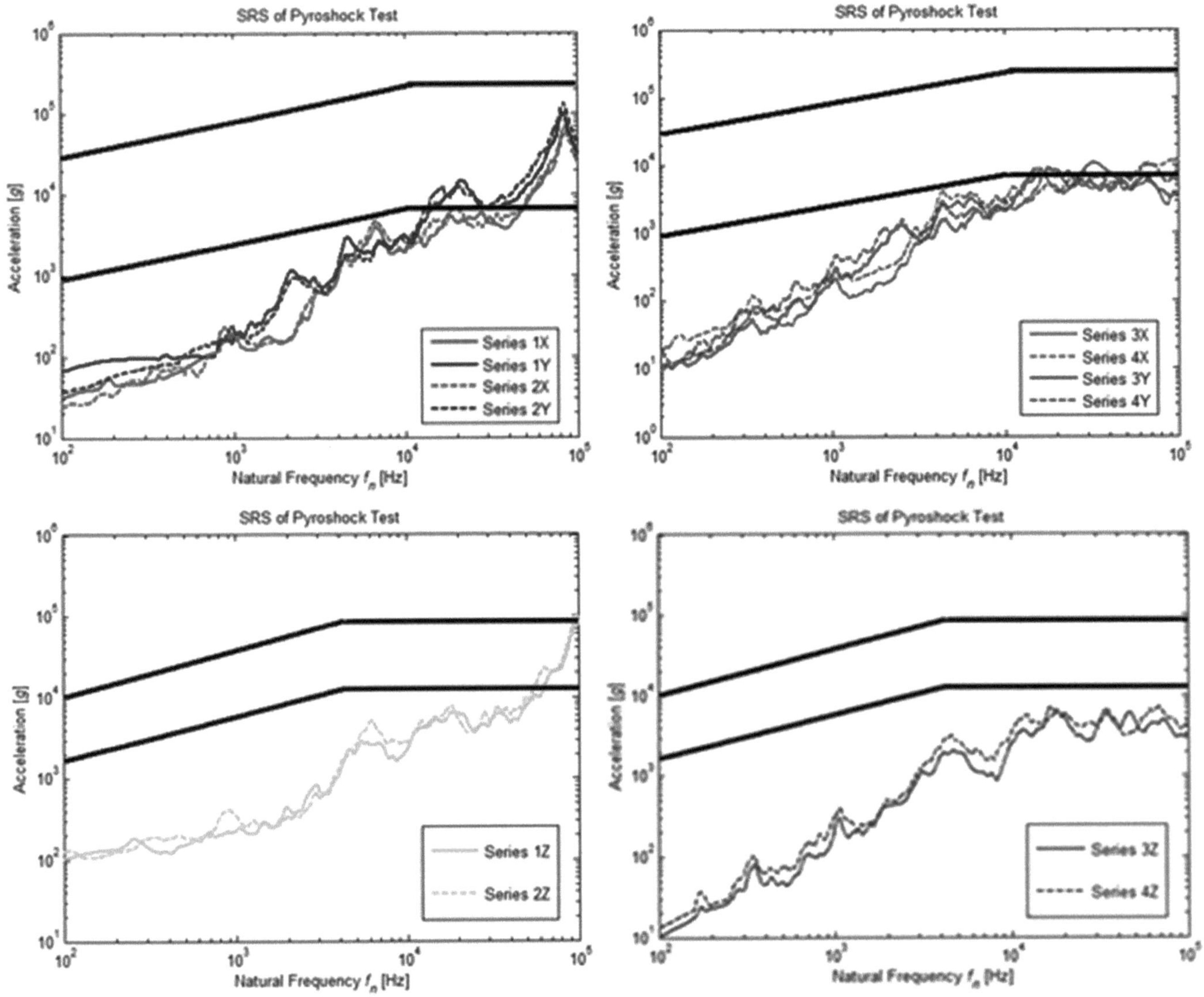

Fig. 12.14 SRS for single and multiple input responses

of interesting phenomena, specifically the consistency in SRS behavior between tests. While this is expected when applying a mechanical shock, it is not as expected when using live explosives to apply energy to the system. It verifies that our test apparatus stays in the elastic regime and is not damaged, or altered, between tests. Additionally, the plots show that in all directions the tests were generally unsuccessful in achieving the minimum SRS band. The SRS does reach the minimum bands with very high frequencies in Series 1 and 2. The simultaneous inputs test does not reach the minimum SRS band, which contradicts the idea of higher amplitudes with multiple inputs.

12.6 Summary

Previous sub-scale laboratory testing showed that by providing multiple impacts on a plate, at specific locations, the additional input can generally increases the magnitude of the Frequency Response Functions. Additionally, when looking at the Shock Response Spectra there is a significant increase in the damage potential at the output point due to the simultaneous impacts. In general, after initial testing with the full scale test setup, simultaneous inputs didn't have as significant effect on the magnitude of the shock response. Whether looking at overall time-domain vs acceleration data or frequency data, there is nothing that consistently portrays a drastic magnitude change between single and multiple inputs. For the SRS analyses, only at very high frequencies were we able to see any peak acceleration inside the necessary bands. For future testing, further emphasis should be given to sensors with similar plate locations on the MAPP rather than all the focus being on the bookshelf. There is far too much variability between configurations due to edge boundary condition, shock wave interactions, and the actual true magnitude of your input. When presenting multiple inputs into the MAPP, more understanding is needed to characterize the shock waves and how they interact. New frequencies and wave properties are introduced to the problem that weren't present in the single input tests. If some of these issues can be addressed then computational simulation for response prediction at certain locations on the MAPP may be in the realm of possibilities. Finally, the 2011 modal data that was never processed should be compared to this pyroshock data for analysis of possible mapping functions.

References

1. Wolfson J et al (2010) Modal testing of complex hardened structures. In: IMAC XXVIII A conference and exposition on structural dynamics, Jacksonville
2. Wolfson J (2011) Turning a pyroshock input to a SRS. In: Proceedings of 82nd symposium and shock vibration, SAVIAC, Baltimore
3. Irvine T (2002) An introduction to the shock response spectrum
4. Scavuzzo RJ, Pusey HC (1995) Principles and techniques of shock data analysis, 1995th edn. SAVIAC, Arlington

Chapter 13
Experimental Study of Glass Fiber Reinforced Polyester Under Repeated Impacts

Ahmed M. Elmahdy, Abdelhalim M. Elhabak, Mahmoud A. Adly, and Mohamed M. Elbawab

Abstract The results of an experimental study on the repeated impact behavior of woven glass fiber reinforced polyesters are presented. Tests were carried out using a specially designed repeated impact testing machine, which utilizes a pneumatic system to apply the loads, and a data acquisition system to directly measure the number of impacts and deflections of the specimens. Calibration results of the device are reported and a comparison is carried out between experimental and analytical impact forces. Dynamic deflections and impact fatigue trend of composite specimens under repeated impacts are also discussed.

Keywords Impact • Polymeric composites • Glass fibers • Delamination • Low cycle fatigue

13.1 Introduction

Fibrous polymeric composite materials have increasingly been used in various industries such as aerospace, automotive and sports industries. Their importance comes from the fact that composite materials can be tailored to produce specific mechanical properties for various applications. However, due to the lack of through-the-thickness reinforcement in fibrous composites, they can be subjected to damage and failure when transverse loads are applied on the laminates. Such transverse loads can be static loads or dynamic loads such as impact and shock loads. Impacts can range from small stones hitting a composite car bumper or small tools dropped on a composite surface, to a high speed bullet fired on a composite panel.

The importance and complexity of the subject has led to numerous research in order to understand the response of composite materials under impact loads. Researchers studied different parameters such as residual strength and stiffness after impact [1–3], residual flexure properties after impact [4], fiber content [5], hybridization [6], in-plane dimensions [7], and testing temperature [8]. While extensive research was carried out covering the response of composite materials subjected to single impact events, relatively little research was done on the response of polymeric composites to repeated impacts or impact fatigue events. Low velocity impacts can develop internal microcracks that are barely visible in the composite structures [9]. With the accumulation of microcracks due to repeated impacts, these structures can be subjected to severe failures without prior notice.

Several testing machines and setups were used to apply repeated impacts on the tested specimens. Falling weight impact testing machines were used by Sevak [10], Icten [11] and De Morais [12], while others like Bora [13], Çoban [14] and Roy [15] used swinging pendulum machines. New devices were also introduced by Azouaoui [16], which transforms rotary motion into linear impact motion for high cycle impact fatigue, and a device by Mittelman [17], which takes advantage of stored energy in a spring to apply impacts with a spring cam mechanism. Several parameters were also studies by some researchers such as testing temperature [11], hybridization [10], number of impacts [13–15], residual strength [16], and contact time [18].

The aim of this study is to design a new compact device, which can be used to apply repeated impacts at low cycle fatigue rates, and studying the dynamic behavior and impact fatigue trend up to failure of glass fiber reinforced polyester composites.

A.M. Elmahdy (✉) • A.M. Elhabak (✉) • M.A. Adly (✉)
Department of Mechanical Design and Production, Faculty of Engineering, Cairo University, Giza 12613, Egypt
e-mail: ahmedelmahdy@eng.cu.edu.eg

M.M. Elbawab
Rexam Beverage Cans, 6th of October City, Egypt

A. Wicks (ed.), *Shock & Vibration, Aircraft/Aerospace, and Energy Harvesting, Volume 9*,
Conference Proceedings of the Society for Experimental Mechanics Series, DOI 10.1007/978-3-319-15233-2_13

13.2 Experimental Procedures

13.2.1 Specimen Description

The composite specimens used are glass fiber reinforced polyesters. Glass fibers are bidirectional weave cloth of volume fiber 53.3 %. The plate used is made up of eight layers of thickness 0.25 mm each. Specimen strips are cut into dimensions of 120 mm by 30 mm with a total thickness of 4 mm.

13.2.2 Impact Device Description

The repeated impact testing device utilizes a pneumatic system to apply the impacts, as shown in Fig. 13.1. It consists of an air compressor (1) of power 1.5 hp which has a working pressure range of 1 bar up to 5 bars. The capacity of the attached air tank is 24 liters. The compressed air passes through a hose to an air filter (2) and a solenoid valve (3). The solenoid valve is connected to a double action air cylinder (4), which uses the air pressure from the compressor to actuate and provide the force needed for impact. The air cylinder diameter is 70 mm and the maximum stroke of its piston is 100 mm. The solenoid valve controls the repeated action of the impact. It takes its signal and power from a control unit (5), which is programmed to provide specific fixed timing signals to open and close the outputs of the valve. The force of the air cylinder's piston is transferred to a system of connecting rods (6), which is attached to the impactor (7). The impactor hemispherical nose is 12 mm in diameter and the distance between the impactor nose and the specimen – referred to as 'distance X'- ranges from 5 to 60 mm. Different specimen dimensions and materials can be used and fixed in the specimen fixation (8), up to 120 mm in length and 100 mm in width. A photoelectric switch sensor (9) and a linear variable displacement transducer (10) are attached to the device to enable automated measurements of the number of impacts and the deflections of the specimens respectively. Both sensors are controlled via a data acquisition card (11) NI-USB 6212, which is connected to a personal computer (12) to display and record the measurements by a means of an interface designed using LabVIEW program.

When the pressurized air flows into the air cylinder, the control unit sends intermittent signals to the solenoid valve which allows the air to flow in each side of the air cylinder, one at a time, causing the repeated impact force on the specimen. The impact begins with a forward stroke, causing a sudden impact force, then the impactor stays in contact with the specimen for a certain contact time while applying force in the same time. Finally, the impact cycle ends with the return stroke, which releases the impactor away from the specimen. Figure 13.2 illustrates the load cycle of the used impact device. The impact load cycle takes approximately 8.5 seconds to be applied, making the device capable of delivering approximately seven impacts per minute.

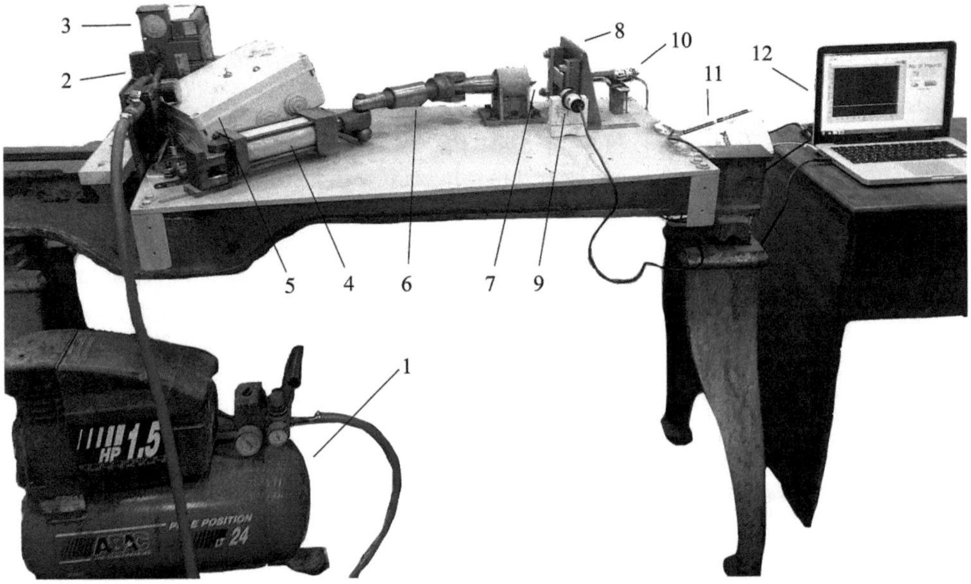

Fig. 13.1 Components of the repeated impact testing device

Fig. 13.2 Load cycle of the repeated impact

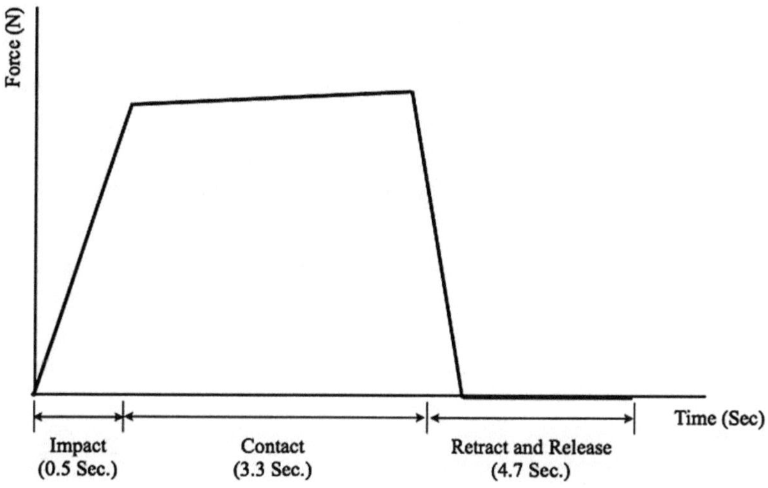

Fig. 13.3 Elastic and plastic behaviors of the specimens under impact

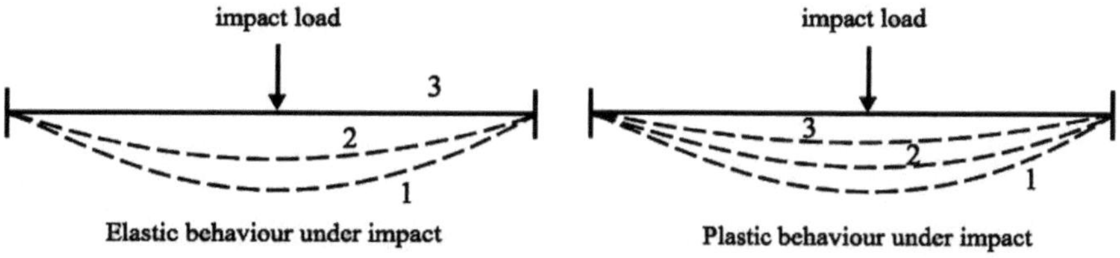

Fig. 13.4 Analytical impact forces vs. distance X between the impactor and the specimen

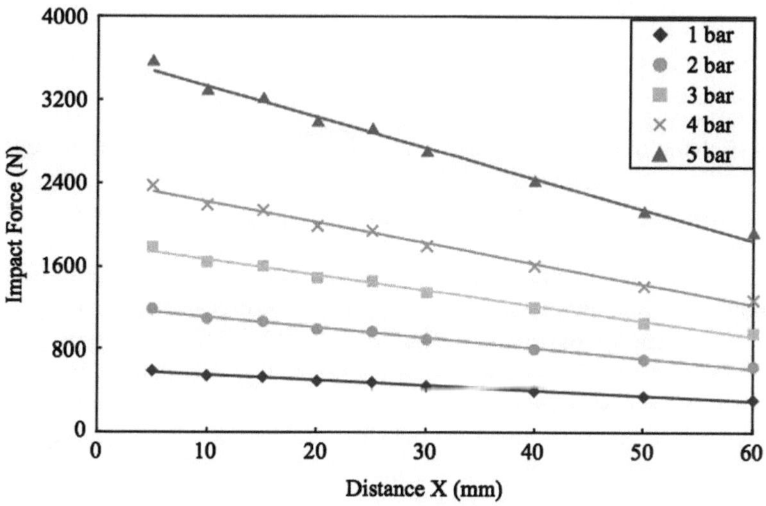

The behavior of specimen material when they are subjected to impact using the repeated impact device is illustrated in Fig. 13.3, where materials can behave either elastically or plastically. When the impactor strikes the specimen, the specimen develops an initial deflection as a result of the impact, denoted by (1). The impactor then stays in contact with the specimen while applying force, causing the specimen to develop an intermediate deflection, denoted by (2), which is less than the initial deflection (1). Upon the release of the load, the specimen will either return to its original position without any plastic deformations if it behaves elastically, or will develop a final permanent deformation, donated by (3), if it behaves plastically. Through the rest of this paper, the initial deflection (1) will be referred to as 'dynamic deflection', while the final deflection (3) will be referred to as 'permanent deflection'.

Different values of impact forces can be obtained by changing the values of the applied pressure and the distance between the impactor and the specimen. Figure 13.4 shows the different values of the calculated impact forces through force analysis techniques [19]. As seen from the figure, the impact forces increase with the increase of the pressure for the same distance X. On the other hand, for the same pressure, the impact forces decrease with the increase of the distance between the impactor nose and the specimen.

13.2.3 Impact Device Calibration

A dummy specimen is used to determine the experimental values of the impact forces. A quarter bridge strain gage system is attached to the dummy specimen and the strain is measured using a KYOWA PCD 320A strain meter. The dummy specimen is made of low carbon steel with a modulus of elasticity (*E*) 200 GPa, and dimensions 120 × 35 × 5 mm.

A stress analysis relation is deduced to determine the value of the impact force through the measured strain, which is derived as follows:

1. The dummy specimen is fixed at both ends, so that the impact force acts as sudden bending force, causing bending stresses to the specimen as shown in Fig. 13.5, where (*P*) is the impact load, (*l*) is the length of the dummy specimen, (*b*) and (*t*) are the width and thickness of the specimen respectively.
2. From the bending moment diagram, the maximum bending moment applied is:

$$M_{\max} = \frac{Pl}{8} \tag{13.1}$$

3. Using stress analysis, and knowing the section modulus of the specimen (rectangular section), the maximum impact bending stress is:

$$\sigma_{\max} = \frac{3Pl}{4bt^2} \tag{13.2}$$

4. Since the calibration experiment is carried out in the elastic limit, Hook's law can be applied, and hence, the impact bending force will be:

$$P = \frac{4bt^2E}{3l}\varepsilon \tag{13.3}$$

5. Substituting the values of *b*, *t*, *E* and *l* in Eq. (13.3), the relation becomes as:

$$P = 1746031.74\varepsilon \tag{13.4}$$

The measurements are recorded within pressures ranges of 1 bar up to 4 bars at the corresponding distances X from impactor tip to the specimen surface. Values of forces corresponding to pressures greater than 4 bars can be extrapolated.

Strain gages used are KYOWA KFG-10-120-C1 general purpose gages of dimensions 16 × 5.2 mm, excitation voltage 2–4 V, resistance 120 Ω, gauge factor approximately 2.1, and measurement accuracy of approximately ±0.05 %. The surfaces of the dummy specimen were ground and polished to a near mirror surface, then carefully cleaned using acetone. Gages are then pasted on the center of the polished surface opposite to the impactor by a means of a cyanoacrylate adhesive.

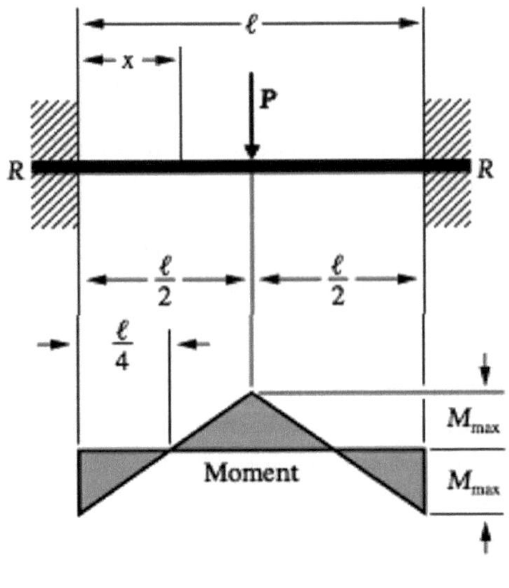

Fig. 13.5 Bending moment diagram of the dummy specimen with fixed-fixed supports

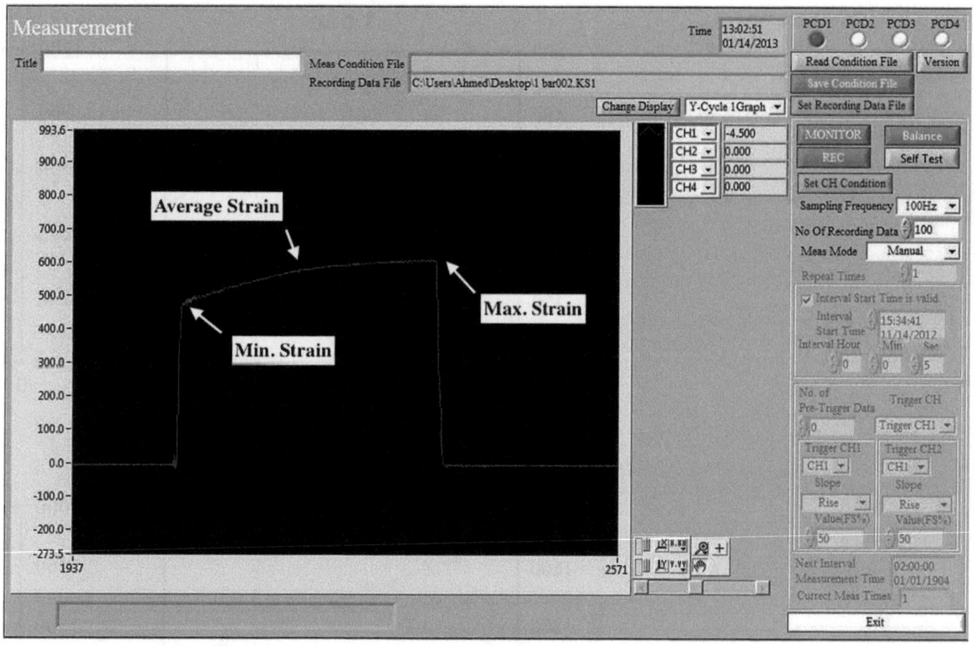

Fig. 13.6 Strain measurement screen

Table 13.1 Repeated impact testing procedures for composite specimens and the corresponding testing parameters

Parameter	Group 1			Group 2				
Pressure (bar)	3	4	5	1	2	3	4	5
Distance X (mm)	45			45				
Number of impacts N	315			Up to failure				

Strain is measured in microstrains ($\mu\varepsilon$) using KYOWA PCD 320 strain meter. According to the behavior of the impact device, there is an initial value of strain at the impact stroke (min. strain). The strain then increases while the impactor is in contact with the specimen till it reaches its maximum value (max. strain). Finally, it drops to zero with the backward stroke. Figure 13.6 shows a strain measuring curve on the PC screen using the strain meter at pressure 4 bar and distance 30 mm. Strains were measured four times at each pressure and distance X. First, the average of the minimum and the maximum strains was calculated four times then the average of the four average measures was determined and substituted in Eq. (13.4).

13.2.4 Repeated Impact Procedures

Repeated impact tests were divided into two groups; the first group of impact tests were carried out to measure the permanent and dynamic deflections of the composite specimens (in millimeters) after a certain number of impacts, while the second group of impact tests were carried out to measure the number of impacts up to the failure of the composite specimen in order to estimate its fatigue trend, as indicated in Table 13.1.

13.3 Results and Discussion

13.3.1 Device Calibration Results

Figure 13.7 shows a comparison between experimental and analytical impact forces at different pressures. A drop in the value of the experimental impact force can be seen at distance 5 mm from the specimen, as compared to the analytical impact force, where the ratio between the analytical and the experimental forces is approximately two. This ratio decreases to approximately 1.8 at distance 30 mm from the specimen, and further decrease to approximately 1.4 at distance 60 mm

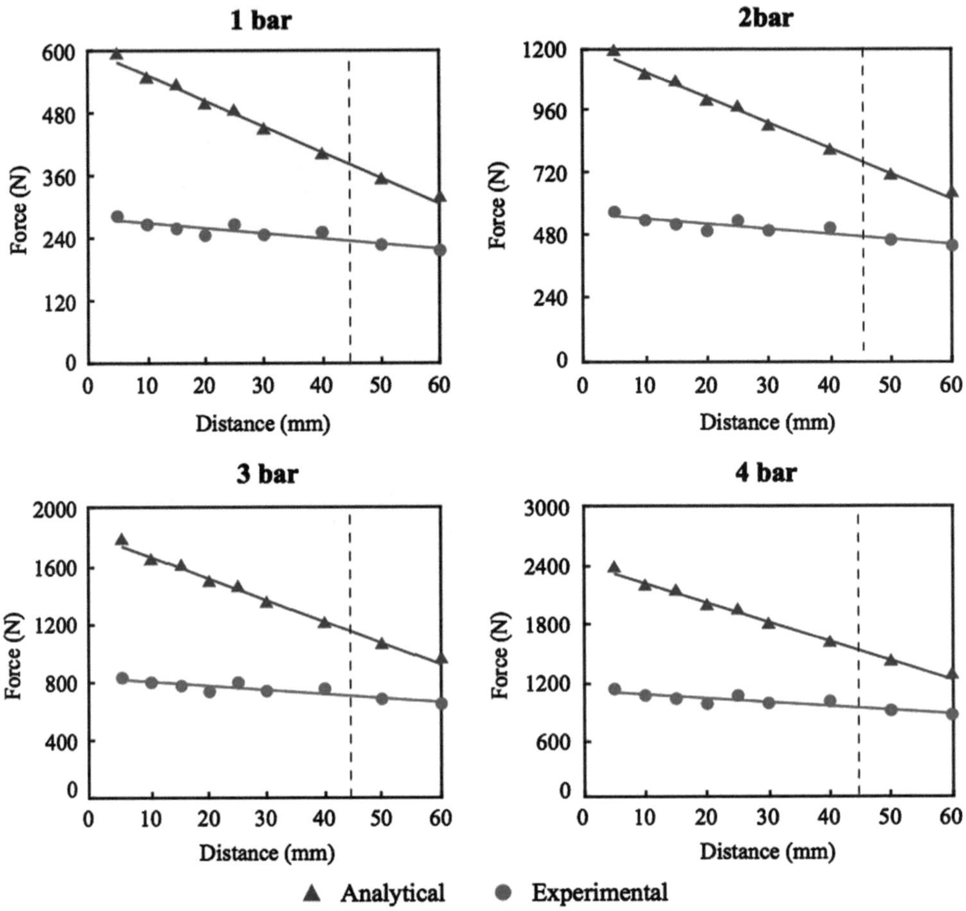

Fig. 13.7 Analytical and experimental impact forces at different distances X

from the specimen. This behavior is constant with the increase of the pressure despite a slight tendency of instability in the system at higher pressures. As a result, the distance between the impactor and the specimen is chosen to be 45 mm for the repeated impact tests, where the difference between the experimental and analytical forces is minimum.

The drop in the experimental force compared to the analytical force is due to the sudden drop in pressure of the air in the air cylinder. As mentioned in Sect. 13.2.2, the device utilizes an air compressor, which compresses the air to a certain desired pressure and the compressed air then expands in the air cylinder causing it to impact the specimen. This sudden expansion is accompanied by a sudden drop in pressure according to ideal gas laws. The pressure then increases gradually, trying to compress the air to the same volume it had before the expansion. This happens while the impactor is in contact with the specimen and stops in the backward stroke. The same behavior is also observed on the pressure gauge fitted in the air filter.

13.3.2 *Deflection of Composite Specimens*

Figure 13.8 summarizes the results of the permanent and dynamic deflections across the 315 impact cycles for the glass fiber reinforced polyester.

Glass fiber reinforced polyester composites do not undergo significant plastic deformations. Permanent deflections reaches their maximum values at 315 impact cycles, ranging from 0.22 mm at force 720 N to 0.4 mm at force 1,200 N, while maintaining a constant behavior up to 315 impacts. The specimens clearly did not suffer any sever delimitations or accumulation of matrix cracks at this early stage of impacts.

Moreover, dynamic deflection increased nearly five times compared to permanent deflection values, and the ratio remains constant with the increase of both the impact forces and the number of impacts. This indicates that dynamic forces have a significant effect on the deflection of the composite specimens. In addition, by increasing the value of the dynamic force by 66 %, the dynamic deflection increase by only 26 %, which indicates a slight effect of the dynamic forces on the dynamic deflections.

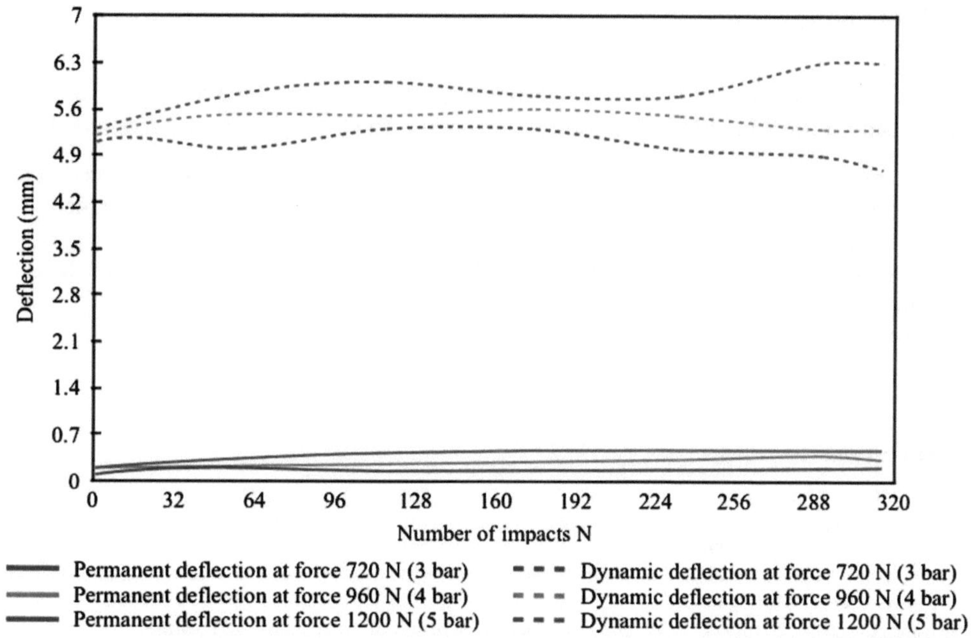

Fig. 13.8 Permanent and dynamic deflections of composite specimens after different number of impacts

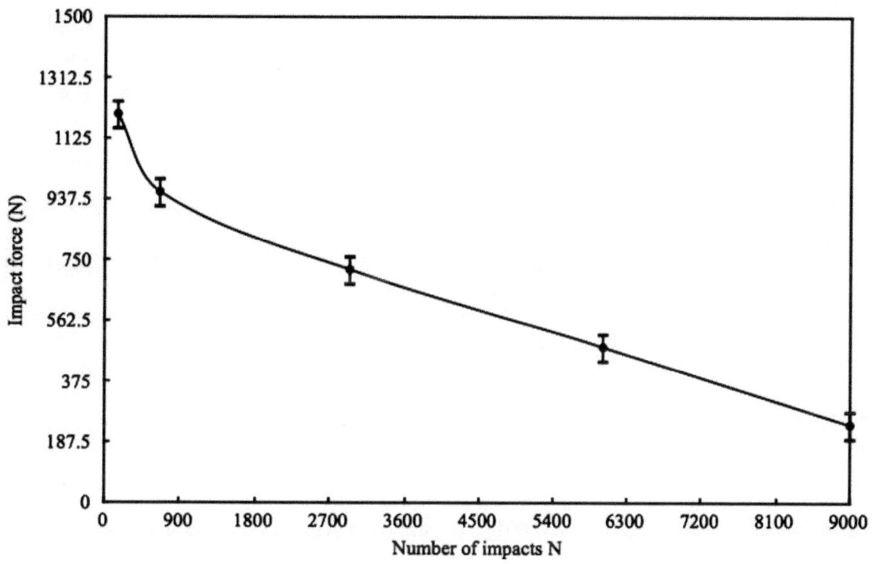

Fig. 13.9 Impact fatigue trend of composite specimens

13.3.3 Impact Fatigue Trend of Composite Specimens

Composite specimens were impacted up to failure at different forces 241, 481, 720, 960 and 1,200 N, corresponding to pressures 1, 2, 3, 4 and 5 bars. Number of impacts to failure was recorded using the photoelectric switch. Figure 13.9 shows the impact fatigue trend of the glass fiber reinforced polyester, where the experimental errors varied between 2 % and 3 %.

At higher impact forces, the rate of delamination and matrix cracks increase very rapidly, causing early and fast failure of the specimens. This can be seen at forces higher than 800 N, where specimens fail at approximately 850 impacts. On the other hand, at lower impact forces, the rate of delamination and matrix cracks tends to be less, leading to a longer fatigue life up to 9,000 impacts at 374 N. It can then be considered that the force level of 800 N is a critical force level, above which a higher rate of delamination and failure occurs, and below it, lower rate of failure occurs.

13.4 Conclusion

A special repeated impact testing device was designed and calibrated to test the behavior and the impact fatigue trend of glass fiber reinforced polyester composite specimens. The following can be concluded from this study:

1. Dynamic forces have a significant effect on the deflection of glass fiber reinforced composite specimens.
2. Dynamic deflection values were five times the permanent deflection values, which also showed constant behavior up to 315 impact cycles.
3. At impact forces greater than 800 N, glass fiber reinforced polyester composites tend to fail at a high rate. Below 800 N, composites show better resistance to impact fatigue. Thus, at the present conditions, the force level at 800 N can be considered a critical force level.

References

1. Davies G, Hitchings D, Zhou G (1996) Impact damage and residual strengths of woven fabric glass/polyester laminates. Compos A Appl Sci Manuf 27(12):1147–1156
2. Santiuste C, Sánchez-Sáez S, Barbero E (2009) Residual flexural strength after low-velocity impact in glass/polyester composite beams. Compos Struct 92(1):25–30
3. Barcikowski M, Semczyszyn B (2011) Impact damage in polyester-matrix glass fibre-reinforced composites. Part II. Residual load bearing abilities. Kompozyty 11(3):235–239
4. Liu Q, Guo O, Ju Y, Lin Y, Li Q (2014) Composite structures. Compos Struct 111(C):332–339
5. Romanzini D, Lavoratti A, Ornaghi HL Jr, Amico SC, Zattera AJ (2013) Influence of fiber content on the mechanical and dynamic mechanical properties of glass/ramie polymer composites. Mater Des 47:9–15
6. Sarasini F, Tirillò J, Ferrante L, Valente M, Valente T, Lampani L, Gaudenzi P, Cioffi S, Iannace S, Sorrentino L (2014) Composites: part B. Compos B Eng 59(C):204–220
7. Aslan Z, Karakul R, Octane B (2002) The response of laminated composite plates under low-velocity impact loading. Compos Struct 59(1):119–127
8. Yang L, Yan Y, Kuang N (2013) Experimental and numerical investigation of aramid fiber reinforced laminates subjected to low velocity impact. Polym Test 32(7):1163–1173
9. Polemic U, Meo M, Almond DP, Angioni SL (2010) Detecting low velocity impact damage in composite plate using nonlinear acoustic/ultrasound methods. Appl Compos Mater 17(5):481–488
10. Sevkat L, Delale R (2010) Effect of repeated impacts on the response of plain-woven hybrid composites. Compos B 41(5):11–11
11. Icten BM (2008) Repeated impact behavior of glass/epoxy laminates. Polym Compos 9999(9999):1562–1569
12. De Morais WA, Montessori SN, d'Almeida JRM (2005) Evaluation of repeated low energy impact damage in carbon–epoxy composite materials. Compos Struct 67(3):307–315
13. Bora MÖ, Çoban O, Sinmazçelik T, Cürgül Î, Günay V (2009) On the life time prediction of repeatedly impacted thermoplastic matrix composites. Mater Des 30(1):145–153
14. Çoban O, Bora MÖ, Sinmazçelik T, Cürgül İ, Günay V (2009) Fracture morphology and deformation characteristics of repeatedly impacted thermoplastic matrix composites. Mater Des 30(3):628–634
15. Azouaoui K, Rechak S, Azari Z (2001) Modelling of damage and failure of glass/epoxy composite plates subject to impact fatigue. Int J Fatigue 23(10):877–885
16. Roy R, Sarkar BK, Bose NR (2001) Impact fatigue of glass fibre–vinylester resin composites. Compos A Appl Sci Manuf 32(6):871–876
17. Mittelman A (1992) Low-energy repetitive impact in carbon-epoxy composite. J Mater Sci 27:2458–2462
18. Kawaguchi T, Nishimura H, Ito K, Sorimachi H, Kuriyama T, Narisawa I (2004) Impact fatigue properties of glass fiber-reinforced thermoplastics. Compos Sci Technol 64(7):1057–1067
19. Elbawab M (2009) Dynamic behaviour of composite materials under impact repetitive loads. M.Sc. thesis, Faculty of Engineering, Cairo University, Egypt

Chapter 14
FE Modeling of Paperboard Material Using Sandwich Structure Method

W. Yang, M.W. Allin, and C.J. Dehenau

Abstract Paperboard materials have been widely used in industrial applications, especially in packaging. It is very important and necessary to model the material behaviors of paperboard in order to accelerate product development processes and optimize the product designs.

In this paper, the sandwich structure method is applied to describe the mechanical behavior of the paperboard, which is modeled by two shell surfaces and one solid core. The two shell surfaces represent the in-plane (machine direction [MD] and cross direction [CD]) properties while the core represents the out-of-plane (thickness direction [TD]) properties. The anisotropic elastic plastic material and isotropic elastic material models within ABAQUS are employed for the two shell surfaces and core, respectively. The two material models are validated based on the tests.

Then the two material models are implemented into a paperboard beverage package to predict the packaging performance under different loading conditions. The quasi-static handle test is also performed to validate the Finite Element Simulation. A good agreement between the FE result and tests is achieved. This method will play important role in packaging products development.

Keywords Paperboard • Packaging • Sandwich structure • ABAQUS

14.1 Introduction

Paperboard is one of the most commonly used materials in every industry. Paperboard is a paper material which generally consists of several pulp fiber sheets bonded by adhesive materials, and is usually a multilayered structure. Schematic of typical paperboard's micro and macro-structure are shown in Fig. 14.1 [1], in which three orthogonal directions are depicted for paperboard. MD refers to the machine direction and CD refers to the cross or transverse direction. The machine and cross directions generate the plane of the structure, and ZD refers to the out of plane or thickness direction. The fibers are mainly oriented in the plane and along machine direction. So the paperboard in-plane properties are dominated by the fibers.

Paperboard materials generally exhibit complex anisotropic and nonlinear mechanical behavior due to fiber orientation distributions, which is also highly affected by the moisture and temperature, and paperboard also exhibits viscosity and has different properties between compression and tension. So in order to have the practical use of the material model, the material model should be as simple as admitted and the material parameters can be obtained easily by the material testing.

There are many material models proposed and developed for paperboard by previous researchers. Some of these material models are developed by the theory of micromechanics and used to predict paperboard mechanical properties based on the properties of fibers and fiber-fiber interfaces. Perkins and Sinha created a based network constitutive model for the in-plane properties of paper [2, 3], in which a representative meso-scale element was built for fibrous paper microstructure. The mechanical behavior of the representative element depends on the properties of fiber and fiber interfaces, and the fiber interfaces play a key role in the overall in-plane inelastic behavior of paper. The material models of paperboard were developed based on classic laminate theory by Page and Schulgasser [4, 5], which can be only used to predict elastic behavior. Gunderson used Tsai–Wu quadratic yield condition to model the failure loci they obtained experimentally [6, 7]. Xia developed an anisotropic elastic-plastic constitutive mode for paper and paperboard [1], in which a multi-surface yield function, anisotropic hardening, different mechanical behaviors between tension and compression, and an out of plane nonlinear elastic description were incorporated. But this complex model requires a lot of experimental work to validate it.

W. Yang (✉) • M.W. Allin • C.J. Dehenau
Senior R&D Engineer, Food & Beverage, MeadWestVaco. 501 S 5th Street Richmond, VA 23219, USA
e-mail: wenong.yang@mwv.com

A. Wicks (ed.), *Shock & Vibration, Aircraft/Aerospace, and Energy Harvesting, Volume 9*,
Conference Proceedings of the Society for Experimental Mechanics Series, DOI 10.1007/978-3-319-15233-2_14

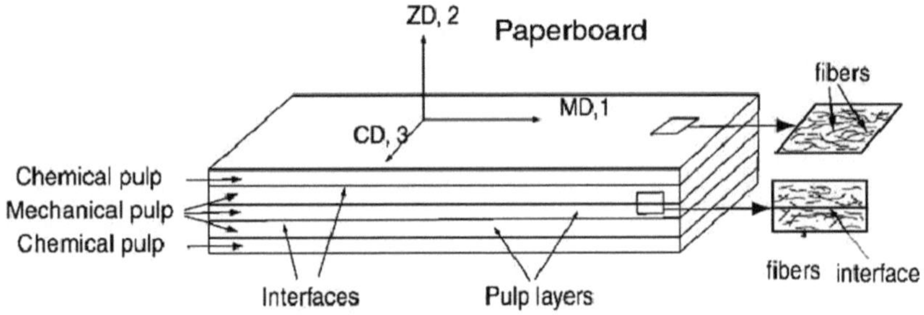

Fig. 14.1 Schematics of paperboard's micro and macro-structure

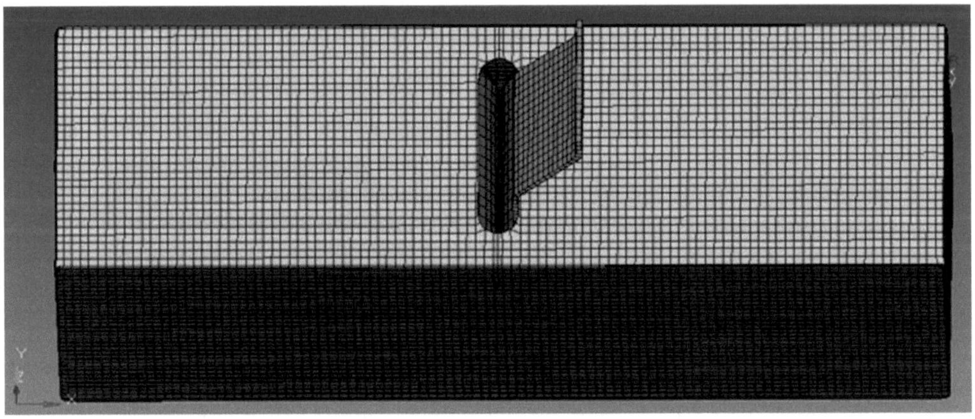

Fig. 14.2 Finite element model of paperboard beverage package handle test

In this paper, the mechanical behavior of paperboard is modeled by the sandwich structure method, which includes two surfaces and one core representing the in-plane and out-of-plane properties of paperboard, respectively. The anisotropic elastic-plastic material model is applied for two shell surfaces and elastic material model is applied for core.

14.2 Finite Element Model

The FE simulation is performed in three main steps, the first is building model using ABAQUS CAE. Then, the nonlinear analysis is carried using ABAQUS v6.13. Finally the results are assessed and presented using ABAQUS CAE, and validated by the experimental tests. The geometry of the paperboard beverage package is of a hollow rectangular solid with cross-section (123.0 mm × 135.0 mm) and 402.0 mm in length. The thickness of the package is 0.45 mm. The handle testing was selected to validate the sandwich structure method for modeling the material behavior of paperboard.

The mesh scheme of the beverage package handle test is shown in Fig. 14.2. The model consists of the paperboard package, aluminum beverage cans and handle. The element types used in this model are eight-node, six-node solid elements and three-node, four-node shell elements. There are 18,242 solid elements and 51,079 shell elements, and 54,177 nodes in the FE model.

Two ABAQUS material models are used for the FE analysis of the handle testing, which are anisotropic elastic-plastic and elastic. The moving handle and aluminum beverage cans are modeled as rigid body. The anisotropic elastic-plastic with potential function is chosen for the two shell surfaces in the beverage handle testing [8], while the elastic material is selected for the core. The mechanical properties are determined by the experimental tests including Instron Machine and 3D Ultrasonic Tester. The potential option is used to define stress ratios for anisotropic yield behavior, which can be used for paperboard that exhibits different yield behaviors in different directions.

Anisotropic yield behavior is modeled by the use of yield stress ratios, R_{ij}. In the case of anisotropic yield the yield ratios are defined with respect to a reference σ^0, which is the metal plasticity definition. Such that if σ_{ij} is applied as the only nonzero stress, the corresponding yield stress is $R_{ij}\,\sigma^0$.

The Hill potential function is a simple extension of the Mises function, which can be described in terms of rectangular Cartesian stress components as:

$$f(\sigma) = \sqrt{F(\sigma_{22}-\sigma_{33})^2 + G(\sigma_{33}-\sigma_{11})^2 + H(\sigma_{11}-\sigma_{22})^2 + 2L{\sigma_{23}}^2 + 2M{\sigma_{31}}^2 + 2N{\sigma_{12}}^2} \tag{14.1}$$

Where F, G, H, L, M and N are constants obtained by material tests in different orientations. They are defined as:

$$F = \frac{1}{2}\left(\frac{1}{{R_{22}}^2} + \frac{1}{{R_{33}}^2} - \frac{1}{{R_{11}}^2}\right)$$

$$G = \frac{1}{2}\left(\frac{1}{{R_{33}}^2} + \frac{1}{{R_{11}}^2} - \frac{1}{{R_{22}}^2}\right)$$

$$H = \frac{1}{2}\left(\frac{1}{{R_{11}}^2} + \frac{1}{{R_{22}}^2} - \frac{1}{{R_{33}}^2}\right)$$

$$L = \frac{3}{2}\frac{1}{{R_{23}}^2}$$

$$M = \frac{3}{2}\frac{1}{{R_{31}}^2}$$

$$N = \frac{3}{2}\frac{1}{{R_{12}}^2} \tag{14.2}$$

Where R_{11} , R_{22} , R_{33} , R_{12} , R_{23} , R_{31} are anisotropic yield stress ratios.

The contact interface type surface to surface is used to define the contact among aluminum beverage cans, handle and paperboard. The boundary conditions refer to the moving handle. During the handle test, the moving handle is modeled as a rigid body at a quasi-static velocity.

14.3 Simulation Results

The force displacement curve is selected as output to validate the sandwich structure method for modeling the material behavior of paperboard. There is a good agreement achieved between the simulation result and experimental tests shown in the Fig. 14.3.

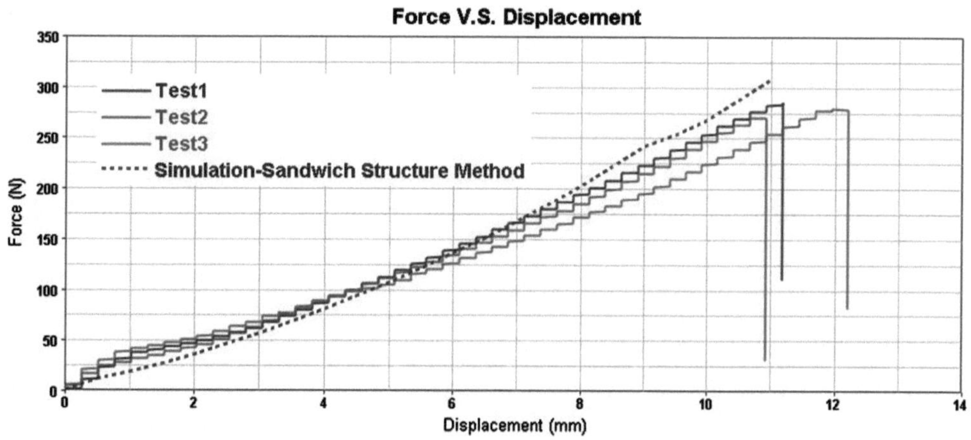

Fig. 14.3 Plot of force displacement curves of simulation result and experimental tests

14.4 Conclusions

Paperboard is a very complex material, In order to accelerate the packaging product development process, it is very necessary to model the paperboard material, so the sandwich structure method is utilized in this paper, which provides very practical application in modeling and simulation of paperboard packaging performance and help optimize the paperboard packaging designs. Based on the above work performed on the paperboard structural modeling, the sandwich structure method with anisotropic elastic plastic material model provided a good correlations between simulation result and experimental test (see Fig. 14.3), which shows about 90 % correlation has been achieved. This material model can now be used as a basis for continued work correlating simulated and experimental packaging performance, and then provide recommendations for packaging designs and solutions.

References

1. Xia QS, Boyce MC, Parks DM (2002) A constitutive model for the anisotropic elastic-plastic of paper and paperboard. Int J Solids Struct 39:4053–4071
2. Perkins R, Sinha S (1992) A micromechanics plasticity model for the uniaxial loading of paper materials. In: Plastic flow and creep. ASME Applied Mechanics Division, New York
3. Sinha S, Perkins R (1995) A micromechanics constitutive mode for use in finite element analysis. In: Mechanics of cellulosic materials. ASME Applied Mechanics Division, New York
4. Page D, Schulgasser K (1989) Evidence for a laminate model for paper. In: Mechanics of cellulose and polymeric materials. ASME Applied Mechanics Division, New York
5. Schulgasser K, Page D (1988) The influence of transverse fiber properties on the in-plane elastic behavior of paper. Compos Sci Technol 32:279–292
6. Gunderson DE (1983) Determining paperboard strength. In: International paper physics conference
7. Gunderson DE, Bendtsen LA, Rowlands RE (1986) A mechanistic perspective of the biaxial strength of paperboard. ASME J Eng Mater Technol 108:135–140
8. ABAQUS user manual

Chapter 15
Evaluation of Seismic Performance of an Excavation Support System

Omer F. Usluogullari, Yalcin Bulut, and Ahmet Temugan

Abstract In this study a braced excavation support system of a subway station is evaluated under two cases: (1) Lateral soil pressure caused by excavation and overburden load (2) Seismic load. The substructure is instrumented and the lateral deformations and settlements are observed during excavation and construction of substructure.

A three dimensional finite element model of the substructure is developed and verified using field observations. The structural model is used to calculate the seismic response of the substructure. A bracing system which consists of secant piles and tie-back anchors is introduced to the substructure and seismic performance is examined. Results of three dimensional finite element analysis showed that predicted lateral deformations and vertical movements have similar patterns with measured results and these movements is almost three times under seismic load comparing to static case.

Keywords Seismic load • Excavation support • Tie-back anchors • 3D finite element analysis • Cut-cover tunnel

15.1 Introduction

Rapid population growth and the exponential increment at the needs of the people in Turkey, increased the importance of city planning. Especially in the capital city of Ankara (Fig. 15.1a) population is densely conglomerated in several locations and resulted in limited space for transportation. Due to these reasons underground metro projects gain importance and deep excavations adjacent to buildings need to be evaluated more carefully.

Deep excavations possess too much risk considering the dense inner-city traffic and too many buildings near to the metro lines. Settlements and lateral deformations of supporting systems might cause high damage to overlay pavements or adjacent buildings. Several methods have been used to evaluate lateral deformations and settlements by many researchers. Some researchers used case studies and field observations [1–5] some of them used numerical and analytical studies [6, 7] and few of them used both field observations and numerical analysis [8, 9, 10]. Also the effect of seismic movement was reported by [11, 12, 13]. In this study static and seismic performance of supporting system was evaluated by investigating lateral deformations and ground settlements in the excavation area located between Mecidiye-Belediye stations, which is a part of Tandogan-Kecioren (M4) line (Fig. 15.1b).

15.2 Soil Properties and Site Conditions

A comprehensive field exploration program was performed to obtain soil parameters. The subsurface consisted of two uniform layers of soil. Top layer consists of soft to stiff alluvial soil with an approximate uniform depth of 9 m overlying stiff sand stone. The rivers pass through this Metro line and alluvial deposits located at tributaries generally consist of the mix of greenish brown, medium plastic, medium to very stiff sandy silty clay, clayey sand and clayey sandy gravels. Groundwater table was observed at 3 m depth. The soil properties were characterized by in situ standard penetration tests, pressuremeter and piezometer tests and a series of laboratory tests including unconfined compression tests, tri-axial tests and other soil characterization tests. Geological strength index (GSI) was used to determine strength of soft rock sand stone. Idealized soil profile with geotechnical parameters was shown in Table 15.1.

O.F. Usluogullari (✉) • Y. Bulut • A. Temugan
Civil Engineering Department, Turgut Ozal University, Ankara, Turkey
e-mail: ofusluogullari@turgutozal.edu.tr

A. Wicks (ed.), *Shock & Vibration, Aircraft/Aerospace, and Energy Harvesting, Volume 9*,
Conference Proceedings of the Society for Experimental Mechanics Series, DOI 10.1007/978-3-319-15233-2_15

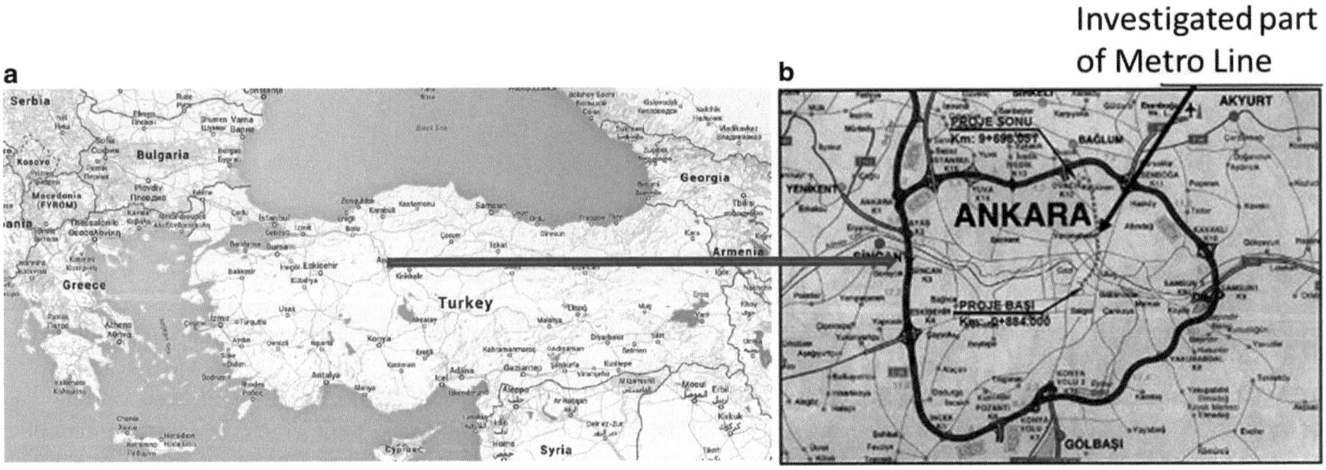

Fig. 15.1 (**a**) Location of Ankara in Turkey, (**b**) location of metro line

Table 15.1 Geotechnical properties of soil layers

Layers	γ_{unsat} (kN/m^3)	Cohesion c (kPa)	Friction angle (ϕ^o)	Elasticity modulus (MPa)	Shear modulus (MPa)	Shear velocity (m/sn)
0–9 m	20	2	30	19.5	8.125	63.1
9–30 m	24	100	36	2,700	1,038	651.2

Fig. 15.2 An illustration of the excavation area

In this case history, excavation area is closely located between residential and commercial buildings. Due to existence of these buildings pile retaining walls with post-tension tieback anchors was used as bracing system and aimed to have a lateral soil movement limited to maximum 1–1.5 cm.

Optical deformation points and inclinometers were used to measure settlements and lateral movement of soil, respectively (Fig. 15.2). Figure 15.3 shows the locations of the instrumentation points to measure the lateral deformations and settlements of retaining pile walls.

The excavation area had an approximate width of 16 m and depth at maximum level of excavation was around 9 m. Generally adjacent buildings consist of four-story reinforced concrete buildings. In this study the section between 5 + 724 km and 6 + 164 km was investigated which closely represents the whole section.

15.3 Construction Phases and Numerical Modeling

Finite element analysis, by the software PLAXIS 3D, was used to investigate the static and seismic behavior of pile retaining walls, under lateral loads caused by the excavation of soil and earthquake loads using ground motion acceleration records of El-Centro 1940 earthquake in order to construct cut-cover tunnel, respectively. Excavated area was braced with secant

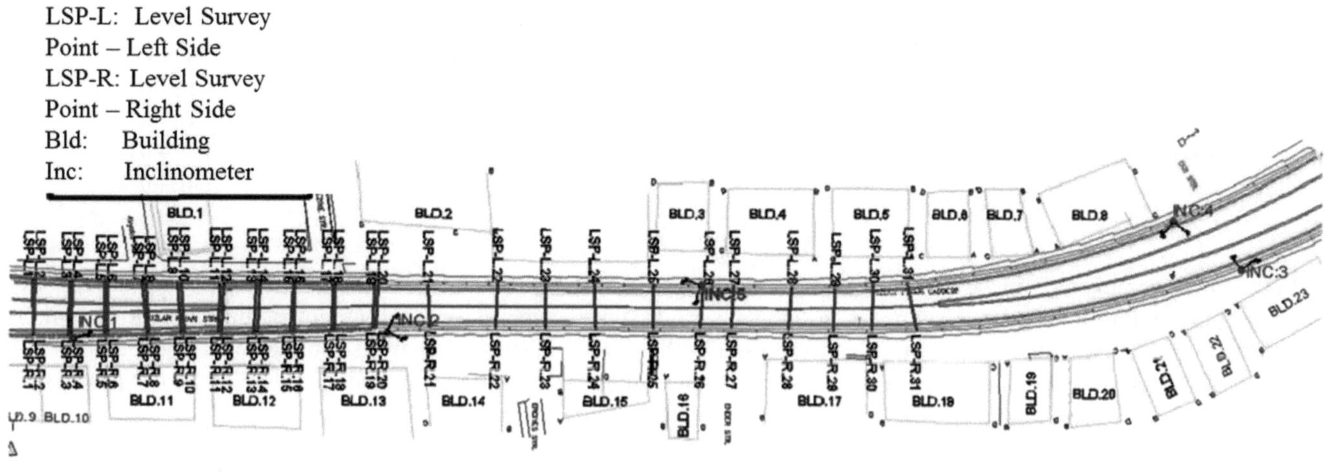

Fig. 15.3 Instrumentation plan of cut and cover tunnel construction between Belediye and Mecidiye stations

Fig. 15.4 Geometry and structural elements of bracing system

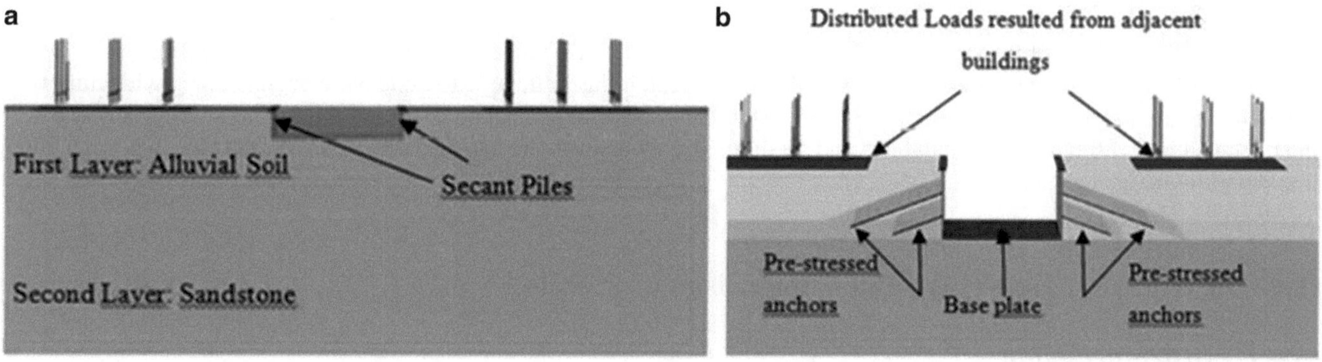

Fig. 15.5 3D finite element model of cut-cover tunnel excavation (**a**) first stage and (**b**) last stage

piles and tie-back anchors (Fig. 15.4). A typical finite element model showing excavation area, structures of support system, adjacent buildings is shown in Fig. 15.5. Settlements and lateral movements were investigated to evaluate the performance of support system, in order to be in the desired deformation range under static and seismic loads. Static loads are consisted of lateral earth pressure and overburden load caused by adjacent buildings.

To model the soil, ten node wedge elements were used. Various element sizes ranging from very fine to very coarse were used during calculations. Coarse size elements were selected due to efficiency of calculation of time in regards to change at results. The soil was modelled with Mohr-Coulomb failure criterion approximated to be a linear elastic-perfectly plastic. The constitutive model used to simulate the soil behavior required five basic input parameters and two seismic parameters shown in Table 15.1: the Elastic modulus (E), Poisson's ratio (υ) (0,2), cohesion (c), friction angle (ϕ), unit weight (γ), shear modulus (G) and shear velocity (V_s).

Table 15.2 Material properties of structural elements

Structural elements	Thickness/diameter (cm)	Elastic modulus (kN/m^2)	EA (kN)
Secant piles (embedded piles)	0.80	2.19E+07	1.1E+07
Free section of anchors (post tension steel tendons)	–	–	1.1E+05
Grouted section of anchors (embedded piles)	0.14	2.83E+07	4.4E+04
50 cm basement floor	0.5	2.85E+07	1.4E+07

In the finite element analysis as-built construction drawings have been used to develop model. Excavation was supported by secant piles with 80 cm radius having 11 m length were driven into rock layer with an embedment length of 2 m. To model the pile retaining wall, embedded piles were used in the three dimensional analysis. Studies on embedded pile [14] validated the accuracy by comparing with volume piles. That study also showed that embedded piles can be used to analyze the actual behavior of real piles.

The post-tensioned tie-back anchors with steel tendons, 80 cm grouted body and the anchor spacing of 2.6 m were installed to support soil during excavation. The first free parts of anchors without grouting material were modeled as node-to- node anchor and the second sections having 8 m grouted length were modeled as embedded beam elements providing the interface simulated the grout-soil interaction. Finally the basement slab was modeled as 50 cm thick plate element (Fig. 15.5a, b). The material parameters of structural elements used in this study were summarized in Table 15.2.

The closest buildings adjacent to the excavation area were 10 m away from site. Those four story buildings have approximately 1.5 m foundation height and the resulting load due to dead weight of one of those building was estimated approximately 50 kN/m^2 and applied to surface as distributed surcharge load (Fig. 15.5b).

The construction stages used in the finite element model and corresponding actual construction steps can be summarized as follows:

1. Installment of secant piles and excavation of 3.5 m top soil till 0.5 m below of first row of anchors (15.08.2012–19.10.2012),
2. Installation of anchors at −3 m depth having 21 m length (19.10.2012–24.10.2012),
3. Excavation of the soil 0.5 m below second row anchors (−6.5 m) (24.10.2012–3.11.2012),
4. Installation of anchors with a length of 15 m at −6 m depth (3.11.2012–11.11.2012),
5. Excavation to final depth at −9 m level and construction of base slab (11.11.2012–08.12.2012).

The predicted values using PLAXIS 3D was compared with measured lateral movements and settlements obtained by inclinometers and level instrument, respectively. A typical finite element model used with the appropriate boundary conditions in the analysis of the staged excavation is shown in Fig. 15.5a (first stage) and b (final stage). Fifteen node wedge elements to model the soil, embedded piles to model the pile retaining wall, node to node bar elements along with embedded pile elements to model the tie-back anchors were used in the three dimensional analysis. Various element sizes between very fine to coarse were used during calculations and medium size elements were selected due to having efficiency at calculation time regarding to change at results.

15.4 Instrumentation System and Comparison of Results

15.4.1 Instrumentation

To evaluate effects of excavation on adjacent buildings and road, vertical ground movements and lateral deformations were monitored with an instrumentation system consists of level measurements and inclinometer readings, respectively (Fig. 15.3).

Tan and Li [2] suggested that unless soil under the excavated area does not bear to hard stratum and does not have consolidation issues, significant retaining wall settlement is not expected. The results of level survey measurements of the points at pile heads agree with that suggestion and support system kept settlements in the range of desired limits which are 1–2 mm. In order to measure ground settlements near to buildings, level measurement points were chosen at four sides of 23 buildings 10 m away from excavation area (Fig. 15.3). Settlements of piles located at right and left hand sides of excavation area were monitored by level instruments at 31 survey points (Fig. 15.3).

In addition to ground movement, the performance of supporting system was evaluated with monitoring the lateral soil movements during excavation and with the construction of structural elements. The deflections of piles were intended to be observed by 5 inclinometer tubes located at both sides of excavation area. However, due to construction issues, data were able to collect from three of them. Inclinometer readings were collected every 7 days after the construction of anchors at

−3 m level till the end of excavation, which is around 3 months. The retrieval of settlement readings were started with the time of excavation and continued to end of excavation, which is around 5 months.

15.4.2 Ground and Pile Settlements

The results of level survey measurements of the points near to buildings and pile heads are presented in Figs. 15.6 and 15.7, respectively.

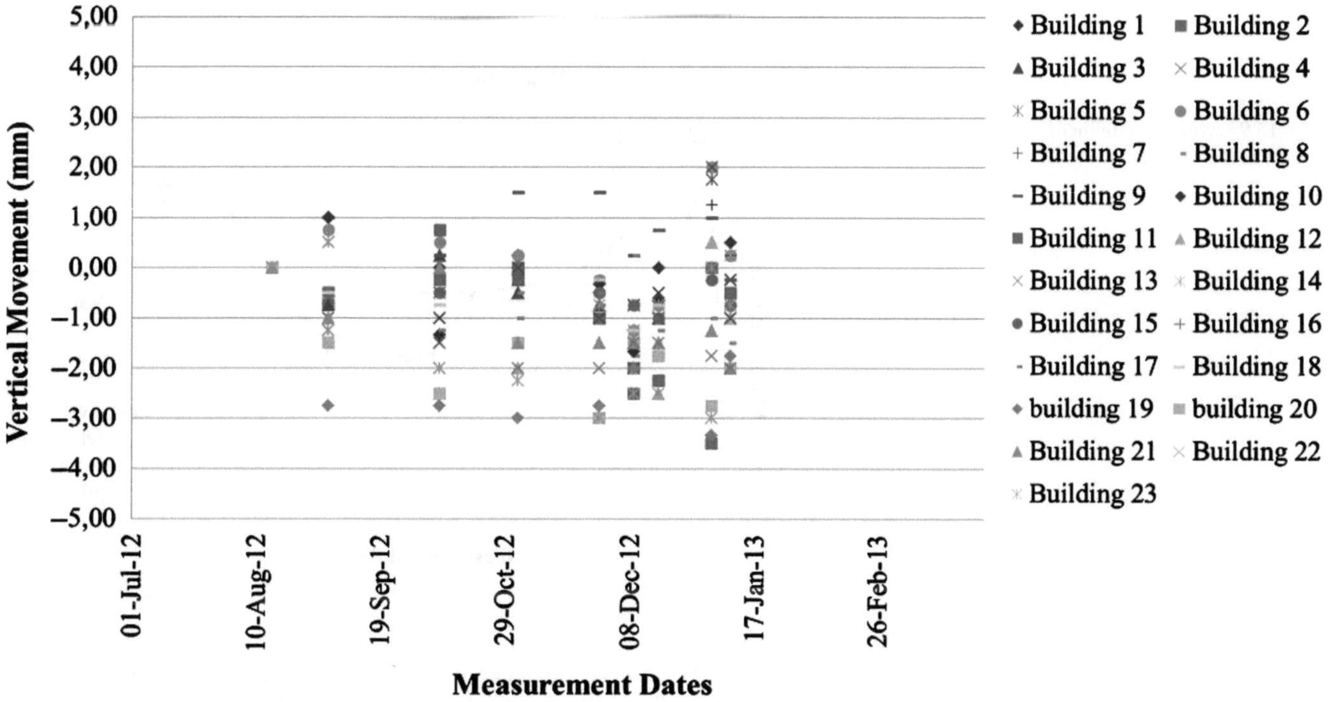

Fig. 15.6 Ground settlements near to adjacent buildings during

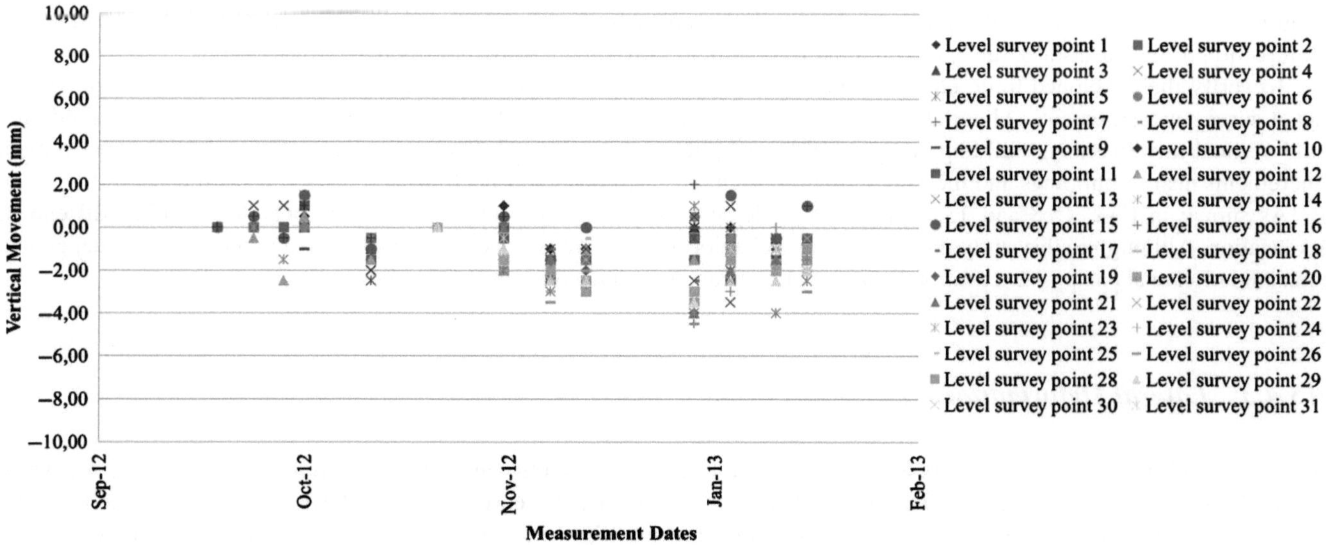

Fig. 15.7 Settlements at top of piles

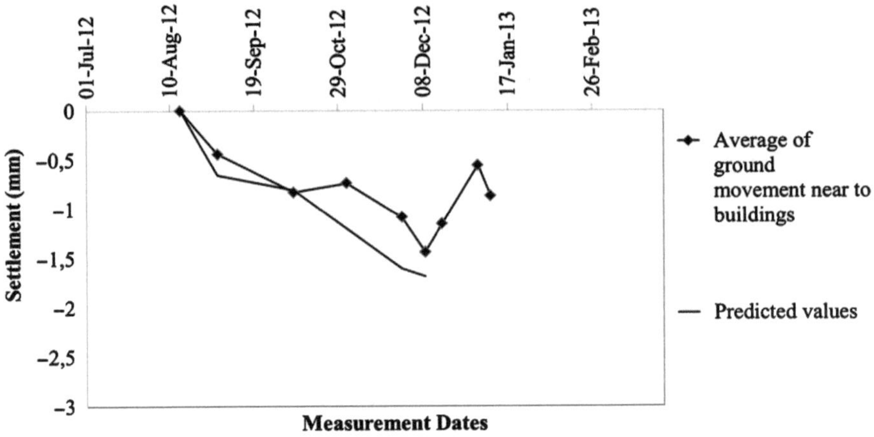

Fig. 15.8 Average ground settlements near to adjacent buildings during excavation

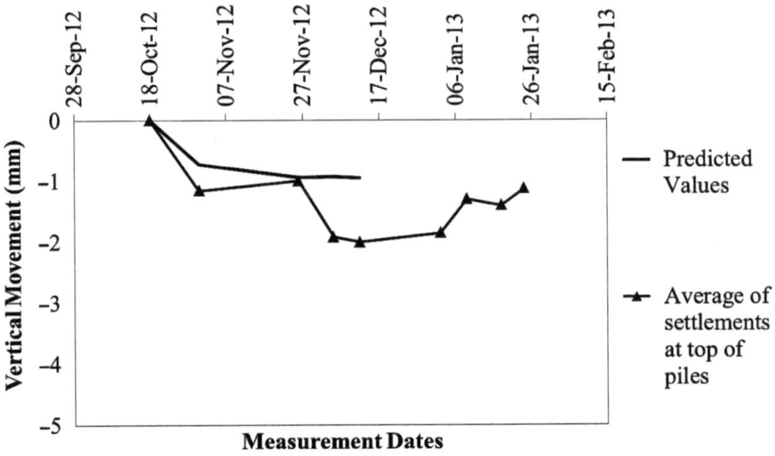

Fig. 15.9 Average settlement results at top of piles

At top of piles and near to adjacent building corners, a wavy vertical movement was observed at soil when all survey points are investigated in one graph (Figs. 15.6 and 15.7). To observe the generalized behavior of soil movement the average results on relevant dates were selected and compared to three dimensional static finite element analysis as shown in Figs. 15.8 and 15.9.

According to these figures as excavation proceeded to lower levels, the ground settlements near to buildings increased continuously (Fig. 15.8) and the pile retaining walls began to settle (Fig. 15.9) when excavation was completed. After completion of the base slabs, the excavation induced vertical movements tended to stabilize around 1 mm.

To verify the developed finite element model, the predicted values were compared with the inclinometer readings and level survey measurements. Material properties shown in Tables 15.1 and 15.2 were used in three dimensional modeling and vertical movements at building corners and top of piles were estimated. Comparison of estimated vertical movement increments near to buildings and measured in-situ settlements are shown in Fig. 15.8.

As shown in Figs. 15.8 and 15.9, predicted settlements successfully simulate the actual behavior of piles till end of excavation. Although in this study post excavation behavior of soil have not been observed, with a numerical time-dependent consolidation analysis post-excavation behavior of soil can be evaluated by further studies.

15.4.3 Lateral Deformations

In excavation induced movements major concern is the lateral movements caused by stress relief of soil next to excavated area. Those soils tend to replace the excavated soil and result with an active earth pressure. The averages of three inclinometer readings regarding to associated measurement dates are shown in Fig. 15.10. The results illustrated in Fig. 15.10 shows that after excavation measurements reaches to −6.5 m levels lateral movement tend to increase and reaches two times more than previous stages, then increment of the movement reduce with adding anchors. This may be the result of, after curing of

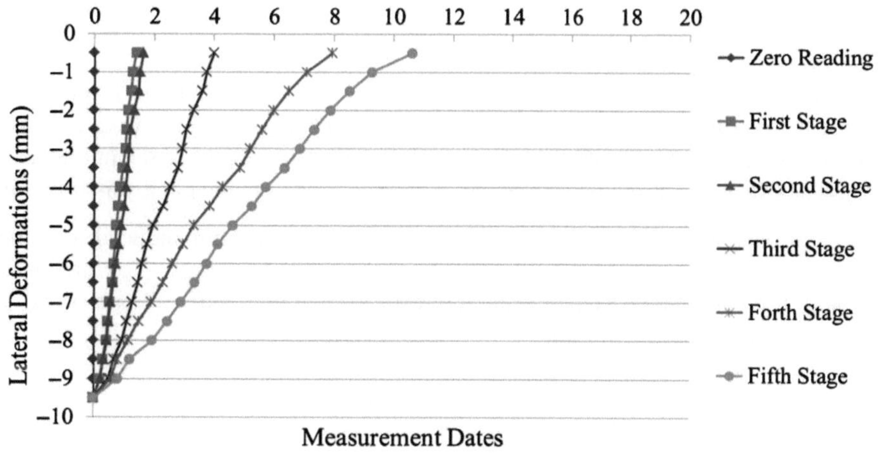

Fig. 15.10 Average pile deflections measured at different stages

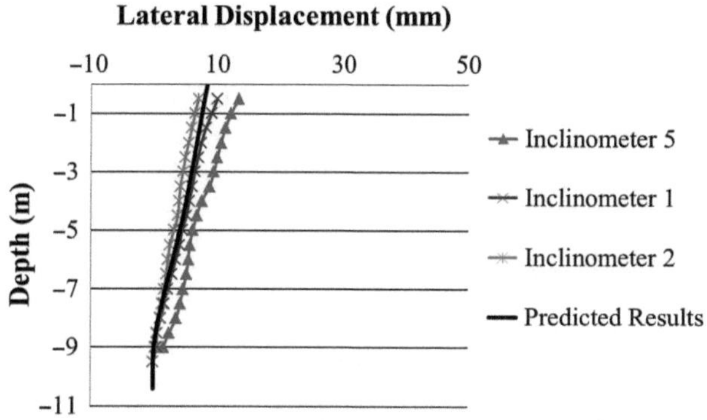

Fig. 15.11 Comparison of measured and numerically predicted lateral deformations

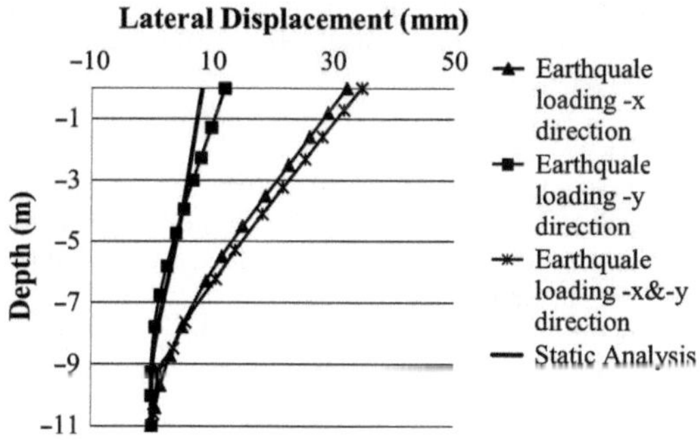

Fig. 15.12 Comparison of results of seismic and static analysis

concrete completes in anchors and basement slab, they provide sufficient strength to suppress pile deflections till end of excavation and post excavation.

The relationship between the estimated lateral deformation of the piles using three dimensional finite element model and measured results obtained by three inclinometers after completing excavation and installation of the anchors are presented in Fig. 15.11.

After obtaining a relationship between predicted numerical analysis and measured in-situ results, a series of seismic analysis was performed. Effects of an earthquake loading on excavation induced lateral deformation of piles were investigated in different analysis by applying El-Centro earthquake loads as prescribed displacement multiplying with acceleration data in three components, −x direction (direction of anchors), −y direction (throughout the piles) and combination of −x an −y directions. In the seismic analysis; soil conditions, material properties and the excavation geometry, were assumed to be same with actual site conditions presented above. Figure 15.12 shows the comparison of seismic and static analysis.

15.5 Conclusion

In this case study the excavation induced lateral deformations and vertical movements were investigated during the construction of a cut and cover tunnel in the Capital City of Turkey, Ankara. This study focused on, comparison of predicted and measured results of lateral deformation of piles and effects of seismic loads developed by an earthquake. Based on the results of the excavation case study and numerical finite element analysis following conclusions can be advanced:

1. In the Mecidiye-Belediye part of Ankara metro the excavation induced settlements at soil near to adjacent buildings and at top of piles and maximum lateral deformations were at 1.5 mm, −2.0 mm and −11.0 mm levels, relatively. This relatively less numbers were assumed to be the result of having a stiff soil and using secant piles which makes the piles behaves similar to a diaphragm wall.
2. The predicted lateral deformations and vertical movements show similar patterns with measured results.
3. Performed seismic analysis comparing to static analysis the support system presented in this study, shows that in seismic loading maximum lateral deformation increase approximately three times when the seismic load applied in the direction of excavation.
4. The ratio between lateral deformations and settlements were around 8–10, which indicates that lateral deformations are major concern at excavation induced movements.
5. Having a database about tunnel excavations in Turkey is a neglected issue, this work aims to be a sample for future studies in this area.

References

1. Finno RJ, Atmatzidis DK, Perkins SB (1989) Observed performance of a deep excavation in clay. J Geotech Eng 115(8):1045–1064
2. Tan Y, Li MW (2011) Measured performance of a 26 m deep topdown excavation in downtown Shanghai. Can Geotech J 48(5):704–719
3. Ou CY, Chiou DC (1993) Three-dimensional finite element analysis of deep excavation. In: Proceedings of 11th Southeast Asia geotechnical conference, Institution of Engineers Malaysia, Kuala Lumpur, pp 769–774
4. Wong IH, Poh TY, Chuah HL (1997) Performance of excavations for depressed expressway in Singapore. J Geotech Geoenviron Eng 123(7):617–625
5. Ou CY, Liao JT, Lin HD (1998) Performance of diaphragm wall constructed using the top-down method. J Geotech Geoenviron Eng 124(9):798–808
6. Finno RJ, Harahap IS (1991) Finite element analyses of HDR-4 excavation. J Geotech Eng 117(10):1590–1609
7. Whittle AJ, Hashash YMA, Whitman RV (1993) Analysis of deep excavation in Boston. J Geotech Eng 119(1):69–90
8. Kung GTC (2009) Comparison of excavation-induced wall deflection using top-down and bottom-up construction methods in Taipei silty clay. Comput Geotech 36:373–385
9. Ou CY, Chiou DC, Wu TS (1996) Three-dimensional finite element analysis of deep excavation. J Geotech Geoenviron Eng 122(5):337–345
10. Wang X, Zhao DS, Xu ML (2012) Numerical study on the effect of twin-tube shield excavation on adjacent pipe-lines. Appl Mech Mater 170:1491–1496
11. Sharma S, Judd WR (1991) Underground opening damage from earthquakes. Eng Geol 30(263):276
12. Power MS, Rosidi D, Kaneshiro JY (1998) Seismic vulnerability of tunnels and underground structures revisited. In: Proceedings of North American tunneling '98, Balkema Rotterdam, Newport Beach, pp 243–250
13. Kaneshiro JY, Power M, Rosidi D (2000) Empirical correlations of tunnel performance during earthquakes and aseismic aspects of tunnel design. In: Proceedings of the conference on lessons learned from recent earthquakes on earthquakes in Turkey 1999, 8 Nov 2011
14. Dao TPT (2011) Validation of PLAXIS embedded piles for lateral loading. Master of science thesis, Delft University of Technology, The Netherlands

Chapter 16
Calculating Stress and Strain from Experimental ODS Data

Brian Schwarz, Shawn Richardson, and Mark Richardson

Abstract In this paper it is shown how a Finite Element Analysis (FEA) model can be used together with experimental Operating Deflection Shape (ODS) data to calculate stresses & strains in a machine or mechanical structure. This allows for the on-line monitoring of structural stress & strain, which can be compared with prescribed warning levels to insure that dangerous levels are not exceeded. Examples are included to illustrate how ODS data measured with multiple accelerometers can be used to calculate stress & strain. Also, when this data is displayed together an ODS in animation on a 3D model of the machine or structure, high levels of stress or strain, or ***"hot spots"***, are quickly observed.

Keywords Operating deflection shape (ODS) • Stress & strain • Finite element analysis (FEA) mode shape • Modal assurance criterion (MAC)

16.1 Introduction

In a rotating machine, the dominant forces are applied at multiples of the machine running speed, called orders. An ***order-tracked ODS*** is assembled from the peaks at one of the order frequencies in a set of response frequency spectra of a machine. When displaying in animation on a 3D model [1], an order-tracked ODS is a convenient way to visualize vibration levels, and therefore monitor the health of the machine.

In a companion paper [2], it is shown how ***modes participate*** in an order-tracked ODS of a rotating machine, and how they participate differently at different operating speeds. It is also shown how the ***modal participation*** can be used to ***expand*** an order-tracked ODS so that it is suitable for display on a model of the machine.

It is well known that most rotating machines will exhibit different vibration levels under different loads and speeds. ODSs are conveniently acquired by attaching multiple accelerometers to the machine surfaces, and acquiring vibration data from the machine while it is running. In addition to visualizing the deflection of the machine in real time, an ODS can be used to calculate stress & strain by ***deflecting an FEA model*** of the machine.

16.2 Variable Speed Rotating Machine

Figure 16.1 shows a variable speed rotating machine, instrumented with eight tri-axial accelerometers. An accelerometer is attached to the top of each bearing block, and six accelerometers are attached to the base plate; three on the front edge and three on the back edge. This test setup was used to measure order-tracked ODSs under different machine speeds.

A laser tachometer with its beam pointed at the outer wheel of the machine was used to measure the machine speed, as shown in Fig. 16.1. The outer wheel had reflective tape on it, so the laser measured the once-per-revolution speed of the machine.

Figure 16.2 contains a model of the machine that was used to display ODSs in animation. Each of the numbered test points is displayed as a cube icon.

B. Schwarz • S. Richardson • M. Richardson (✉)
Vibrant Technology, Inc., Scotts Valley, CA, USA
e-mail: mark.richardson@vibetech.com

A. Wicks (ed.), *Shock & Vibration, Aircraft/Aerospace, and Energy Harvesting, Volume 9*,
Conference Proceedings of the Society for Experimental Mechanics Series, DOI 10.1007/978-3-319-15233-2_16

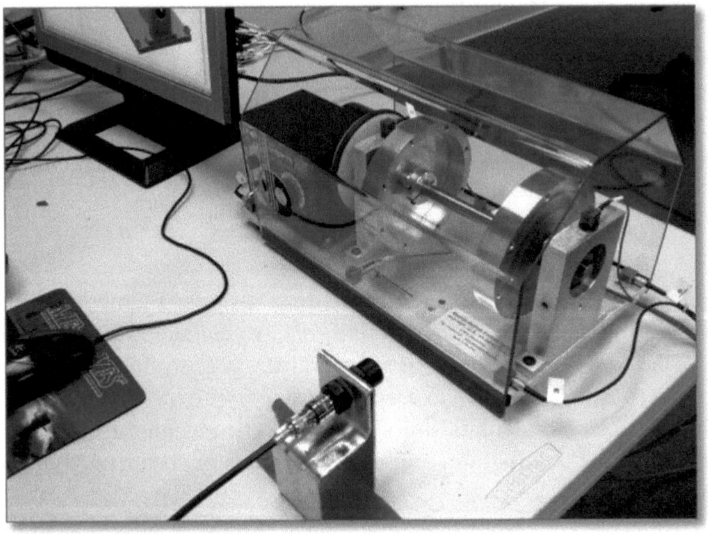

Fig. 16.1 Variable speed rotating machine

Fig. 16.2 Rotating machine model

16.3 Mode Shapes of the Machine

The order-tracked ODSs of the machine in Fig. 16.1 contain participation of both ***"rigid body"*** and ***"flexible"*** mode shapes of the base plate and bearing blocks. Since the machine is resting on four rubber mounts (one under each corner), its rigid body modes will participate significantly in its ODS. The machine has ***six rigid body*** mode shapes. These mode shapes describe the ***free-free motion*** of the machine in space, but they also participate in its ODS since it is resting on ***four soft springs***.

The rigid body and flexible body mode shapes of the machine were obtained from an FEA model of the base plate and one of the bearing blocks. The first 20 mode shapes of the base plate together with 20 modes for each of the bearing blocks were used together with the SDM method [6] to solve for the modes of the combined substructures. Paper [1] contains more details of this procedure.

Some of the mode shapes of the base plate and bearing blocks (60 in all) are listed in Fig. 16.3. Notice that the first six modes are rigid body modes, with frequencies of **"0"**.

16.4 ODS Expansion

ODS data was acquired from the rotating machine in Fig. 16.1 at two different machine speeds. The mode shapes of the base plate and bearing blocks were then used to ***expand*** the experimental ODSs acquired from the eight accelerometers to ODSs for all DOFs on the base plate and bearing blocks [4, 5].

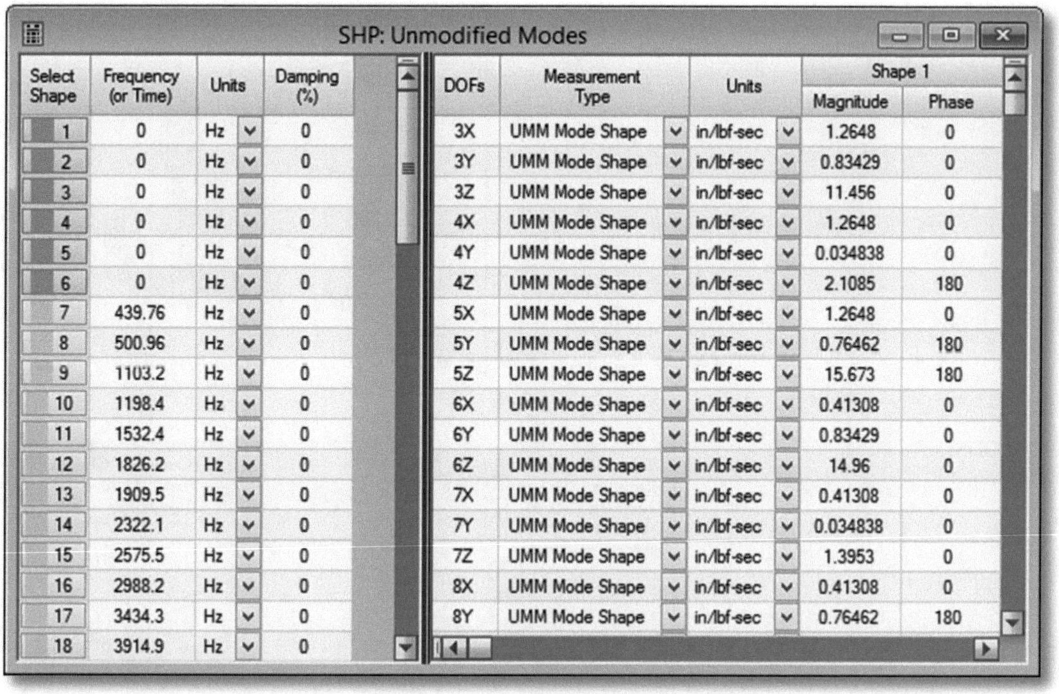

SHP: Unmodified Modes

Select Shape	Frequency (or Time)	Units	Damping (%)
1	0	Hz	0
2	0	Hz	0
3	0	Hz	0
4	0	Hz	0
5	0	Hz	0
6	0	Hz	0
7	439.76	Hz	0
8	500.96	Hz	0
9	1103.2	Hz	0
10	1198.4	Hz	0
11	1532.4	Hz	0
12	1826.2	Hz	0
13	1909.5	Hz	0
14	2322.1	Hz	0
15	2575.5	Hz	0
16	2988.2	Hz	0
17	3434.3	Hz	0
18	3914.9	Hz	0

DOFs	Measurement Type	Units	Shape 1 Magnitude	Shape 1 Phase
3X	UMM Mode Shape	in/lbf-sec	1.2648	0
3Y	UMM Mode Shape	in/lbf-sec	0.83429	0
3Z	UMM Mode Shape	in/lbf-sec	11.456	0
4X	UMM Mode Shape	in/lbf-sec	1.2648	0
4Y	UMM Mode Shape	in/lbf-sec	0.034838	0
4Z	UMM Mode Shape	in/lbf-sec	2.1085	180
5X	UMM Mode Shape	in/lbf-sec	1.2648	0
5Y	UMM Mode Shape	in/lbf-sec	0.76462	180
5Z	UMM Mode Shape	in/lbf-sec	15.673	180
6X	UMM Mode Shape	in/lbf-sec	0.41308	0
6Y	UMM Mode Shape	in/lbf-sec	0.83429	0
6Z	UMM Mode Shape	in/lbf-sec	14.96	0
7X	UMM Mode Shape	in/lbf-sec	0.41308	0
7Y	UMM Mode Shape	in/lbf-sec	0.034838	0
7Z	UMM Mode Shape	in/lbf-sec	1.3953	0
8X	UMM Mode Shape	in/lbf-sec	0.41308	0
8Y	UMM Mode Shape	in/lbf-sec	0.76462	180

Fig. 16.3 FEA mode shapes of the base plate & bearing blocks

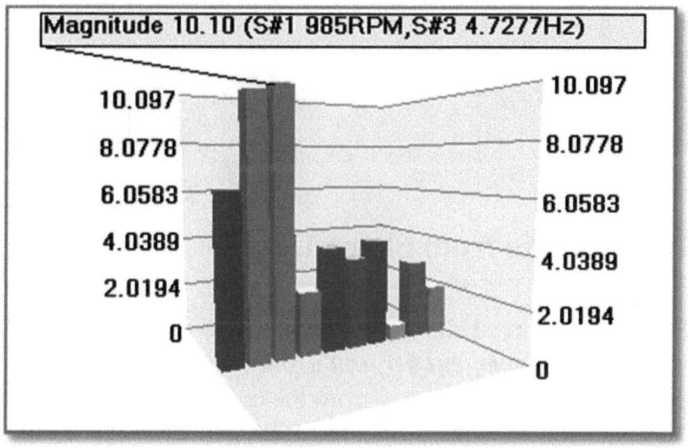

Fig. 16.4 Magnitudes of modal participation at 985 RPM

The modal participation of the first 10 FEA modes in the 985 RPM ODS is shown in Fig. 16.4 [3]. The modal participation factors show that the ***first three modes*** are the ***dominant contributors*** to the 985 RPM ODS. All three of these ***rigid body*** modes are being excited at this speed, and the machine is simply ***"bouncing"*** vertically and sideways on its rubber mounts.

The modal participation of the first 10 FEA modes in the 2280 RPM ODS is shown in Fig. 16.5. The participation factors of modes 2, 5, and 6 indicate that they ***dominate*** the 2280 RPM ODS. At this higher speed, the machine is ***"rocking and twisting"*** on its rubber mounts, with more motion at the outer bearing location.

Figure 16.6 is a comparison display of the two expanded ODSs on a model of the base plate and bearing blocks. Animation of these shapes more clearly shows that they are different, but their ***low MAC value (0.17)*** also indicates that they are different. The 2280 RPM ODS has much more motion at the outboard bearing block. This is due to the greater cyclic force created at the higher speed by the unbalance weight that was added to the outboard wheel.

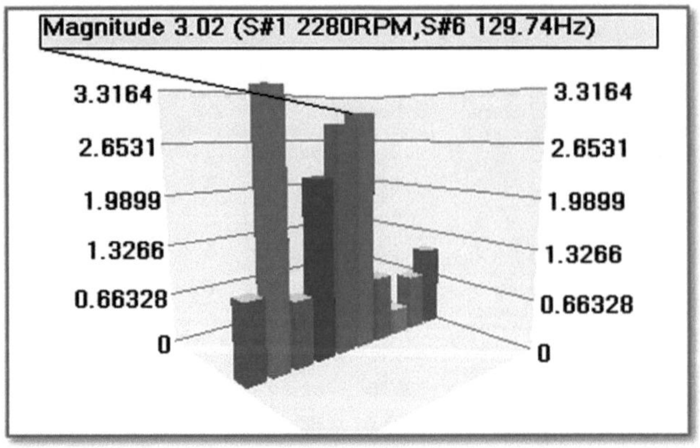

Fig. 16.5 Magnitudes of modal participation at 2280 RPM

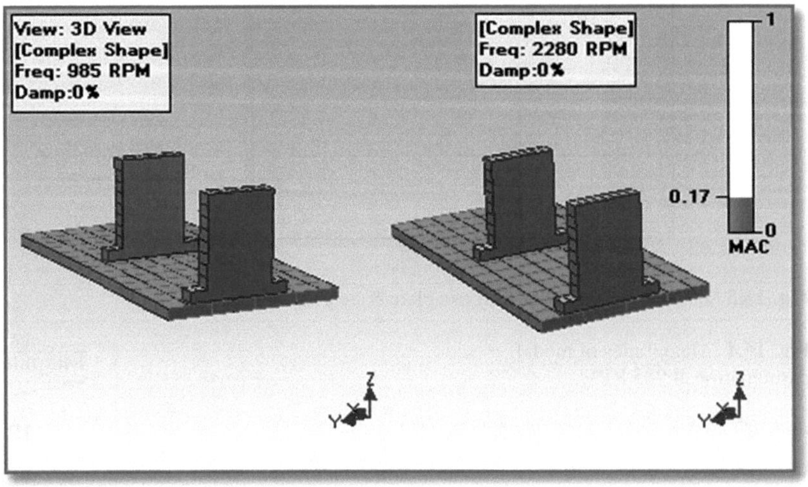

Fig. 16.6 Expanded ODSs of base plate & bearing blocks

16.5 Strain from Shape Data

Strain is the forced change in the dimensions of a structure, measured as a change in ***dimension per unit length***. Its units are typically ***displacement per unit displacement***. Strain in an FEA element is a function of the deflections of its vertices. These deflections can be obtained from the components of an expanded ODS, or from mode shapes.

When the dynamics of a machine can be adequately represented by mode shapes, an ODS can be represented as a ***weighted summation*** of mode shapes. This is the well known ***superposition*** property of modes. In this case, strain can also be calculated for each mode shape.

The equation for calculating the **normal & shear strain** at an FEA element vertex is:

$$\begin{Bmatrix} \varepsilon_x \\ \varepsilon_y \\ \varepsilon_z \\ \gamma_{xy} \\ \gamma_{yz} \\ \gamma_{xz} \end{Bmatrix} = [B] \begin{Bmatrix} u_1 \\ v_1 \\ w_1 \\ \vdots \\ u_n \\ v_n \\ w_n \end{Bmatrix} \tag{16.1}$$

Where:

[B] = displacement strain matrix for an FEA element vertex

$$[B] = \begin{bmatrix} \frac{\partial}{dx} & 0 & 0 & \cdots & \frac{\partial}{dx} & 0 & 0 \\ 0 & \frac{\partial}{dy} & 0 & & 0 & \frac{\partial}{dy} & 0 \\ 0 & 0 & \frac{\partial}{dz} & \cdots & 0 & 0 & \frac{\partial}{dz} \\ \frac{\partial}{dy} & \frac{\partial}{dx} & 0 & & \frac{\partial}{dy} & \frac{\partial}{dx} & 0 \\ 0 & \frac{\partial}{dz} & \frac{\partial}{dy} & \cdots & 0 & \frac{\partial}{dz} & \frac{\partial}{dy} \\ \frac{\partial}{dz} & 0 & \frac{\partial}{dx} & & \frac{\partial}{dz} & 0 & \frac{\partial}{dx} \end{bmatrix} \quad \text{(6 by 3n)}$$

$$\begin{Bmatrix} u_1 \\ v_1 \\ w_1 \\ \vdots \\ u_n \\ v_n \\ w_n \end{Bmatrix} = \text{ODS (or mode shape) components at all vertices}$$

n = number of element vertices

$$\begin{Bmatrix} \varepsilon_x \\ \varepsilon_y \\ \varepsilon_z \\ \gamma_{xy} \\ \gamma_{yz} \\ \gamma_{xz} \end{Bmatrix} = \text{normal \& shear strain at an FEA element vertex}$$

16.6 Stress from Strain

Stress is the amount of ***force acting within a cross sectional area*** of a structure. Its units are typically ***force per unit area***. Stresses are calculated from strains with the following equation:

$$\begin{Bmatrix} \sigma_x \\ \sigma_y \\ \sigma_z \\ \tau_{xy} \\ \tau_{yz} \\ \tau_{xz} \end{Bmatrix} = [C] \begin{Bmatrix} \varepsilon_x \\ \varepsilon_y \\ \varepsilon_z \\ \gamma_{xy} \\ \gamma_{yz} \\ \gamma_{xz} \end{Bmatrix} \quad (16.2)$$

Where:

$$\begin{Bmatrix} \sigma_x \\ \sigma_y \\ \sigma_z \\ \tau_{xy} \\ \tau_{yz} \\ \tau_{xz} \end{Bmatrix} = \text{normal \& shear stress at an FEA element vertex}$$

[C] = stress strain matrix (6 by 6)

$$[C] = \frac{E}{(1+v)(1-2v)} \begin{bmatrix} (1-v) & v & v & & 0 & \\ v & (1-v) & v & & & \\ v & v & (1-v) & & & \\ & & & \frac{(1-2v)}{2} & 0 & 0 \\ & 0 & & 0 & \frac{(1-2v)}{2} & 0 \\ & & & 0 & 0 & \frac{(1-2v)}{2} \end{bmatrix}$$

E = Young's modulus of elasticity
v = Poisson's ratio

16.7 Strain from Mode Shapes

The participation factors shown in Figs. 16.4 and 16.5 indicate that the ***flexible body*** modes 7, 9, and 10 contributed significantly to both the 985 and 2280 RPM ODSs. Figures 16.7, 16.8 and 16.9 is a display of the ODS deflection together with the **normal strain in the x-direction** of the base plate and bearing blocks, for each of these modes.

Fig. 16.7 Normal strain, X-direction, mode #7

Fig. 16.8 Normal strain, X-direction, mode #9

Fig. 16.9 Normal strain, X-direction, mode #10

Fig. 16.10 Normal stress, X-direction, mode #7

Notice that where local bending occurs in each mode shape, positive strain (***tension***) on one side of the base plate equals the same amount of negative strain (***compression***) on the other side. Also, the cross section of the plate transitions from positive to negative strain values.

16.8 Stress from Mode Shapes

Figures 16.10, 16.11 and 16.12 is a display of the ODS deflection together with the **normal stress in the x-direction** of the base plate and bearing blocks, for modes 7, 9, and 10.

Notice again that where local bending occurs, positive stress (***tensile***) on one side of the base plate equals the same amount of negative stress (***compressive***) on the other side of the plate. Also, the cross section of the plate transitions from positive to negative stress values.

Fig. 16.11 Normal stress, X-direction, mode #9

Fig. 16.12 Normal stress, X-direction, mode #10

16.9 Strain in the 985 & 2280 RPM ODS

Figures 16.13 and 16.14 is a display of the **normal strain in the x-direction** calculated for both the 985 RPM and 2280 RPM ODSs. Notice that the strain distributions of the two ODSs are different, primarily because the ODSs themselves are different, as shown in Figure 16.6. Also, the peak strain levels are higher for the 2280 RPM ODS, which is expected. Unlike the mode shape strain values, these values are realistic because they are based on the experimental ODS values in ***inches*** of displacement.

16.10 Stress in the 985 & 2280 RPM ODS

Figures 16.15 and 16.16 is a display of the **normal stress in the x-direction** calculated for both the 985 RPM and 2280 RPM ODSs. Again, the stress distribution is noticeably different between the two ODSs. The peak stress levels are also higher for the 2280 RPM ODS, as expected.

Fig. 16.13 Normal strain, X-direction, 985 RPM ODS

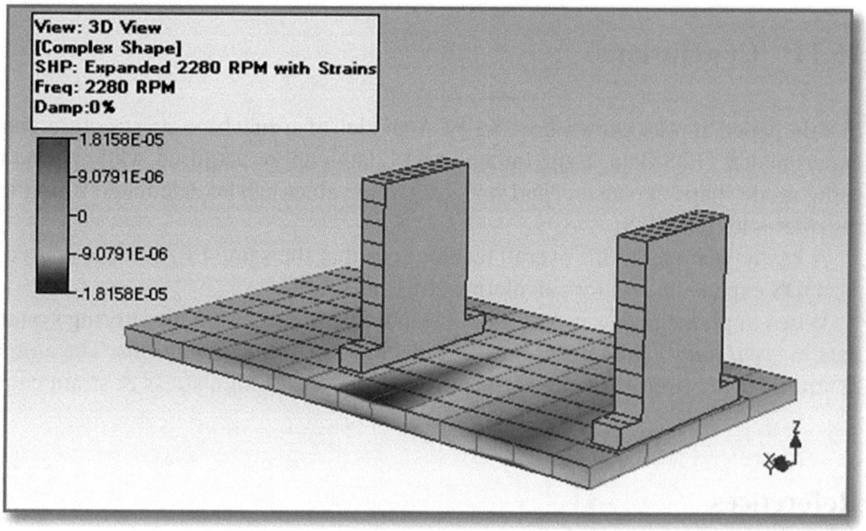

Fig. 16.14 Normal strain, X-direction, 2280 RPM ODS

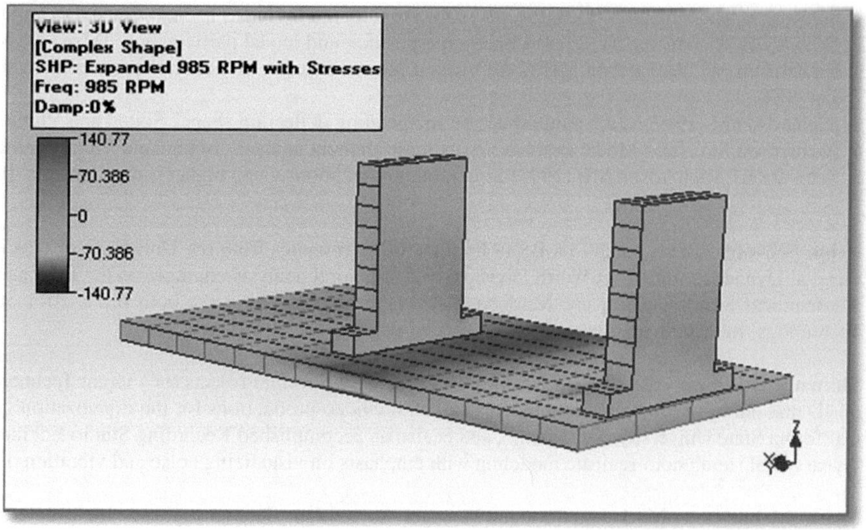

Fig. 16.15 Normal stress, X-direction, 985 RPM ODS

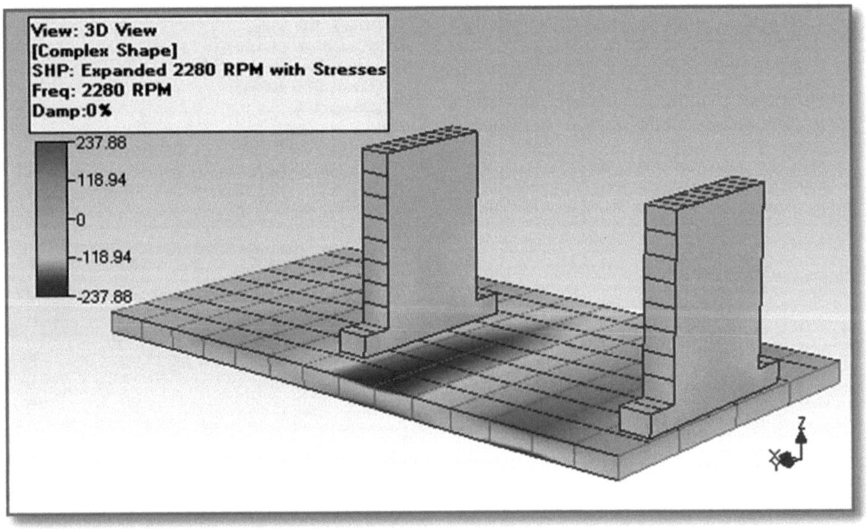

Fig. 16.16 Normal stress, X-direction, 2280 RPM ODS

16.11 Conclusions

In this paper, it was shown how an FEA model of a machine or structure can be used to calculate stress & strain from experimental ODS data. Experimental ODS data can be acquired while a machine is running, expanded into many DOFs using mode shape expansion, and stress & strain values can be calculated from the expanded ODS and displayed in animation together with the ODS.

A key advantage of this overall technique is that the same FEA model is used both to calculate the mode shapes required for ODS expansion and for calculating stress & strain.

When implemented as part of a troubleshooting or long term monitoring system, this technique can provide stress & strain data in ***"real time"***, even under different machine operating conditions. The animated display on an ODS together with stress & strain is very useful for quickly identifying regions of high stress & strain caused by vibration.

References

1. Schwarz B, Richardson S, Richardson M (2015) Using mode shapes for real time ODS animation. IMAC XXXIII, 3–5 Feb 2015
2. Schwarz B, Richardson M (2014) Linear superposition and modal participation. IMAC XXXII, 3–6 Feb 2014
3. Richardson SC, Richardson MH (2008) Using photo modeling to obtain the modes of a structure. In: 16th IMAC conference, Orlando, FL, Jan 2008
4. Richardson M (1997) Is it a mode shape or an operating deflection shape? Sound and Vibration Magazine, Mar 1997
5. Richardson M (2005) Modal analysis versus finite element analysis. Sound and Vibration Magazine, Sept 2005
6. Schwarz BJ, Richardson MH (1997) Structural modifications using higher order elements. In: 15th IMAC conference, Orlando, FL, Feb 1997

Brian Schwarz Brian received a BS in Engineering Mechanics from the University of Cincinnati in 1987. Upon graduation, he began work at General Dynamics, Inc., Fort Worth Division, as a structural analysis engineer on the F-16 and A-12 programs. In 1993, he joined the Lockheed Aeronautical Systems Company, Marietta, GA where he worked on the both the L-1011 and F-22 programs. In 1995, Brian joined Vibrant Technology, Inc where he is currently the Director of Engineering Applications.

Shawn Richardson Shawn Richardson is the Director of Special Projects for Vibrant Technology, Inc., and is widely responsible for the creation of 3D and photo-realistic structure models and audio/video productions for the organization. Shawn has a B.S. in Marketing and a M.B.A. from California State University, Sacramento, and is also an accomplished Recording Studio Engineer and studio owner. Shawn has devoted more than 7 years to 3D and photo-realistic modeling with emphasis on visualizing noise and vibration data.

Mark Richardson Mark is a Product Manager for Vibrant Technology Inc. He oversees the development, testing and marketing for the MechaniCom Machine Surveillance software. Prior to joining Vibrant in 2011, Mark worked for various enterprise software companies in product management and sales. He has a B.S. in Environmental Design from the University of Colorado and graduate studies in Business and Technology.

Dr. Richardson founded Vibrant Technology, Inc., Scotts Valley, California in 1991, and is currently its President and CEO. Vibrant is the developer of the ME'scope Visual Engineering Series™ noise and vibration analysis software, and the MechaniCom Machine Surveillance System™ on-line machine and structural health monitoring software.

During the 1970s as product manager at the Hewlett Packard Co., Mark directed the development of the first commercially available FFT-based modal testing system, and the first dedicated modal testing instrument, the HP 5423A Structural Dynamics Analyzer.

In 1979, Dr. Richardson co-founded Structural Measurement Systems, Inc (SMS) and was its President and CEO. SMS developed and marketed the Star Modal™ software for analyzing structural noise and vibration problems.

Mark has authored over 60 technical papers and magazine articles on structural testing, digital signal processing, modal analysis, and machine fault detection. He has traveled worldwide conducting tests, giving seminars, and teaching short courses.

Chapter 17
Case Study: Modeling Generator Chassis Responses with ODS Data

Sze Kwan Cheah

Abstract Creating a correlated model of an operating generator set is a complicated and expensive endeavor. This case study describes the process of modeling the generator chassis with finite elements and applying generator set measured acceleration values as inputs for a harmonic forced vibration analysis. The underlying idea of this approach is to capture the complex excitation function via experimental data on a subsystem that is relatively linear. Various model responses of the chassis are compared to the measured operating responses.

Keywords Operating deflection shape • Large mass method • Correlation • Generator • Finite element analysis

17.1 Introduction

In this case study, the generator can generally be described as having an engine, alternator, and chassis. The engine and alternator are frequently qualified as individual parts while the chassis is designed to support both components. Both the engine and alternator are fairly complicated systems. The engine has explosive cylinder pressure driving a multi-body dynamic system with internal flowing coolant. The alternator generates and absorbs electromagnetic forces to produce power. The chassis is a relatively simpler component constructed by welded joints of hollow structural section and/or I-beams. The entire system is typically bolted together.

When vibration and strain measurements tests were performed to qualify the generator chassis, vibration measurement locations were selected to capture large deformation and responses of key modes. The strain gage locations were selected from expected high stress concentration locations. The number of available channels and strain gages limits the number of high stress locations that can be measured.

Traditionally, vibration response of an operating generator set [1] can be simulated via commercial code such as AVL EXCITE and LMS Virtual.Lab. Simulating the entire generator set from cylinder forces to responses at the chassis has many challenges such as uncertainties of damping, non-linear joints and deviation from nominal geometry. Correlation depends heavily on the accuracy of various inputs.

To reduce modeling effort, only the chassis is modeled in detail using Finite Element Analysis (FEA). The chassis is a simpler component and little updating is historically necessary to achieve reasonable correlation. The engine and generator are viewed as substructures that are bolted to the chassis. To avoid the complexity of modeling these substructures, the vibration responses or operating deflection shapes (ODS) were measured directly on the chassis near the system interfaces as well as other additional locations. The measured motion is then enforced onto the chassis using the large mass method via harmonic analysis. The dynamic response of the entire chassis can then be computed. In order to evaluate the effectiveness of this approach, a selection of force and strain measurement on the chassis were used to check the accuracy of the predicted response.

S.K. Cheah (✉)
Cummins Power Generation, 1400 73rd AVE NE, 55432 Minneapolis, MN, USA
e-mail: sze.kwan.cheah@cummins.com

A. Wicks (ed.), *Shock & Vibration, Aircraft/Aerospace, and Energy Harvesting, Volume 9*,
Conference Proceedings of the Society for Experimental Mechanics Series, DOI 10.1007/978-3-319-15233-2_17

17.2 Theoretical Background

The majority of the large mass method application has been on base excitation for shaker table or seismic analysis [2, 3]. The theory is applicable to multi-point excitation that are independent of each other. The finite element model equation of motion [4] could be symbolically represented with the following equation:

$$[M]\{\ddot{y}\} + [C]\{\dot{y}\} + [K]\{y\} = \{F\} \tag{17.1}$$

where M, C & K are respectively the mass, damping and stiffness matrix; y vector represents the response of the system and F vector is the external force on the system. The enforced motion is applied by attaching a large mass to the node where enforced motion is desired. The force is computed by Newton's second law:

$$\{F\} = LM\,\{\ddot{u}_b\} \tag{17.2}$$

where LM is the large mass and u_b is the enforced motion at the corresponding FEA node. The scalar value of the large mass should be sufficiently big such that the damping and spring forces adjacent to the node are negligible, i.e. $c_{ij}/LM = k_{ij}/LM = 0$. It was found through numerical experiments [5] that large mass ratio of 10^3 to 10^{10} provides accuracy that matches theoretical results. The large mass method [6] is introduced into Eq. (17.1) by replacing the ith equation with the following:

$$LM_i\ddot{y}_i + \sum_{j\neq i} m_{ij}\ddot{y}_j + \sum_j c_{ij}\dot{y}_j + \sum_j k_{ij}y_j = LM_i\ddot{u}_{bi} \tag{17.3}$$

17.3 Experiment Setup

A prototype diesel generator set shown in Fig. 17.1 was tested for both linear vibration and strain simultaneously. Ten triax consisting of PCB 353B33 accelerometers were used on the chassis. An example accelerometer triax configuration is shown in Fig. 17.2. Vibration data acquisition was taken using LMS SCADAS III & Test.Lab 11B at 2,048 samples/s. Strain measurements were acquired with CEA-06-062UW-350 single element gage. An example strain gage is shown in Fig. 17.3. 5 V DC excitation were applied on all strain gages from the SOMAT e-DAQ acquisition system. Furthermore, dynamic axial loading of a bolt was measured with HBM KML style load washer as shown in Fig. 17.4. While multiple different runs were performed, only the full load steady state 50 Hz power generation at 1,500 RPM operation was considered for this analysis.

Fig. 17.1 Prototype generator set

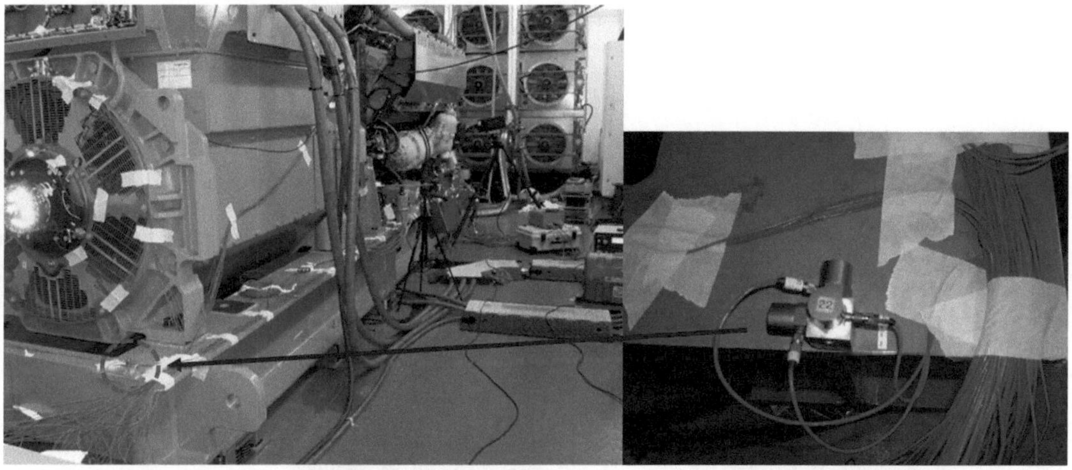

Fig. 17.2 Generator set on spring isolators, PCB 352C33 accelerometers

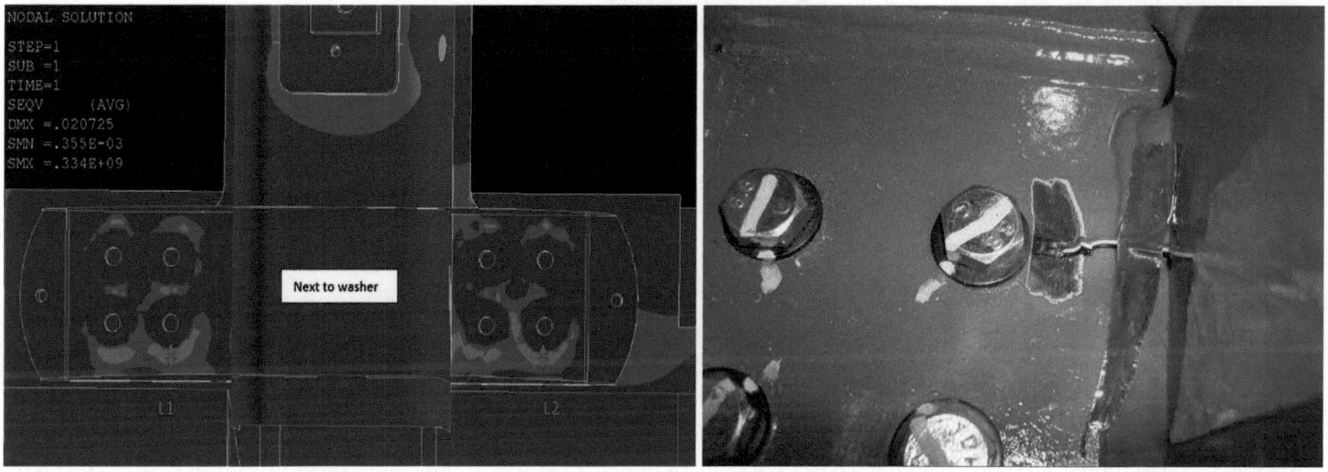

Fig. 17.3 Top view of Trunnion support bracket & strain gage #1

Fig. 17.4 Bolt load washer

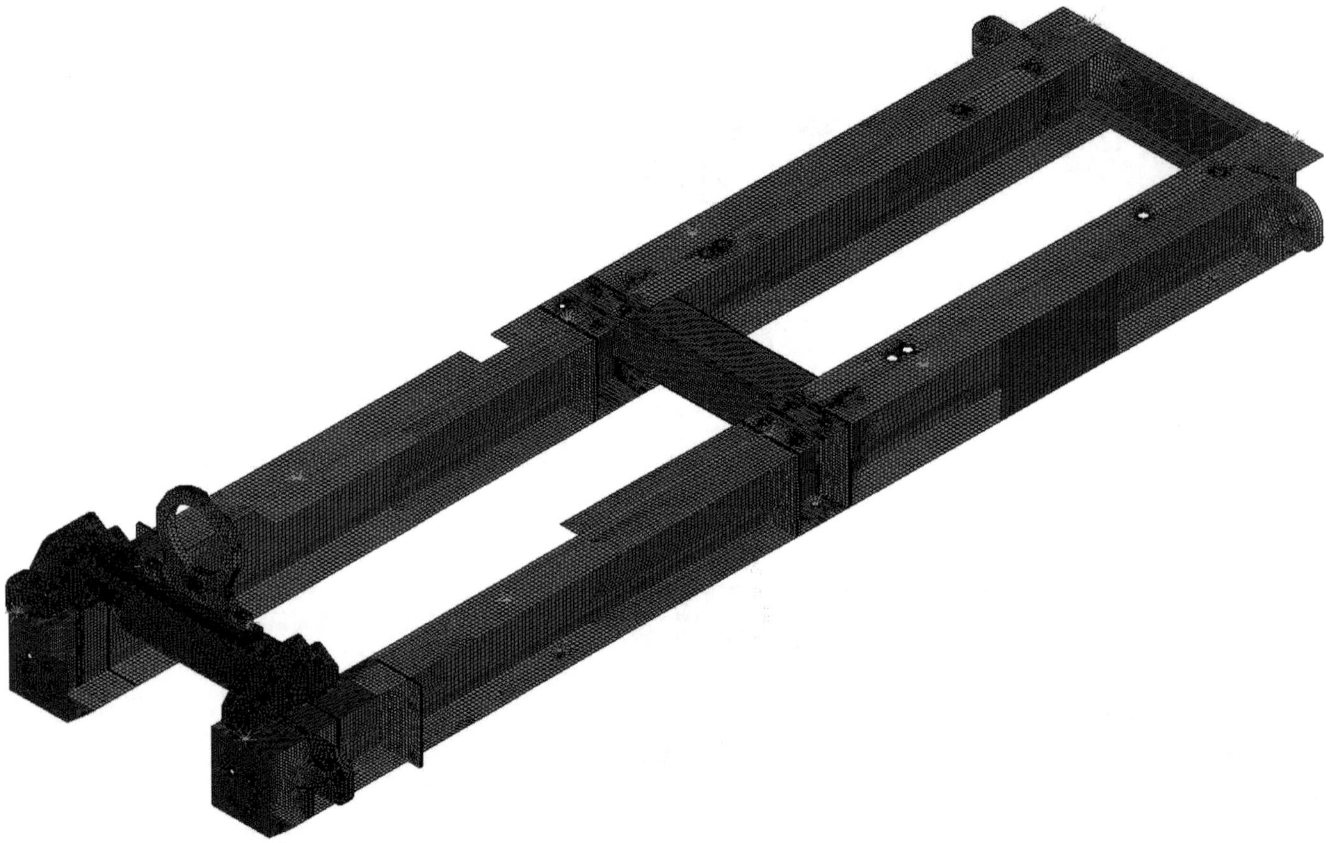

Fig. 17.5 FEA model of generator chassis

17.4 Model Setup

The finite element model was constructed in ANSYS as shown in Fig. 17.5. The model consists of mostly SHELL93 elements with some SOLID95, BEAM4, COMBIN14 & MASS21 elements. Model size is roughly 1.3 million nodes. The bolts were modeled using beam elements and rigid constraint equations (CERIG). Spring elements were modeled under the chassis. Large mass elements (each a thousand times the chassis mass) were added to each measurement location on the chassis. They are represented by a star symbol on the top surface in Fig. 17.5.

The operating deflection shape spectrum data of the chassis (black points in Fig. 17.6) were exported from LMS Test.Lab into ANSYS. Data was only extracted at every half order as these have the major responses. A modal analysis was performed for 30 modes for the use of modal superposition harmonic analysis. It was found that 22 modes would capture up to 200 Hz when the chassis did not have any large mass elements attached to it. Thirty modes was selected to capture higher modes.

The harmonic analysis was performed with modal superposition at every half order up to 200 Hz. Forces consisting of the large mass multiplied by the measured acceleration were applied at frequencies of 12.5 Hz, 25 Hz, 37.5 Hz, 50 Hz... 200 Hz.

Damping ratio of 1 % was assumed.

17.5 Results and Discussion

Analytical results were extracted from the model in the frequency domain at every half order. The response was compared to the test results in both the frequency and time domain for bolt alternating force (Fig. 17.7) as well as strain gage #1 & #2 (Figs. 17.8 and 17.9). A stress concentration of 2 was assumed for the strain gages. The measured strain data plots are amended to provide direct comparisons. All values are shown to be normalized to an arbitrary number.

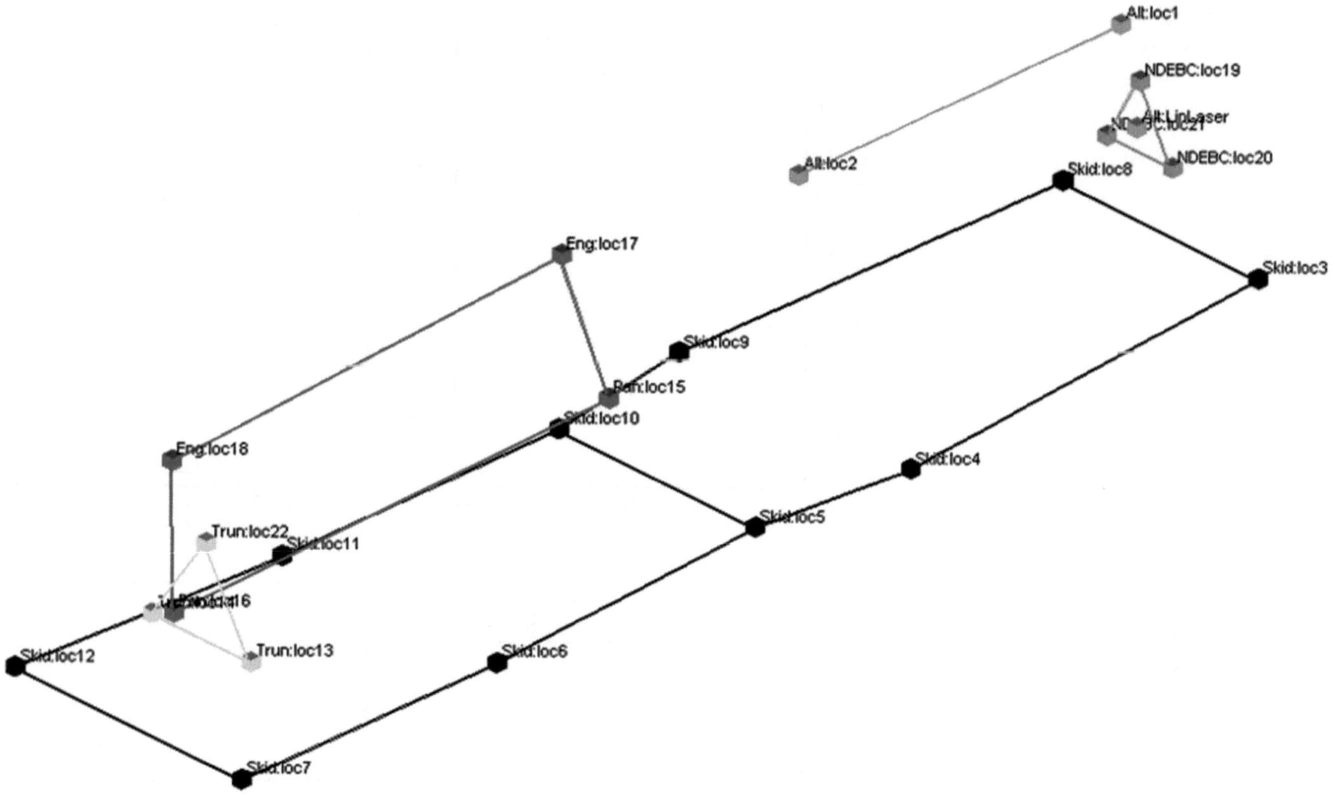

Fig. 17.6 Accelerometer measurement locations

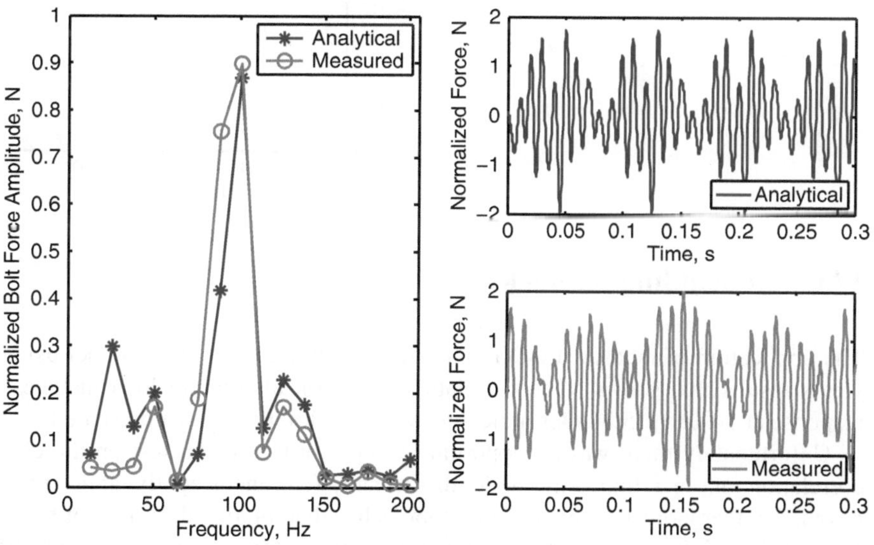

Fig. 17.7 Comparison of dynamic bolted force in frequency & time domain

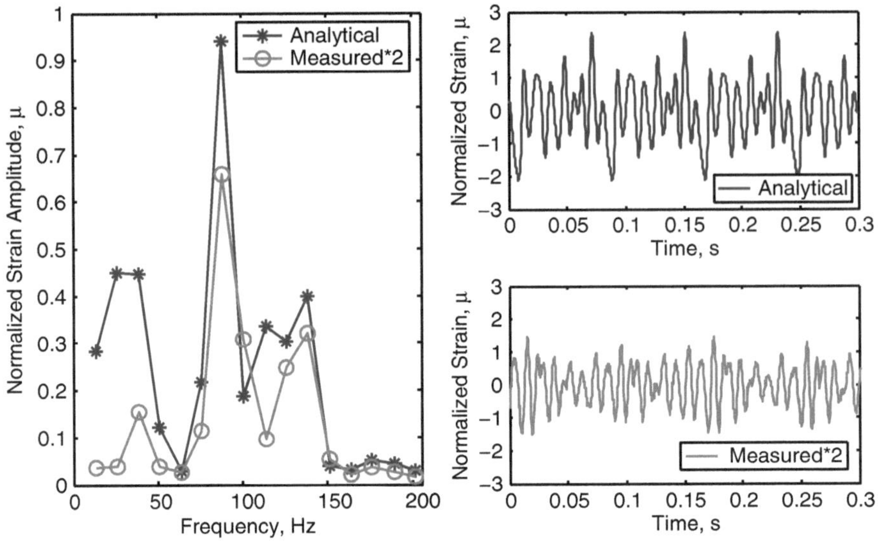

Fig. 17.8 Comparison of strain in gage #1 in frequency & time domain

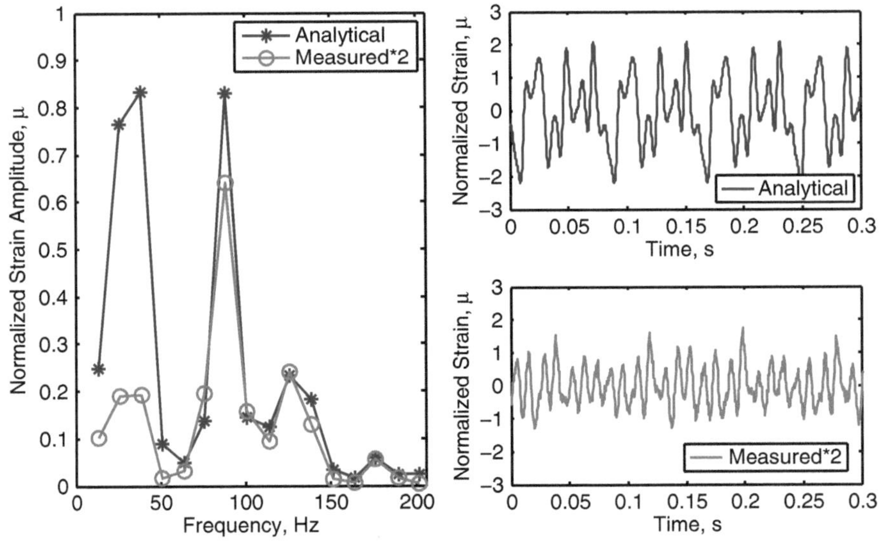

Fig. 17.9 Comparison of strain in gage #2 in frequency & time domain

17.6 Concluding Remarks

The application of measured ODS onto the generator chassis FEA model using large mass method via harmonic analysis yields results that shows some correlation. The bolt alternating load matches well but the strain data overestimated the response at 1st and 1.5th order indicating some modes were overrepresented.

Unfortunately there was no opportunity to perform a modal test on the chassis in a free-free state to allow for model correlation and updating. Any model discrepancy (e.g. incorrect dynamic spring stiffness) is expected to contribute to the errors. Another large deficiency of this approach is that rotational degrees of freedom are not captured in the ODS. Depending on the selected accelerometer measurement locations, this may be important. While the above case study shows reasonable results, confidence in this approach needs to be substantiated for other systems.

Acknowledgements The author would like to thank Steve Seidlitz, Maria Wood and James Hogben for the exceptional vibration test data. Gunjan Maheshwari, Alastair Clifford contributed to the excellent strain measurement. Special thanks to Hari Prasad Konka for creating the initial FEA models. This work was supported by Cummins Power Generation.

References

1. Pan C, Chen XL, Brackett ME, Dokun OD (2006) Electrical power generator package forced response simulation and validation. In: IMAC-XXIV: conference & exposition on structural dynamics, St Louis MO
2. Hale S (2012) Base acceleration harmonic response analysis in workbench. https://caeai.com/sites/default/files/WB_harmonic_shaker_table.pdf
3. Pamidi PR, Reymond M. Improvements to dynamic analysis in MSC. Nastran. http://web.mscsoftware.com/support/library/conf/auto00/p04200.pdf
4. Harris CM (1995) Shock and vibration handbook, 4th edn, McGraw-Hill, New York City
5. Kim Yong-Woo, Jhung Myung Jo (2014) A study of large mass method for dynamic problem of multiple degree-of-freedom system excited by ground acceleration time history. J Mech Sci Technol 28(1):25–41
6. Liu Yuxin, Lu Zhitao (2010) Methods of enforcing earthquake base motions in seismic analysis of structures. Eng Struct 32:2019–2033

Chapter 18
Shock Calibration with Zero Shift Using a Digital Filter Technique

Hideaki Nozato, Wataru Kokuyama, and Akihiro Ota

Abstract In compliance with ISO 16063-13, a shock calibration system for accelerometers ranging from 50 to 10,000 m/s^2 and with pulse widths more than several milliseconds was developed. However, a piezoelectric accelerometer output voltage with zero shift was induced by the low-frequency response of the charge amplifier. So we designed a virtual amplifier with infinite impulse response filters as a digital replica of the charge amplifier. Since the virtual amplifier has the same input–output characteristics as the real charge amplifier, we can obtain the shock sensitivity of a piezoelectric accelerometer by applying the input acceleration to the virtual amplifier. As a result, we obtain flat charge sensitivities from 50 to 10,000 m/s^2, which are comparable to the vibration calibration results.

Keywords Shock calibration • Accelerometer • Digital filter • Zero shift • Charge amplifier

18.1 Introduction

There is a strong demand by Japanese industries for shock calibration for accelerometers at acceleration levels from several hundred m/s^2 to several thousand m/s^2 to improve human safety levels in car crash tests and electric device drop-tests. In general, accelerometers are calibrated using a secondary shock calibration system to compare calibrations, and a high-performance piezoelectric accelerometer is used as the reference accelerometer in various secondary shock calibration systems. Thus, accurate shock calibrations by piezoelectric accelerometers are in high demand. But since a piezoelectric accelerometer is used in combination with a charge amplifier, a zero voltage drift is generated due to the high-pass characteristics of the charge amplifier in the low-frequency region [1]. Although the ISO standards 16063-13 [2] and 16063-22 [3] state that the zero shift shall be within 1 %, zero shift exists as a large error component in shock calibrations with a long pulse. The aim of this study is to reduce the effect of the zero shift to achieve superior shock calibrations for piezoelectric accelerometers. To reduce the zero shift, we developed a virtual amplifier that works on a personal computer in which the virtual amplifier operates with the same input–output characteristic as a real charge amplifier [4]. This paper reports on the use of this virtual amplifier for the shock calibrations of a high-performance piezoelectric accelerometer.

18.2 Shock Calibration System

Figures 18.1 and 18.2 show a schematic of the principles of shock calibration and a photograph of the shock calibration system at the National Metrology Institute of Japan (NMIJ), respectively. The shock calibration system provides primary calibrations with laser interferometry in compliance with ISO 16063-13. This system consists of two parts; the first is a shock exciter which generates a pulsed shock with peak acceleration from 50 to 10,000 m/s^2. The shock is induced by a collision between two rigid bodies supported by an air bearing, and pulse widths vary between 0.5 and 5.0 ms. The second part is a measurement system with two digitizers (ATS660, Alazer Technologies Inc.) and a modified homodyne Michelson laser interferometer. The laser interferometer monitors one position along the sensitive axis of an accelerometer and employs a He-Ne laser wavelength (632.8 m) as the length standard. The two digitizers have a vertical resolution of 16 bits, a signal processing sampling frequency of more than 100 MHz, and they record quadrature signals depending on

H. Nozato (✉) • W. Kokuyama • A. Ota
National Metrology Institute of Japan (NMIJ), National Institute of Advanced Industrial Science and Technology (AIST), 1-1 Umezono, 305-8563 Tsukuba, Ibaraki, Japan
e-mail: hideaki.nozato@aist.go.jp

A. Wicks (ed.), *Shock & Vibration, Aircraft/Aerospace, and Energy Harvesting, Volume 9*,
Conference Proceedings of the Society for Experimental Mechanics Series, DOI 10.1007/978-3-319-15233-2_18

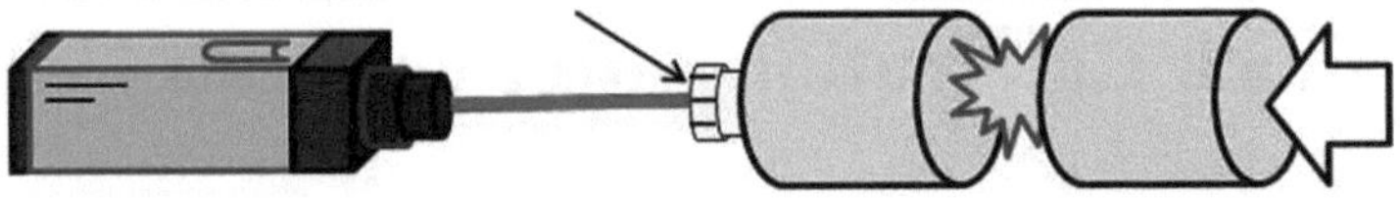

Fig. 18.1 Principles of shock calibration

Fig. 18.2 Shock calibration system

the displacement by the shock. With respect to the demodulation process from the quadrature signal to the acceleration waveform, we implemented phase unwrapping and differentiation twice through a Butterworth digital low-pass filter with a cut-off frequency of 5 kHz [5]. The accelerometer output is also passed through the Butterworth digital low-pass filter with a cut-off frequency of 5 kHz.

18.3 Experimental Results

Figure 18.3a–c and d–f show typical experimental waveforms for two cases with accelerations of 50 and 10,000 m/s^2, respectively. Here Fig. 18.3a, d show the acceleration measured by the He-Ne laser interferometer, and Fig. 18.3b, e show the voltage output from the combined piezoelectric accelerometer and charge amplifier. Figure 18.3c, f show expanded graphs around zero voltage for Fig. 18.3b, e, respectively.

ISO 16063-13 for shock calibration defines the shock sensitivity of an accelerometer as two peak ratios between the input acceleration and the accelerometer output, as shown in Eq. (18.1).

$$S_V = \frac{V_P}{A_P} \tag{18.1}$$

where:

S_V is the shock sensitivity,
A_p is the peak value of the acceleration input to the accelerometer,
V_p is the peak value of the accelerometer output.

However, V_p includes the frequency response effect of not only the piezoelectric accelerometer but also the charge amplifier. Thus, by eliminating the effect of the charge amplifier, the charge sensitivity of the piezoelectric accelerometer can be calibrated.

18.4 Charge Amplifier and Virtual Amplifier

Figure 18.4 shows the frequency response of a Brüel & Kjaer (BK) 2635 charge amplifier which has high-pass characteristics in the low-frequency region. Due to the effect of these high-pass characteristics, the charge amplifier output voltage with zero shift results in a zero voltage drift, which results in a large error component in the shock calibration. To reduce the effect of the

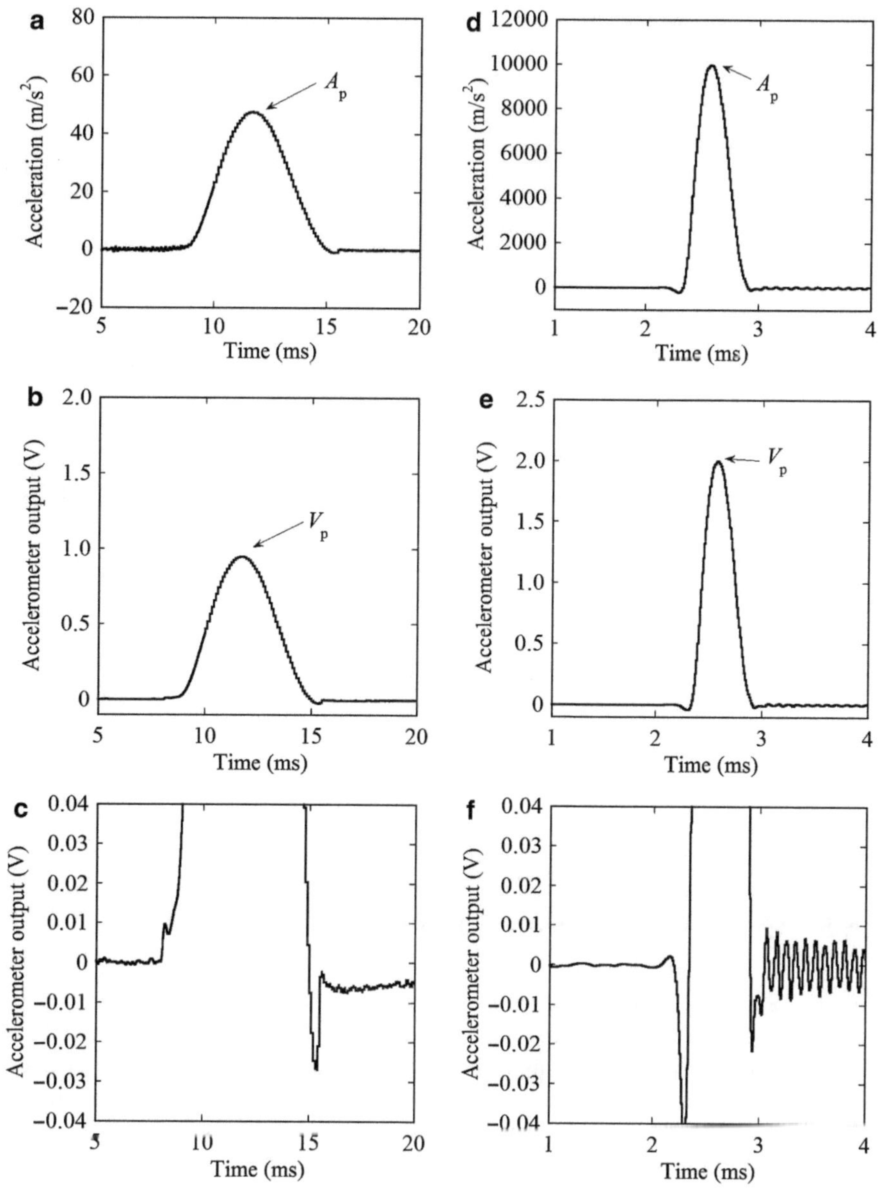

Fig. 18.3 Waveform examples of the acceleration and accelerometer outputs

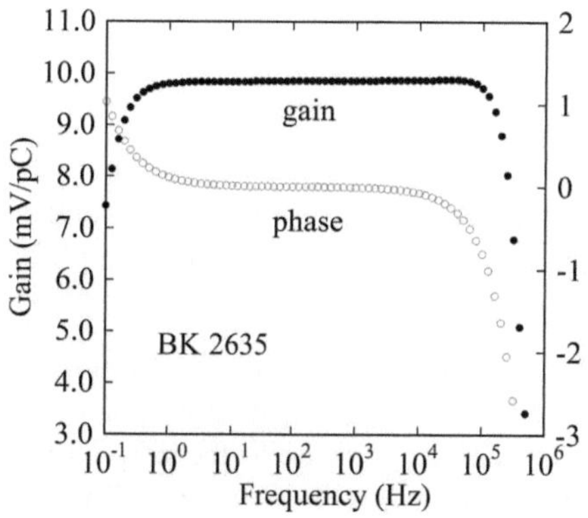

Fig. 18.4 Frequency response of the BK 2635 charge amplifier

zero shift, we developed a virtual amplifier by designing an impulse infinite response digital filter [4]. The virtual amplifier operates with the same input–output characteristics as the charge amplifier in personal computer software. Figure 18.5 shows an example of outputs from charge and virtual amplifiers, and we see that the output of the virtual amplifier agrees with that of the charge amplifier.

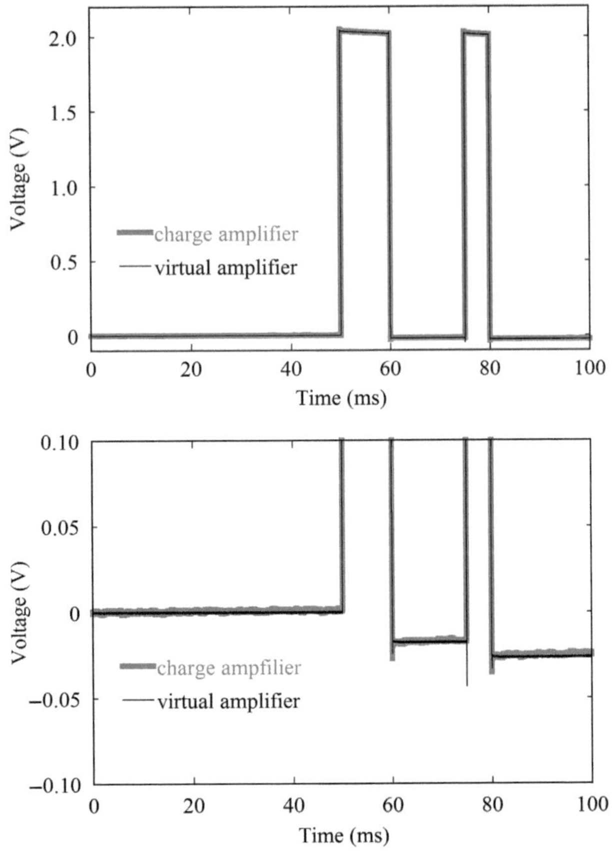

Fig. 18.5 Comparison of results between the charge and virtual amplifiers

18.5 Calibration Procedure and Results

Figure 18.6 shows the procedure for calculating the charge sensitivity of a piezoelectric accelerometer. Here, the three transfer functions of the piezoelectric accelerometer for both the charge and virtual amplifiers are $K(s)$, $H(s)$, and $H^*(s)$. The measured acceleration is input into the virtual amplifier to obtain the peak output, B_p, from the virtual amplifier. If the frequency response of a piezoelectric accelerometer is almost constant in the frequency component of a shock, the charge sensitivity of the piezoelectric accelerometer can be evaluated by dividing the two peak outputs (V_p and B_p) between the charge and virtual amplifiers, as in Eq. (18.2).

$$S_q = \frac{V_P}{B_P} \tag{18.2}$$

where:

S_q is the charge sensitivity for the shock sensitivity,
V_p is the peak value of the charge amplifier output,
B_p is the peak value of the virtual amplifier output.

Figure 18.7 indicates the calibration result of a piezoelectric accelerometer (Meggitt 2270, back-to-back type) based on the above-mentioned calibration procedure using the virtual amplifier. Normally, the charge sensitivity decreases in the low-acceleration region due to the zero shift. However, we obtained a flat calibration result for the charge sensitivities by the advantage of using the virtual amplifier. Also, the charge sensitivities in the shock calibration are comparable to those in the vibration calibration.

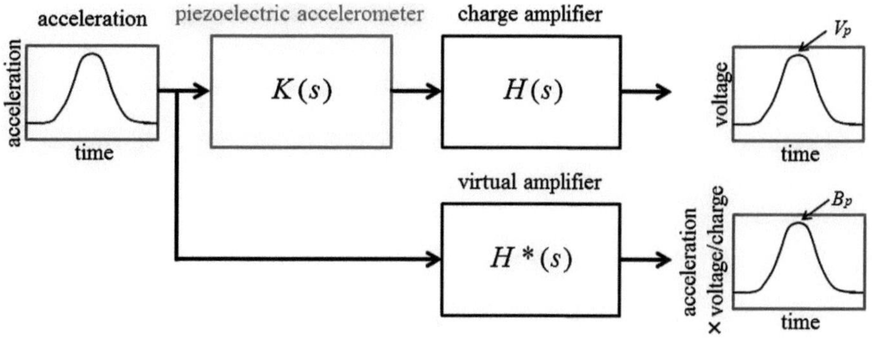

Fig. 18.6 Calibration procedure for a piezoelectric accelerometer using the virtual amplifier

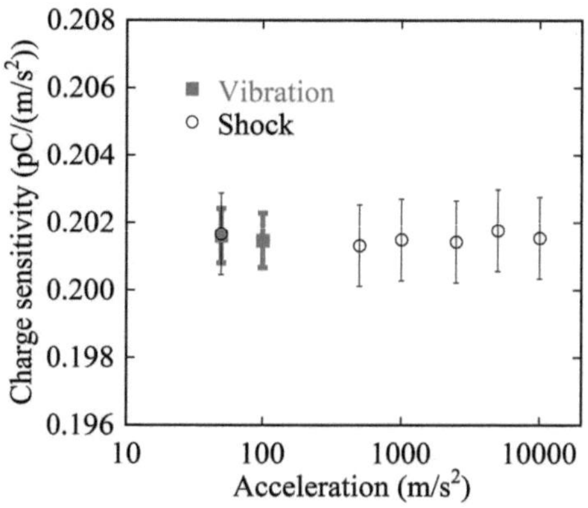

Fig. 18.7 Calibration results for a piezoelectric accelerometer

18.6 Conclusion

By constructing a calibration procedure that uses a virtual amplifier for the shock calibration of piezoelectric accelerometers, we obtained superior calibration results for the charge sensitivities in the acceleration range of 50–10,000 m/s^2 with pulse widths in the order of milliseconds. We investigated the charge sensitivities of a high-performance piezoelectric accelerometer (Meggitt 2270) by implementing this calibration procedure, and achieved flat charge sensitivities in the acceleration range. In addition, a BK 2635 charge amplifier was used as the virtual amplifier, which has superior input–output characteristics as compared with a charge amplifier.

References

1. Chu A (1987) Zeroshift of piezoelectric accelerometers in pyroshock measurements. Shock Vib Bull 57, Part I:71–80
2. ISO (2001) Methods for the calibration of vibration and shock transducers: part 13. Primary shock calibration using laser interferometry ISO 16063-13.
3. ISO (2005) Methods for the calibration of vibration and shock transducers: part 22. Shock calibration by comparison to a reference transducer ISO 16063-22
4. Nozato H, Bruns T, Volkers H, Oota A (2014) Digital filter design with zero shift on charge amplifiers for low shock calibration. Meas Sci Technol 25:035005
5. Nozato H, Usuda T, Oota A, Ishigami T (2010) Calibration of vibration pick-ups with laser interferometry: part IV. Development of a shock acceleration exciter and calibration system. Meas Sci Technol 21:065107

Chapter 19
Mechanical Shock Environment Synthesis for Structural Failure Elicitation

Cassidy L. Fisher, Kaitlyn S. Kliewer, Gregory M. Naranjo, Stuart G. Taylor, and Kendra Van Buren

Abstract Shock Response Spectra (SRS) are commonly used in dynamic testing to describe the mechanical environment in high-energy, non-stationary events, such as impacts or pyrotechnic shocks. Oftentimes, the service environment to which a structure will be exposed is difficult to reproduce in the laboratory, but design engineers desire a laboratory screening test to determine whether the structure will survive an anticipated shock environment. Herein, a combined experimental and numerical study is pursued to evaluate the efficacy of different methods to elicit failure modes of a service shock through destructive shaker tests of a custom-designed test article fabricated using commercially available 3-D printers. Design of the test article is explored through use of finite element modeling, which is found to correlate well to experimentally-obtained natural frequencies. Four techniques to synthesize a service shock are compared: least favorable response, sum of decaying sinusoids, wavelet, and matching temporal moments. Destructive shaker tests of the shock responses are performed using 25 nominally identical test articles to assess the ability of each method to impose similar damage states as those obtained when using the service shock. We find that the method of matching temporal moments best replicates failure modes of the service shock; however, further testing is needed to validate our observations.

Keywords Mechanical shock environment • Shock response spectra (SRS) • Shock synthesis • Qualification testing • Test-analysis correlation

19.1 Introduction

A mechanical shock is an excitation of short duration which induces transitory dynamic stress in structures [1]. The unique characteristics of a given shock make up its environment, for example, the source type, shock wave propagation, and displacement of a structure from the initial shock [1, 2]. Shock environment testing is an essential tool for effectively designing structures subjected to mechanical shock throughout their service lifespan. Examples of shock environments include pyrotechnic shocks from multi-stage rocket detachments, explosions, earthquakes and shocks from military ordnance.

A variety of methods have been used to represent shock environments, but the Shock Response Spectrum (SRS) is the most commonly applied. The SRS describes the peak amplitude response of multiple single-degree-of freedom (SDOF) systems to a base excitation input pulse [3]. This method has been essential to theoretical and experimental shock environment testing but has proven difficult to use when the original time history needs to be returned from the service SRS [3]. Therefore, alternative techniques have been established to synthesize shock time histories. Some of these techniques include: the sum of decaying sinusoids method, matching of temporal moments, and wavelet reconstruction, all of which arise from several measured time signals, as well as least favorable response method which is generated from a measured input and measured

C.L. Fisher
Department of Mechanical Engineering, Montana State University, Bozeman, MT 59717, USA

K.S. Kliewer
Department of Civil Engineering, Princeton University, Princeton, NJ 08544, USA

G.M. Naranjo
Department of Mechanical Engineering, University of New Mexico, Albuquerque, NM 87131, USA

S.G. Taylor (✉) • K. Van Buren
Los Alamos National Laboratory, Engineering Institute, Los Alamos, NM 87545, USA
e-mail: sgtaylor@lanl.gov

A. Wicks (ed.), *Shock & Vibration, Aircraft/Aerospace, and Energy Harvesting, Volume 9*,
Conference Proceedings of the Society for Experimental Mechanics Series, DOI 10.1007/978-3-319-15233-2_19

frequency response function (FRF). Each of the resultant time histories from these various methods and their corresponding SRS can be compared with the service environment's SRS for verification. This paper focuses on determining the most effective method to replicate a service environment for laboratory testing through a comparison of SRS and shock-induced failure, both synthesized and service.

The most effective means of determining structural response to a shock environment is to perform the experimental testing using the real service shock; however, such real-world testing is difficult or impossible to perform in a laboratory setting. An effective method for modeling these environments is then essential. Although the SRS is a strong tool for shock characterization, it is not necessarily the most effective as an infinite number of time histories can share a given SRS. Therefore a test environment defined to match a given SRS could be very different from the service environment. This disparity could result in a structure that is under or over designed for the intended service environment. By correctly identifying the parametric form of the test shock signal, the test shock parameters, design variables, design objectives, and constraints can be properly optimized.

This paper will involve the design and construction of a test article capable of failure under specified testing parameters. After testing a part under the anticipated shock environment, replicate test parts will be tested using four different time histories generated using the following methods: sum of decaying sinusoids, wavelet method, least favorable response, and temporal moments with the sum of decaying sinusoid method. These methods will then be assessed through a comparison to the original structural response under the anticipated shock environment. The test that most accurately reproduces the service shock environment through a series of destructive experiments will be rated based on the ability to elicit anticipated failure modes.

19.2 Theoretical Background

19.2.1 Shock Response Spectra (SRS) Method

Shock Response Spectra (SRS) are commonly used for shock analysis. The SRS looks at the response of a hypothetical series of single degree of freedom systems to a base excitation. This is done assuming a constant damping coefficient across all systems while varying the natural frequency of each system [1]. The peak accelerations are then plotted against their corresponding natural frequencies to generate the SRS. In practice, when a structure is tested under a time history that returns a SRS that matches the SRS of the anticipated service environment, it is assumed to be acceptable. If the experimental SRS sits above that of the anticipated environment SRS the test is conservative [3]. A SRS from a pyrotechnic shock can be seen in Fig. 19.1 with $\pm$3db bands. The SRS method of analysis has some drawbacks as tests typically yield an over-conservative approach and lack the ability to resemble characteristics of underlying excitations. The techniques described below help to accommodate for these disadvantages.

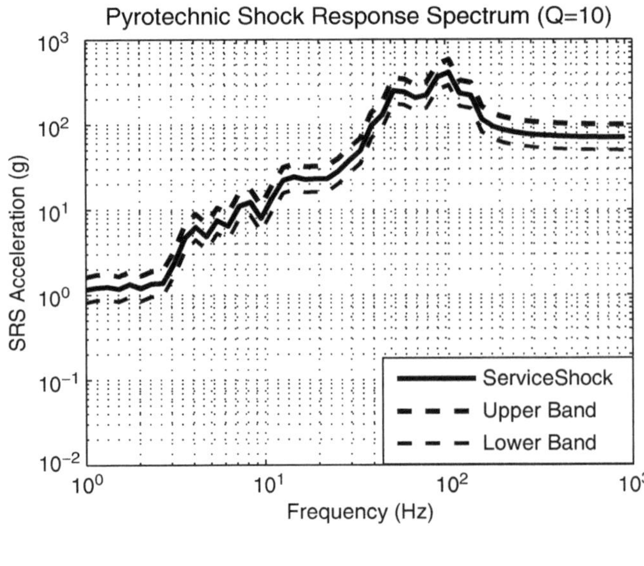

Fig. 19.1 SRS from a pyrotechnic shock

19.2.2 Least Favorable Response

The least favorable response (LFR) provides an alternate method to represent a shock environment; however its formulation is dependent on the frequency response function (FRF) rather than the SRS. This method aims to maximize the response of a structure at a specific point of interest, thus generating the least favorable response. This method tends to be more conservative than other synthesis methods as it is geared to produce the maximum response [4]. To generate the response to an acceleration input, *Y(t),* the LFR is defined as [5]

$$Y(t) = \frac{1}{2\pi}\int_{-\infty}^{\infty} H(\omega)\, X(\omega)\, e^{i\omega t} d\omega,\ t \geq 0 \tag{19.1}$$

where $H(\omega)$ is the FRF at the point of interest and $X(\omega)$ is the power spectral density of the service shock, enveloping the Fourier transform of all possible inputs. To determine the maximum response of the system, the absolute value of the acceleration input response is taken, therefore,

$$|Y(t)|_{max} = \frac{1}{2\pi}\int_{-\infty}^{\infty} |H(\omega)|\, X(\omega)\, d\omega. \tag{19.2}$$

By manipulating Eq. 19.7 to isolate the system input and twice differentiating the expression yields the acceleration input. Therefore the modified expression can be written as

$$\ddot{X}_T(t) = \frac{1}{2\pi}\int_{-\infty}^{\infty} X_e(\omega)\, e^{i(\omega(t-to)-\varnothing(\omega))}\, d\omega. \tag{19.3}$$

19.2.3 Sum of Decaying Sinusoids Method

The sum of decaying sinusoids method involves using a service SRS to determine the parameters to generate a series of decaying sinusoid functions [6]. The summation of these sinusoid waves will form a composite wave with an SRS closely replicating that of the service SRS. The general equation for the sum of decaying sinusoids can be expressed as

$$x(t) = \sum_k A_k e^{\xi_k \omega_k t} \sin(\omega_k t - \phi_k)\ t \geq 0, \tag{19.4}$$

where A is the amplitude, ζ is the decay rate, ω is the frequency, and ϕ is the phase angle. The service shock SRS is broken into 1/12 octaves and for each octave, a decaying sinusoid is generated. The amplitudes of the decaying sinusoids are iterated until the SRS for the composite wave has the least acceptable error with the service SRS [6].

19.2.4 Wavelet Method

Similar to the sum of decaying sinusoids method, the wavelet method uses the service SRS to generate parameters for a series of wavelets. The composite waveform of these wavelets creates the synthesized time history that has an SRS closely resembling the SRS of the original service shock. Wavelet pulses consist of frequency, number of half cycles, amplitude, delay, and time. The general equation for an individual wavelet is expressed as

$$W(t) = Asin\left(\frac{2\pi f}{N}(t-\varphi)\sin\left(2\pi f\,(t-\varphi)\right)\right),\ for\ \varphi \leq t \leq \left(\varphi + \frac{N}{2f}\right), \tag{19.5}$$

where *W(t)* is the wavelet acceleration, *A* is the wavelet amplitude, *f* is the wavelet frequency, *N* is the number of half sines, and φ is the wavelet time delay [7]. The SRS of the service shock is broken into octaves used to generate a random wavelet for each octave. Each of these waves can be combined to generate a single waveform. The total acceleration denoted as $(\ddot{x})$ at time t for a set of n wavelets is

$$(\ddot{x}) = \sum_{m=1}^{n} W_m(t). \tag{19.6}$$

The parameters of each wavelet can be scaled to generate new waves. The compound wavelet SRS can be determined and compared to the original SRS. The parameters of each wavelet are iterated until the acceleration, velocity, and displacement error between the SRS is reduced.

19.2.5 Temporal Moments Methods

The temporal moments of a waveform describe the characteristics of a time history such as the energy distribution that may be lost when converted to an SRS. Both the wavelet method and the sum of decaying sinusoid method are not unique in the possible waveforms generated for each SRS and they lose some of the valuable characteristics of the original service shock by matching only the SRS. By matching the temporal moments in addition to the SRS, some of the crucial characteristics of the original service shock may be maintained to create a waveform that better replicates the original service environment [8, 9]. The mathematical definition of a temporal moment is defined as,

$$m_i(a) = \int_{-\infty}^{\infty} (t-a)^i * [f(t)]^2 dt, \tag{19.7}$$

where t is the location, a is the time shift, and *f(t)* is the time history [3]. The first five moments are the total signal energy, centroid, duration, skewness, and kurtosis:

$$E = m_0 \tag{19.8}$$

$$Centroid\ (\tau) = \frac{m_1(0)}{m_0(0)} = \frac{m_1}{E} \tag{19.9}$$

$$RMS\ Duration\ \left(D_t^2\right) = \frac{m_2(\tau)}{E} \tag{19.10}$$

$$D_S^3 = \frac{m_3(\tau)}{E} \tag{19.11}$$

$$Skewness\ (S_t) = \frac{m_3(\tau)}{E} = \frac{D_S}{D_t} \tag{19.12}$$

$$D_k^4 = \frac{m_4(\tau)}{E} \tag{19.13}$$

$$Kurtosis\ (K_t) = \frac{D_k}{D_t} \tag{19.14}$$

19.3 Test Methods

19.3.1 Test Article Description

To examine and compare the various shock synthesis methods, a test article was designed to elicit desired failure characteristics when tested on an electrodynamic shaker. The goal in designing the test article was to develop a component that experienced different modes of vibration at various frequencies within the allowable test frequency range. The test article was therefore designed as a rigid body cylinder with several cantilever arms, each of different diameter, with masses at each end of the arms, as seen in the right image of Fig. 19.2. The test article was manufactured with ABS plastic using a MakerBot Replicator II 3D printer, as seen in left image of Fig. 19.2. Using the material and geometric properties of the test article, the natural frequency of the center support column was determined to be approximately 14 kHz, well above the test range of 2 kHz. The rigidity of the center support column allows the test article to act as a rigid body and each of the cantilever arms to act independently during testing.

The cantilevered arms had a 0.25 in [0.635 cm] tapered section connecting each of the end masses, 1 in [2.54 cm] spheres, to induce intentional stress concentrations and encourage a shear failure at the base of the spheres. The design of the test article was refined through several iterations using both experimental and numerical analysis. Because 3-D printing creates a non-homogenous material, the finite element analysis was performed to guide the design process with final design iterations performed using experimental testing. As seen in Fig. 19.2, a horizontal plane was located on each sphere to allow for attachment of an accelerometer during testing if necessary. The finalized test article dimensions are shown in Table 19.1, with labels of the test arms shown in the right image of Fig. 19.2.

19.3.2 Numerical Modeling and Experimental Testing

In order to assist in the design process of the test article and to gain a better understanding of its dynamic behavior, a finite element model was developed using ANSYS. The test article was modeled using solid quadratic elements and simplifying the material to an isotropic, homogeneous material with ABS plastic material properties. Initially, a mesh refinement analysis was performed to determine an appropriate mesh size for the computational modeling. Based on the mesh refinement analysis, a mesh size of approximately 200,000 elements was chosen, as seen in the plot in Fig. 19.3. The resulting mesh for the test article is shown in Fig. 19.3.

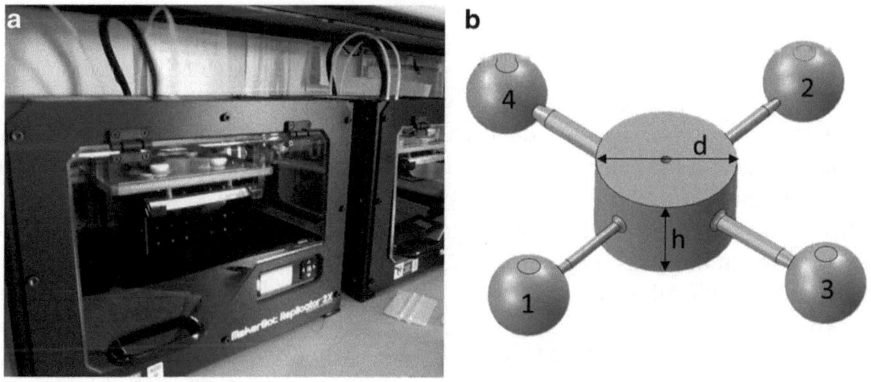

Fig. 19.2 Test article manufacturing with MakerBot 3D printer (*left*) and the final test article design (*right*)

Table 19.1 Test article dimensions

Test article dimensions	
Arm 1 diameter	0.318 cm [0.125 in]
Arm 2 diameter	0.381 cm [0.15 in]
Arm 3 diameter	0.508 cm [0.2 in]
Arm 4 diameter	0.635 cm [0.25 in]
Support diameter (d)	4.445 cm [1.75 in]
Height (h)	4.572 cm [1.8 in]

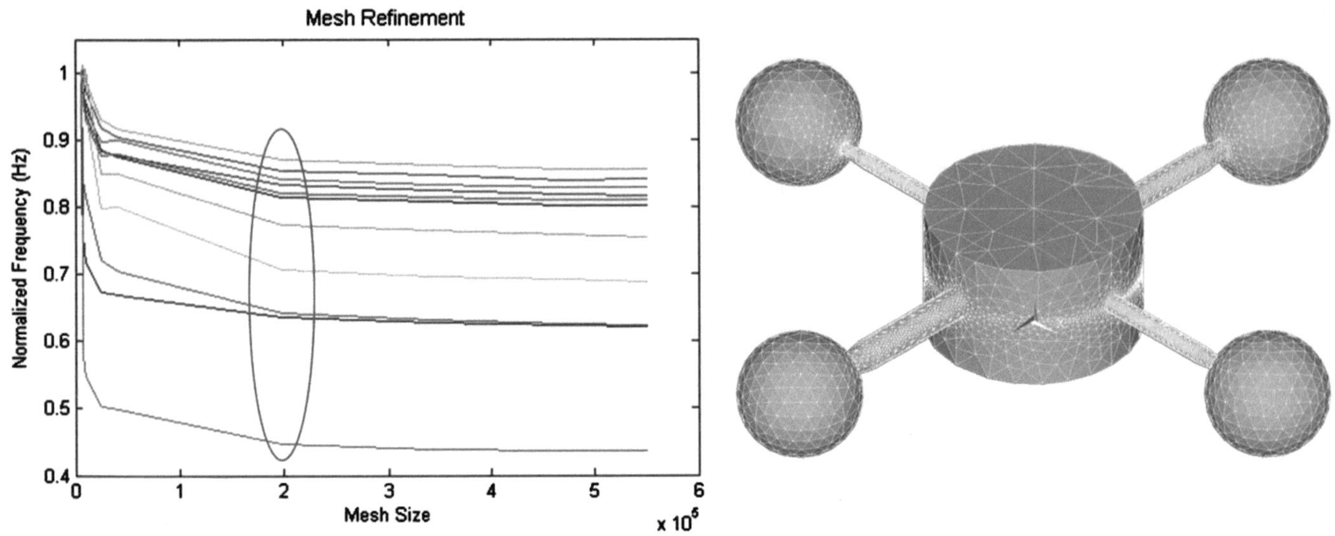

Fig. 19.3 Mesh refinement analysis (*left*) and resulting mesh discretization used for numerical analysis (*right*)

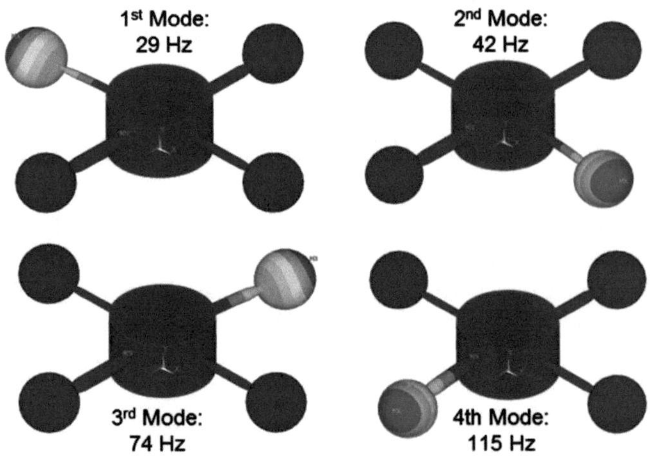

Fig. 19.4 FEA modal analysis displacements

Using this model, a modal analysis was performed in ANSYS assuming a rigid, fixed base. The modes of the structure were evaluated up to 4.6 kHz, which was well beyond the 2 kHz frequency range of the electrodynamic shaker used in physical testing. This analysis demonstrated the first modes of the test article contained no coupling between the arms or modes located in the center support cylinder in the frequency range analyzed. Because of this, it was expected that the arms of the structure would act independently from one another as desired. The results of the first 4 modes of the test article are illustrated in Fig. 19.4 with the color gradient representing displacement. As desired, these results show a range of natural frequencies in the arms ranging from 29 Hz located in the arm with the smallest diameter increasing sequentially through the arms with the 4th mode at 115 Hz located in the arm with the largest diameter.

To better understand the experimental dynamic behavior of the test article, the part was subjected to a random transient wave using a 25 lbf. [111.2 N] electrodynamic shaker. Accelerometers were placed on each cantilever arm, the support column, and the base of the structure, as seen in Fig. 19.5. The acceleration input from the accelerometer located at the base of the test article was used as the base accelerations for the analysis of the results. Six random vibration tests were performed for 19 different test article for a total of 114 vibration tests. Test repeats were performed to help account for the manufacturing and material property variation that could have occurred from test article to test article to allow for characterization of the experimental uncertainty. From these results, a power spectral density was calculated for each arm with the results demonstrated in the left image of Fig. 19.6. The natural frequencies for each arm were taken as the frequency with the maximum peak and demonstrate a distinct frequency value for each cantilever arm, allowing the experimental tests to be tailored towards this frequency range. Unlike the FEA modal analysis, the PSD experimental results indicated potential coupling between some of the arms. This may not have been seen in the FEA analysis as the boundary conditions were modeled as a perfectly rigid, fixed condition whereas no connection is perfect in physical experiments. The experimental

Fig. 19.5 Test article random shock testing set up

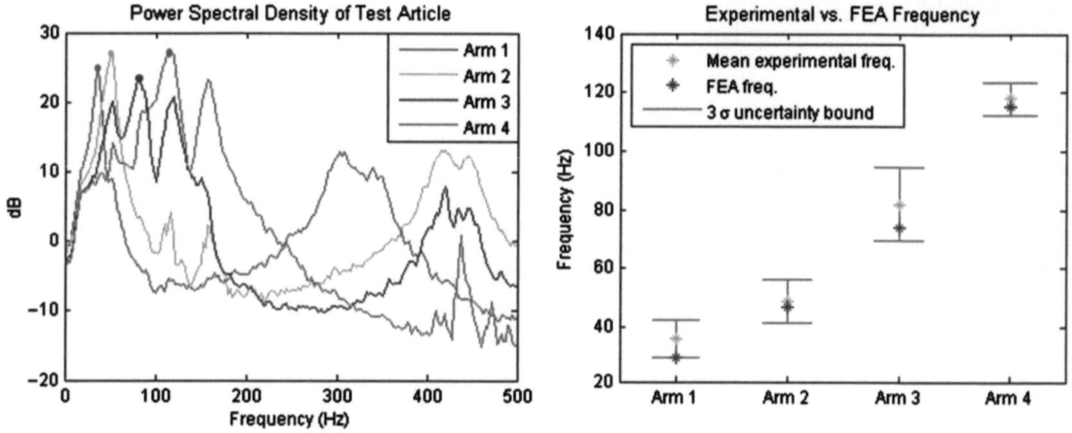

Fig. 19.6 Experimental power spectral density results (*left*) and comparison of the corresponding experimental and numerical analysis frequencies (*right*)

frequency values were compared to the FEA modal analysis frequency values for the first modes of each arm, as shown in the right image of Fig. 19.6. When experimental uncertainty bounds of $\pm 3\sigma$ were taken into account, this comparison demonstrated the agreement of the FEA model with the experimental testing. The good test-analysis correlation suggests that the FEA model was a reasonable design guide.

19.3.3 Test Hardware and Experimental Setup

To perform the proposed failure tests, the hardware in the form of: two high voltage amplifiers, a 100 lbf and 25 lbf electrodynamic shakers, PCB accelerometers, and a shaker controller were utilized. Two types of experiments were set up for testing. The first experimental setup involved using a PA-138 linear amplifier with a maximum output of 500 V, a 25 lbf. electrodynamic shaker, and a Laser USB controller. This setup was mainly used to generate a series of random shock impulses for correlation to the solutions obtained through FEA, and to ensure that each of the accelerometers were properly working, and that each of the cantilevered arms was in fact vibrating at its resonate frequency according to the modal analysis. Eight total accelerometers were mounted on the test article, one on each of the four end masses, three on top of the test article, and one at the bottom of the structure to measure the base excitation. The failure testing experiments were conducted using a larger amplifier, Laser USB controller, and a 100 lbf. electrodynamic shaker. Comet USB Vibration control software was selected for data acquisition for the failure testing as the software included a closed loop control to ensure the base accelerations corresponded to the time-history we input into the system. The test setup is shown in Fig. 19.7.

Initial testing of the test article informed us that additional mass was required at the end of each arm to ensure failure of the arms during the failure tests. This resulted in a change in the natural frequencies which decreased the frequencies of the test articles into the frequency range of interset. This shift in natural frequencies is seen in Table 19.2.

For the experimental testing, a pyrotechnic shock was used as the service shock. The pyrotechnic shock was modified to ensure the test article designed was capable of failure when subjected to this hypothetical service shock. The frequency range

Fig. 19.7 Experimental set up of test article

Table 19.2 Natural frequency shifts due to mass increase of the cantilever arms

Arm	Freq. w/o Mass (Hz)	Freq. w/Mass (Hz)
1	29	17
2	42	24
3	74	42.8
4	115	66.5

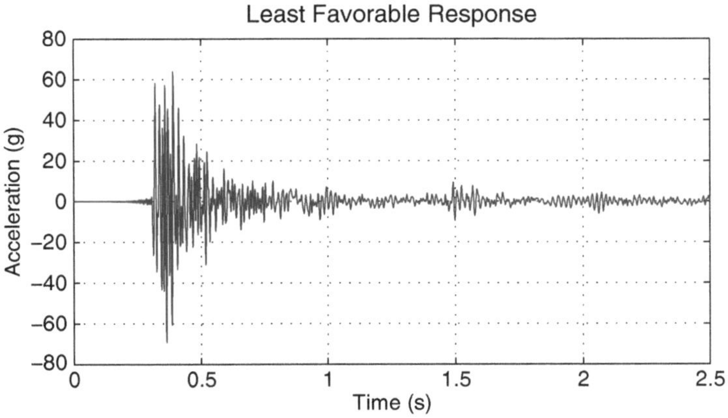

Fig. 19.8 Slowed service shock used in experimental testing

Table 19.3 Temporal moments of synthesized wave compared to service shock

	Temporal moments		
	Synth. wave	Service shock	Percent error
Root mean energy (g)	24.35	23.60	3.14
Centroid (s)	0.20	0.16	28.32
Duration (s)	0.25	0.27	6.49
Skewness	1.69	1.64	3.42

of the original pyrotechnic shock was far higher than the natural frequency range of the test article requiring its slowing to lower the frequency range to that of the test article. This resulted in the final pyrotechnic shock, Fig. 19.8, having a duration of over 2 s which is longer than typically observed in shock environments.

From the pyrotechnic shock seen in Fig. 19.8, a total of four additional shock time histories were generated using the four synthesis methods introduced previously in Sect.19.2, with the resulting time histories provided in Fig. 19.9. These four time histories and the service shock were the five shock environments used in the experimental failure tests. The LFR time history was generated focusing on arm 2 with a frequency response function generated from the random vibration test of the weighted test article. The temporal moments time history was generated by further limiting the decaying sinusoid synthesis method to match the first four temporal moments of the service shock. As seen in Table 19.3, three of the four temporal moments, root mean energy, kurtosis, and skewness, were matched with a reasonable error. However, it was challenging to closely match the centroid of the synthesized wave to the centroid of the service shock, which can be seen with the larger margin of error compared to the other temporal moments matched.

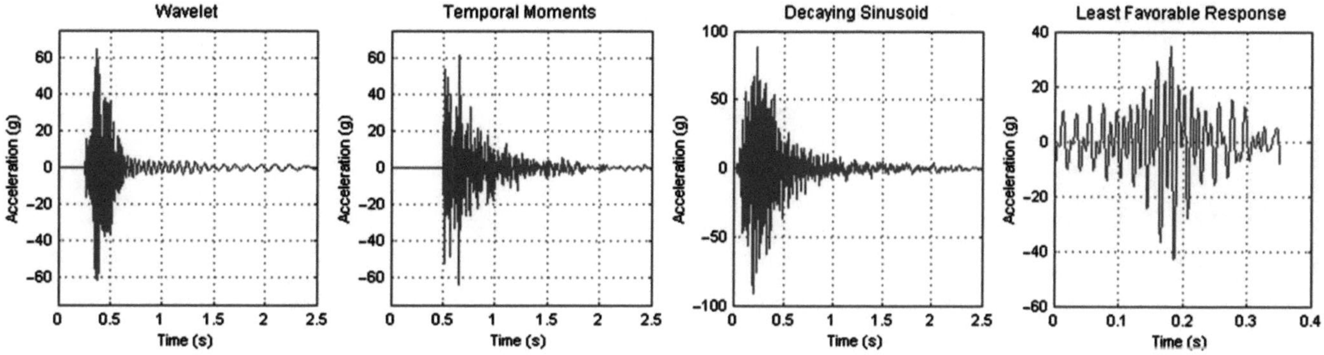

Fig. 19.9 Synthesized time histories from pyrotechnic shock

19.4 Experimental Results

For each time history, 5 tests were performed during which the failure behaviors of the test article were recorded. Additionally, the base accelerations during the testing were recorded to maintain a closed loop control and calculate the SRS for each of the failure tests. These SRSs are shown in Fig. 19.10 as solid lines with the desired SRS shown with ±3 dB error bands in dashed lines. Because of the limitations of the testing equipment, a high pass filter was applied to the time history to limit the magnitude of the displacements resulting from the acceleration input. As expected, this caused the tested service shock SRS curves to not match the desired SRS for frequencies below the frequency of the filter. However, in the frequency range of concern which included the natural frequencies of the cantilevered test article arms, the SRS curves aligned very well and fell well within the 3db bands. The only case where this did not occur was for the LFR wave, which is expected because the LFR method was not developed based on the SRS of the service shock.

For all failures during testing, the test articles failed at the base of the spheres as anticipated in the design. To compare the failures between the time histories, the total number of failures in each arm for all five tests was recorded. The results of these tests can be seen in Fig. 19.11. The results demonstrate that while the decaying sinusoid and wavelet synthesis methods are able to capture some of the same failure patterns of the original service shock, the temporal moments method (indicated by the white bar in Fig. 19.11) performs the closest to that of the service shock. This can be expected as the temporal moments method was applied to further refine the decaying sinusoid method. The LFR method was expected to be a more conservative synthesis method, however, during the failure testing no failures were observed in the LFR tests even in arm 2 for which the synthesis was optimized for. However, there were also no failures observed in arm 2 for the service shock tests. To form a more conclusive analysis for LFR, it is recommended that the method be repeated for another arm of the test article.

To gain a greater insight into the dynamic behavior of our test article, a transient dynamic analysis was performed in ANSYS with the service shock. Because it is computationally expensive to perform a transient dynamic analysis, only one of the time histories was analyzed. The transient analysis showed the highest stress concentrations in arms 1 and 3, as seen in Fig. 19.12. This is similar to what was observed in the experimental testing with the highest number of failures occurring in arms 1 and 3 during the failure testing. The FEA analysis also showed the stress concentrations located at the connection points of the cantilevered arms, with the highest stresses at the base of the sphere. Again, this agrees with what was observed in the experimental testing with all failures occurring at the base of the spheres.

19.5 Conclusion

This paper has demonstrated the feasibility of designing a large number of nominally identical test articles to assess the ability of various synthesis methods to replicate the damage-causing potential of the service shock environment. From a given service shock, waveforms were successfully synthesized using the following four methods: wavelet method, decaying sinusoid method, least favorable response method, and temporal moments of the decaying sinusoid method. The failure modes of the test article were observed and compared with the service shock test article performance. From these results, the temporal moments method with the damped sinusoid methods appeared to best impose the same damage-causing potential

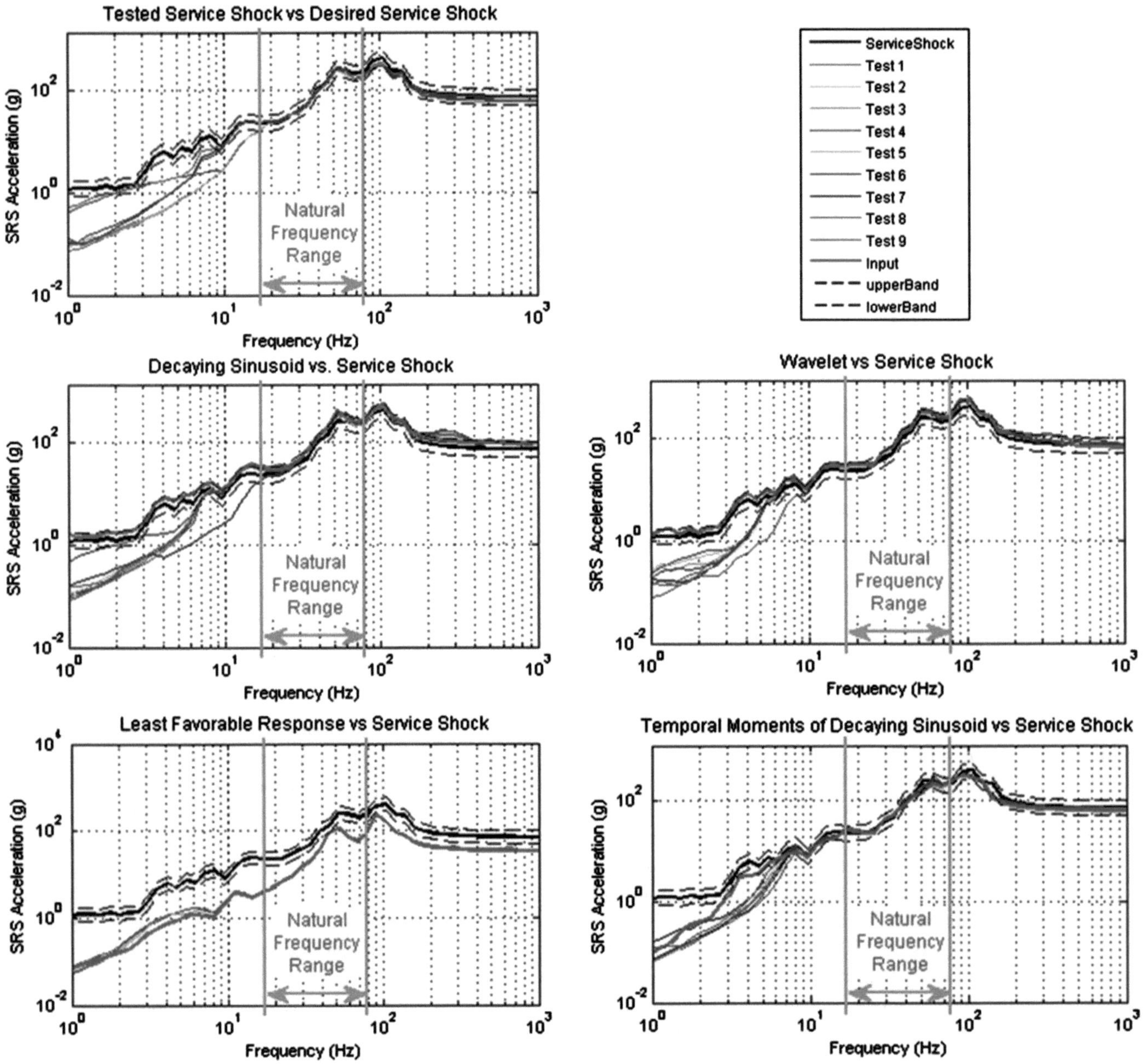

Fig. 19.10 Synthesized SRS compared with the experimental SRS

as the service environment. The wavelet method and decaying sinusoid were able to replicate similar failure patterns of the service shock environment and appear to be acceptable methods to use for laboratory testing of the predicted shock environment.

Recommended future work includes further testing of the LFR method with waveforms targeted at other locations on the test article to better determine the effectiveness of this synthesis method. As the use of temporal moments was effective with the sum of decaying sinusoids, it would be useful to apply this method to the wavelet-generated waves for further analysis. Additionally, it is recommended that further tests be performed for each of the methods to increase the numbers of samples and enhance the comparison amongst the various methods. Given the promising results of the FEA models, a better insight into the performance of the test articles may be obtained by performing a transient dynamic analysis for each of the four synthesis methods. This will allow for a comparison of the dynamic behavior predicted in the computational analysis by comparing the peak stress values and the locations of the peak values.

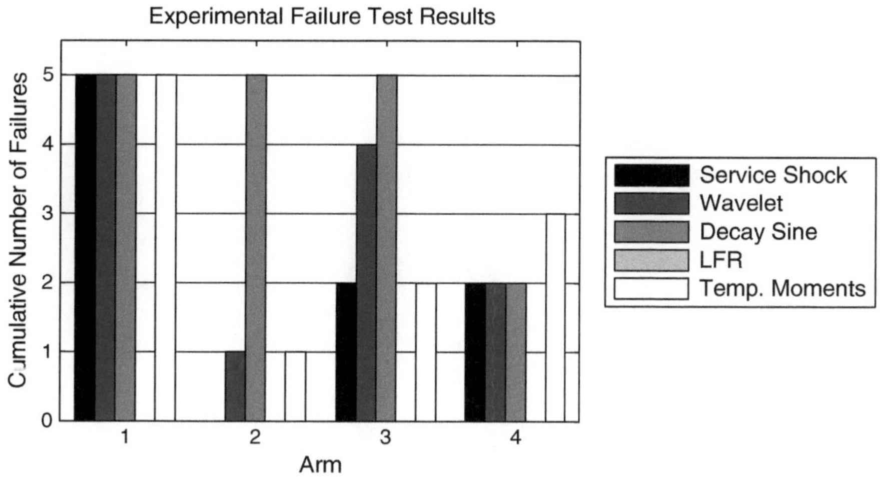

Fig. 19.11 Number of failure tests for each time history

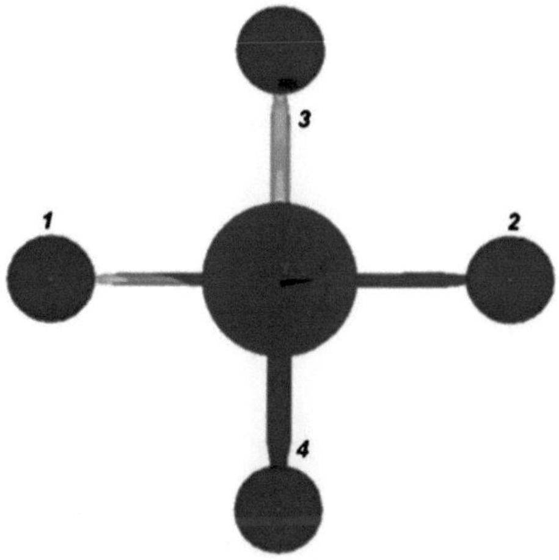

Fig. 19.12 FEA transient analysis stresses

Acknowledgements This work is performed under the auspices of the Los Alamos National Laboratory (LANL) as part of the Los Alamos Dynamics Summer School (LADSS). The authors are grateful to Dr. Charles Farrar, The Engineering Institute, for organizing the LADSS. The authors also wish to express their gratitude to James Thompson and Dr. Tom Irvine, for their assistance and guidance in the experimental testing. The companies Simulia and Vibrant Technology, Inc., graciously contributed Abaqus™ and ME'Scope software licenses to the LADSS, without which this work would not have been possible. LANL is operated by the Los Alamos National Security, L.L.C., for the National Nuclear Security Administration of the U.S. Department of Energy under Contract DE-AC52-06NA25396. This publication has been assigned the release number LA-UR-14-28444.

References

1. Lalanne C (2009) Mechanical shock, 2nd edn. ISTE Ltd, London, pp 51–53. Print
2. Filippi E et al (1999) Pyroshock simulation using the alcatel etca test facility. In: Proceedings of the first European conference on launcher technology, 1999, Toulouse, France
3. Smallwood DO (1994) Characterization and simulation of transient vibrations using band limited temporal moments. Shock Vib 1(6): 507–527. Print
4. Harris CM, Piersol AG (2002) Chapter 26, part II. Harris' shock and vibration handbook, 5th edn. McGraw-Hill, New York, pp 20–21. Print
5. Paez TL (1980) Conservatism in least favorable response analysis and testing (No. SAND-80-2093C; CONF-801033-2). New Mexico Univ., Albuquerque (USA). Dept. of Civil Engineering. Shock and vibration symposium, San Diego, CA, USA, 21 Oct 1980
6. Smallwood DO, Nord AR (1974) Matching shock spectra with sums of decaying sinusoids compensated for shaker velocity and displacement limitations. Shock Vib Bull 44(3): 43–56. Print
7. NASA Center for Aerospace Information (2008) An alternative method of specifying shock test criteria. By Ferebee RC, Clayton J, Alldredge D, Irvine T. NASA/TM 215253. MSFC, Alabama: National Aeronautics and Space Administration, April 2008. pp 10–11. Print
8. Cap J, Smallwood DO (1997) A methodology for defining shock tests based on shock response spectra and temporal moments. In: Proceedings of the 68th shock and vibration symposium, 1997, Baltimore, MD, United States
9. Hemez F, Doebling S (2003) From shock response spectrum to temporal moments and vice-versa. In: 21st international modal analysis conference, Kissimmee, 2003

Chapter 20
Natural Frequencies of Layered Beams Using a Continuous Variation Model

Arnaldo J. Mazzei and Richard A. Scott

Abstract This work involves the determination of the bending natural frequencies of beams whose properties vary along the length. Of interest are beams with different materials and varying cross-sections, which are layered in cells. These can be uniform or not, leading to a configuration of stacked cells of distinct materials and size. Here the focus is on cases with two, or three cells, and shape variations that include smooth (tapering) and sudden (block type) change in cross-sectional area. Euler-Bernoulli theory is employed. The variations are modeled using approximations to unit step functions, here logistic functions. The approach leads to a single differential equation with variable coefficients. A forced motion strategy is employed in which resonances are monitored to determine the natural frequencies. Forcing frequencies are changed until large motions and sign changes are observed. Solutions are obtained using MAPLE®'s differential equation solvers. The overall strategy avoids the cumbersome and lengthy Transfer Matrix method. Pin-pin and clamp-clamp boundary conditions are treated. Accuracy is partially assessed using a Rayleigh-Ritz method and, for completeness, FEM. Results indicate that the forced motion approach works well for a two-cell beam, three-cell beam and a beam with a sinusoidal profile. For example, in the case of a uniform two-cell beam, with pin-pin boundary conditions, results differ less than 1 %.

Keywords Beams with layered cells • Layered structures resonances

Nomenclature

A	Area of the beam cross section (A_i area values for different cells)
E	Young's modulus (E_i Young's modulus values for different cells)
F	Axial compressive force acting on the shaft
f	External transverse force per unit length acting on the shaft
f_1, f_2, f_3, f_4	Non-dimensional functions for material/geometrical properties
F_1, F_2	Non-dimensional parameters $\left(F_1 = \frac{F}{\rho_0 A_0 L^2 \Omega_0^2},\ F_2 = \frac{f}{\rho_0 A_0 L \Omega_0^2}\right)$
$H(x)$	Logistic function
I	Area moment of inertia of the shaft cross section (I_i moment of inertia values for different cells)
K	Real constant
K_1	Non-dimensional stiffness $\left(K_1 = \frac{E_0 I_0}{\rho_0 A_0 L^4 \Omega_0^2}\right)$
L	Length of the beam
R	Cylindrical beam radius
$R_{midspan}$	Sinusoidal shaft radius at mid-length
R_0	Sinusoidal shaft end radius
$S(\xi)$	Non-dimensional spatial function
s_1	Numerical parameter $\left(\sqrt{2\pi \Big/ (\beta^2\pi + 2\pi + 8\beta)}\right)$
t	Time

A.J. Mazzei (✉)
Department of Mechanical Engineering, C. S. Mott Engineering and Science Center, Kettering University, 1700 University Avenue, Flint, MI 48504, USA
e-mail: amazzei@kettering.edu

R.A. Scott
University of Michigan, Herbert H. Dow Building, 2300 Hayward, Ann Arbor, MI 48109, USA

A. Wicks (ed.), *Shock & Vibration, Aircraft/Aerospace, and Energy Harvesting, Volume 9*,
Conference Proceedings of the Society for Experimental Mechanics Series, DOI 10.1007/978-3-319-15233-2_20

w	Shaft displacement in the y direction
xyz	Inertial reference system (coordinates x, y, z)
Y	Non-dimensional shaft displacement in the y direction
β	Numerical parameter $((R_{midspan}/R_0))$
ν	Non-dimensional frequency
ρ	Mass density (ρ_i, density values for different cells)
ξ	Non-dimensional spatial coordinate
τ	Non-dimensional time
Ω	Frequency of the shaft
Ω_0	Reference frequency

20.1 Introduction

This work is an extension of one given in reference [1] in which the determination of the natural frequencies and mode shapes of longitudinal vibrations of layered rods was addressed. These issues were approached using transfer matrix methods, where the solutions for each cell are found and then boundary and interface continuity conditions are applied yielding transcendental equations for the determination of the frequencies and mode shapes. This method is lengthy and can be quite cumbersome. In reference [1] an alternate method was given, in which the discrete cell properties were modeled by continuously varying functions, specifically logistic functions (Heaviside step functions could have been used, but they lead to severe numerical difficulties because of their discontinuous derivatives). This approach has the considerable advantage of working with a single differential equation (albeit one with variable coefficients). Natural frequencies were obtained using a forced motion strategy (monitoring resonances) together with MAPLE®'s ODE[1] solver. In [1] validation was achieved by comparing the results with those of the transfer matrix approach, for certain limiting cases. The transfer matrix method is very cumbersome for beams and the thrust of the current work is the development of alternate strategies for finding the bending frequencies of longitudinally layered beams.

There is an extensive body of work on the bending vibrations of layered beams. Numerous references can be found, for example reference [2], where free vibrations of stepped Timoshenko beams were treated. The problem was tackled via a Lagrange multiplier formalism and results compared well with values obtained using other analytical methods. See also reference [3], in which Euler-Bernoulli stepped beams were studied via exact and FEM approaches. FEM results using non-integer polynomials shape functions (see reference [4]) compared well with exact solutions. General studies on media with discrete layers have been discussed, for example, in references [5–8]. Note that finite difference approaches to the dynamics of non-homogeneous media can be found in reference [9].

The vibration problem discussed here involves stepped beam with layers of distinct materials bonded together. The beams can be non-uniform and shape variations treated include smooth changes (tapering) and step changes (sudden variation). The variations are modeled using approximations to unit step functions, here logistic functions. This leads to a single differential equation with variable coefficients. Solutions are obtained using MAPLE®'s differential equation solvers. A forced motion strategy is employed in which resonances are monitored to determine the natural frequencies. Forcing frequencies are changed until large motions and sign changes are observed. Accuracy is assessed by comparing results with those obtained via Rayleigh-Ritz and FEM approaches.

20.2 Modeling

In this exploratory study, Euler-Bernoulli theory is used. A detailed derivation of the governing equation of motion, based on a Newton-Euler approach to a differential beam element (see Fig. 20.1, which exhibits the underlying variables) was given by the authors in reference [9]. See Eq. (20.1).

$$\frac{\partial^2}{\partial x^2}\left(EI\frac{\partial^2 w}{\partial x^2}\right) + F\frac{\partial^2 w}{\partial x^2} + \rho A\frac{\partial^2 w}{\partial t^2} = f \tag{20.1}$$

[1] www.maplesoft.com

Fig. 20.1 Beam configuration element

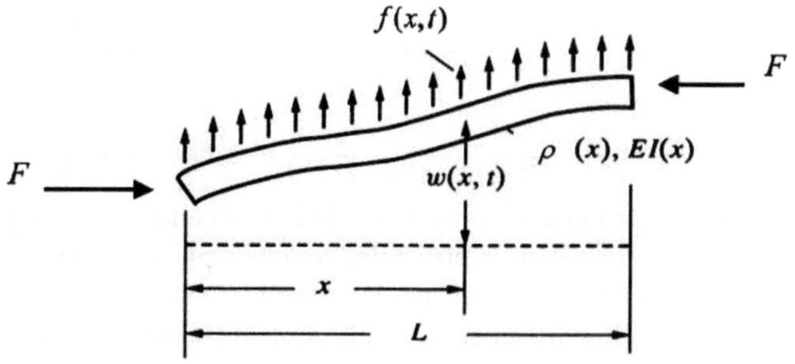

No assumption was made in relation to the material type in the derivation, therefore it can be either a homogeneous or non-homogeneous material.

Configurations studied below include a uniform 2-cell beam, a non-uniform 2-cell beam, a non-uniform 3-cell beam and a sinusoidal 2-cell beam. Circumstances may dictate the use, and shape, of tapers, such as conical (linear) and sinusoidal (following a model of Lee et al. [10]). The sinusoidal profile is used here since it is considered a more challenging shape. For the sinusoidal case the relations controlling the variation in area and area moment of inertia are (see reference [11]):

$$A(x) = A_0 {s_1}^2 \left[1 + \beta \sin\left(\pi \frac{x}{L}\right)\right]^2, \; I(x) = I_0 {s_1}^4 \left[1 + \beta \sin\left(\pi \frac{x}{L}\right)\right]^4 \tag{20.2}$$

where $\beta = R_{midspan}/R_0$ is the ratio between the mid-span radius to one end radius (radii end values are the same), and $s_1 = \sqrt{2\pi \Big/ (\beta^2 \pi + 2\pi + 8\beta)}$. In the example below the circular cross section at the center of the beam is chosen to have a radius which is twice the value of the one at the ends of the beam.

The transitions from one material to another are approximated via logistic functions (step functions are not used since they would generate numerical complications):

$$H(x) \approx \frac{1}{2} + \frac{1}{2}\tanh(Kx) = \frac{1}{1 + e^{-2Kx}} \tag{20.3}$$

In Eq. (20.3) a larger K corresponds to a sharper transition at $x = 0$.

A non-dimensional version of the equation of motion can be obtained by taking:

$$\nu = \Omega/\Omega_0, \; \xi = x/L, \; Y = w/L$$

$$I(x) = I_0 s^4 [\alpha + (1-\alpha)\,\xi]^4 = I_0 f_2(\xi) \text{ or } I(x) = I_0 {s_1}^4 [1 + \beta \sin(\pi \xi)]^4 = I_0 f_2(\xi),$$

$$A(x) = A_0 s^2 [\alpha + (1-\alpha)\,\xi]^2 = A_0 f_4(\xi) \text{ or } A(x) = A_0 {s_1}^2 [1 + \beta \sin(\pi \xi)]^2 = A_0 f_4(\xi),$$

$$E(x) = E_0 H(K\xi) = E_0 f_1(\xi), \; \rho(x) = \rho_0 H(K\xi) = \rho_0 f_3(\xi)$$

Substituting these into Eq. (20.1), one obtains:

$$\left(\frac{E_0 I_0}{\rho_0 A_0 L^4 \Omega_0^2}\right) \frac{\partial^2}{\partial \xi^2}\left(f_1(\xi)\, f_2(\xi) \frac{\partial^2 Y}{\partial \xi^2}\right) + \left(\frac{F}{\rho_0 A_0 L^2 \Omega_0^2}\right) \frac{\partial^2 Y}{\partial \xi^2} + f_3(\xi)\, f_4(\xi) \frac{\partial^2 Y}{\partial \tau^2} = \frac{f}{\rho_0 A_0 L \Omega_0^2} \tag{20.4}$$

where Ω_0 is a reference frequency (here $\frac{\Pi^2}{L^2}\sqrt{\frac{EI}{\rho A}}$).

In Eq. (20.4), seeking vibration frequencies, on can assume $Y(\xi, \tau) = S(\xi) \sin(\nu\tau)$ and harmonic external forcing $f = f \sin(\nu\tau)$. This leads to:

$$K_1 \frac{\partial^2}{\partial \xi^2}\left(f_1(\xi) f_2(\xi) \frac{\partial^2 S(\xi)}{\partial \xi^2} \right) + F_1 \frac{\partial^2 S(\xi)}{\partial \xi^2} - \nu^2 f_3(\xi) f_4(\xi) S(\xi) = F_2 \tag{20.5}$$

where $K_1 = \frac{E_0 I_0}{\rho_0 A_0 L^4 \Omega_0^2}$, $F_1 = \frac{F}{\rho_0 A_0 L^2 \Omega_0^2}$ and $F_2 = \frac{f}{\rho_0 A_0 L \Omega_0^2}$.

Given the material layout (stacked cells) and cross section variation, i.e., $f_1(\xi), f_3(\xi), f_2(\xi)$ and $f_4(\xi)$, numerical solutions for the equation can, in principle, be obtained. Here, neglecting the axial force ($F_1 = 0$), Eq. (20.5) would have to be solved subjected to a set of boundary conditions. (Note that, in general, this problem does not have analytic solutions.)

In the following an approach described by the authors in reference [11] is utilized for extracting resonances. The strategy employed is to use MAPLE®'s two-point boundary value solver to solve a forced motion problem. It consists of assuming a constant value for the forcing function F_2 and varying the frequency ν . By observing the mid-span deflection of the beam, the resonant frequency can be found on noting where an abrupt change in sign occurs. Note that higher modes could be obtained by extending the search range.

For some examples an assumed mode approach is used to confirm results obtained using the forced motion method. The procedure is described by one of the authors in reference [12].

20.3 Numerical Examples

Consider the beam shown in Fig. 20.2, which is composed by a number of n stacked cells of different materials.

20.3.1 Two-Cell Beams

For the first example, consider two cells made of the following materials: Aluminum ($E_1 = 71$ GPa, $\rho_1 = 2{,}710$ Kg/m^3) and Silicon Carbide ($E_2 = 210$ GPa, $\rho_2 = 3{,}100$ Kg/m^3).

Taking a uniform, rectangular section beam (0.01 m by 0.01 m and $L = 1$ m), the non-dimensional logistic functions and cross-section functions can be written as (see Fig. 20.3):

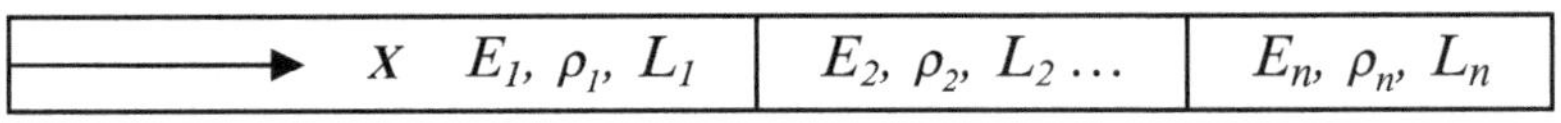

Fig. 20.2 Layered beam

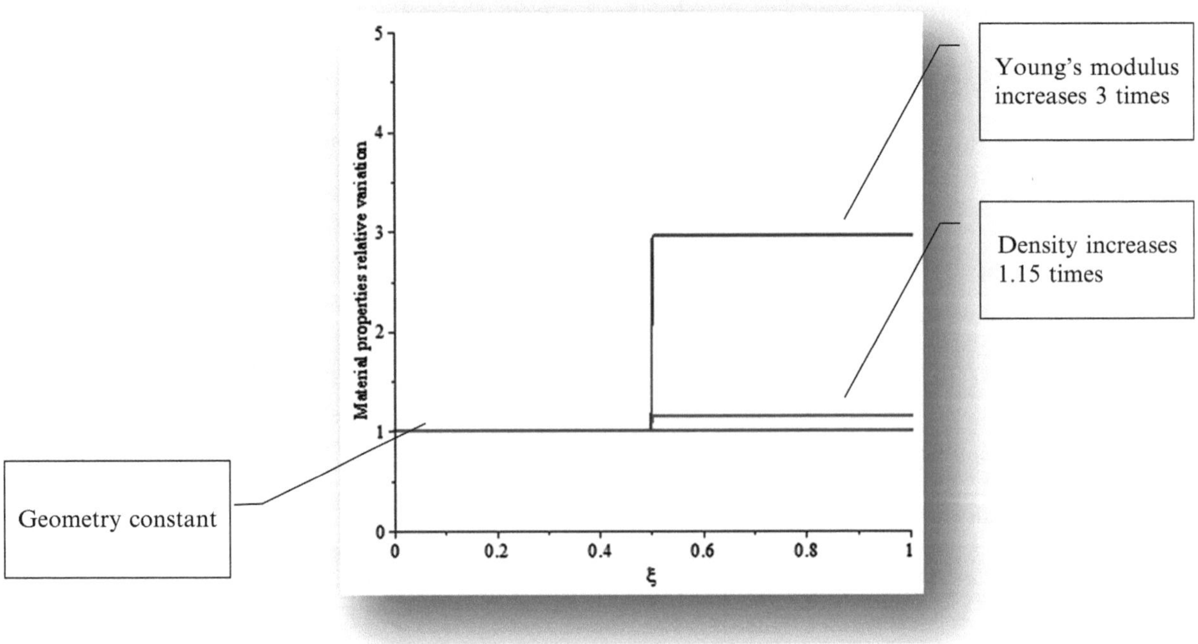

Fig. 20.3 Relative properties variation for two-cell beam

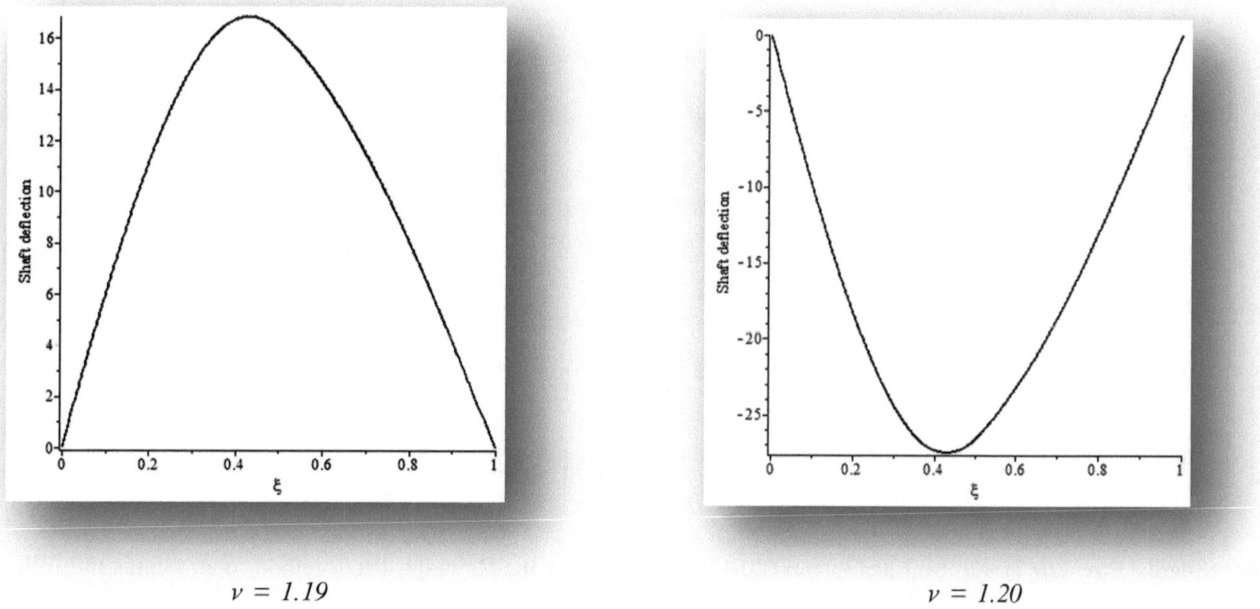

Fig. 20.4 Pin-pin beam deflections as a function of ν

$$
\begin{aligned}
f_1(\xi) &= 1 + \tfrac{E_2 - E_1}{E_1}\left(\tfrac{1}{2} + \tfrac{1}{2}\tanh\left(1000\left(\xi - \tfrac{1}{2}\right)\right)\right), \ f_2(\xi) = 1, \\
f_3(\xi) &= 1 + \tfrac{\rho_2 - \rho_1}{\rho_1}\left(\tfrac{1}{2} + \tfrac{1}{2}\tanh\left(1000\left(\xi - \tfrac{1}{2}\right)\right)\right), \ f_4(\xi) = 1
\end{aligned}
\tag{20.6}
$$

Assuming a value of *1* for the external forcing and using the numerical method described above, the resultant deflections are plotted in Fig. 20.4 for two distinct values of the frequency ν. The boundary conditions used are pin-pin.

Resonance is taken to occur at $\nu = 1.20$. Next a second approach is pursued to confirm this value.

For the assumed mode method, the following function is used for the beam elastic form (harmonic motion):

$$w(x, t) = \varphi(x)\sin(\lambda t) \tag{20.7}$$

where:

$$\varphi(x) = \sum_{1}^{k} c_i \phi_i(x) \tag{20.8}$$

The shape function $\varphi(x)$ is taken to be a linear combination of k beam characteristic orthogonal polynomials ϕ_i where the c_i are arbitrary constants to be determined. Each polynomial satisfies the geometric boundary conditions at the ends of the beam.

The set of orthogonal polynomials is generated by the Gram-Schmidt process [13] as demonstrated by Bhat [14].

For the present problem, the first polynomial is taken to satisfy the two geometrical boundary conditions at the ends and to have the following form (pin-pin non-dimensional beam static deflection under distributed load):

$$\phi_1(x) = s_1\left(x^4 - 2x^3 + x\right) \tag{20.9}$$

The normalization constant s_1 is chosen such that:

$$\int_0^L (\phi_k(x))^2 dx = 1 \tag{20.10}$$

Next the Lagrangian can be written as:

$$\mathcal{L} = T - V \tag{20.11}$$

Where T and V are kinetic and potential energy, respectively. The kinetic and potential energy for the i-segment of the beam (n cells) can be written as:

$$T = \frac{1}{2}\int_{\frac{i-1}{n}L}^{\frac{i}{n}L} \rho_i A_i \left(\frac{\partial}{\partial x} w(x,t)\right)^2 dx,\ V = \frac{1}{2}\int_{\frac{i-1}{n}L}^{\frac{i}{n}L} E_i I_i \left(\frac{\partial^2}{\partial x^2} w(x,t)\right)^2 dx \tag{20.12}$$

The time dependence in Eq. (20.7) can be suppressed by using the time average value of the Lagrangian (see reference [14]):

$$\overline{\mathcal{L}} = \int_0^{\frac{2\pi}{\lambda}} \mathcal{L}\,\mathrm{dt} \tag{20.13}$$

Then utilizing Eqs. (20.7), (20.8), (20.11) and (20.13) and applying the Rayleigh-Ritz method, one obtains the following homogeneous simultaneous equations:

$$\frac{\partial \overline{\mathcal{L}}}{\partial \mathrm{c_i}} = 0,\ \mathrm{i} = 1,\ 2,\ 3 \ldots \mathrm{k} \tag{20.14}$$

which can be cast into a standard eigenvalue problem form, from which the natural frequencies can be calculated.

Using ten orthogonal polynomials the procedure produces the following values for the first five resonances: *1.21, 5.20, 11.23, 15.92,* and *17.21*. Note that the first one is very close to the resonance estimated by the forced motion approach.

Another set of boundary conditions is analyzed next. Taking the same beam with clamp-clamp boundary conditions leads to the following results.

The deflections can be seen in Fig. 20.5 for two distinct values of the frequency ν.

Thus the forced motion approach gives $\nu = 2.80$.

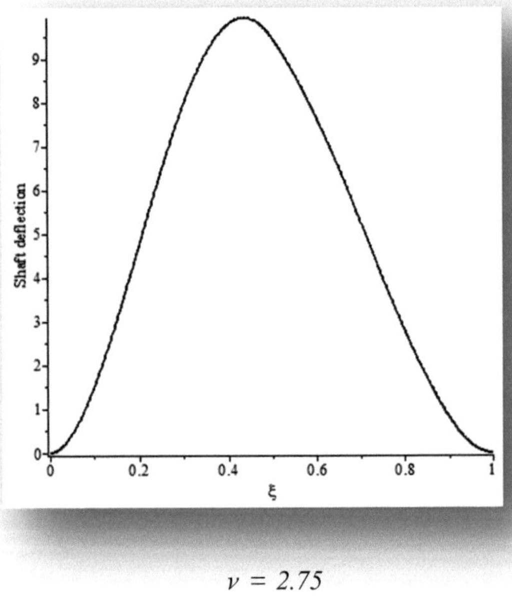

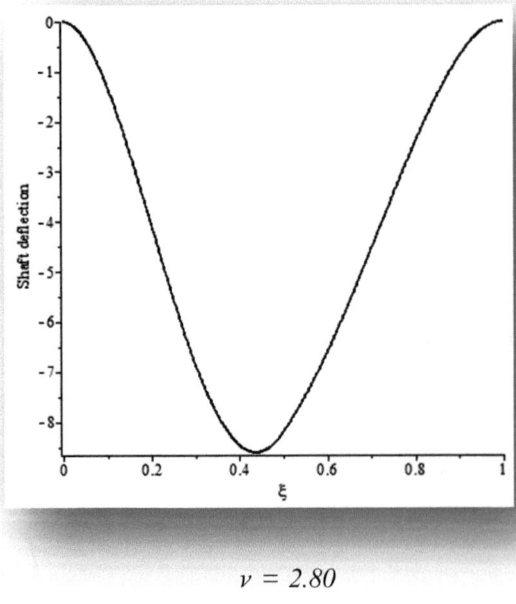

Fig. 20.5 Clamp-clamp beam deflections as a function of ν

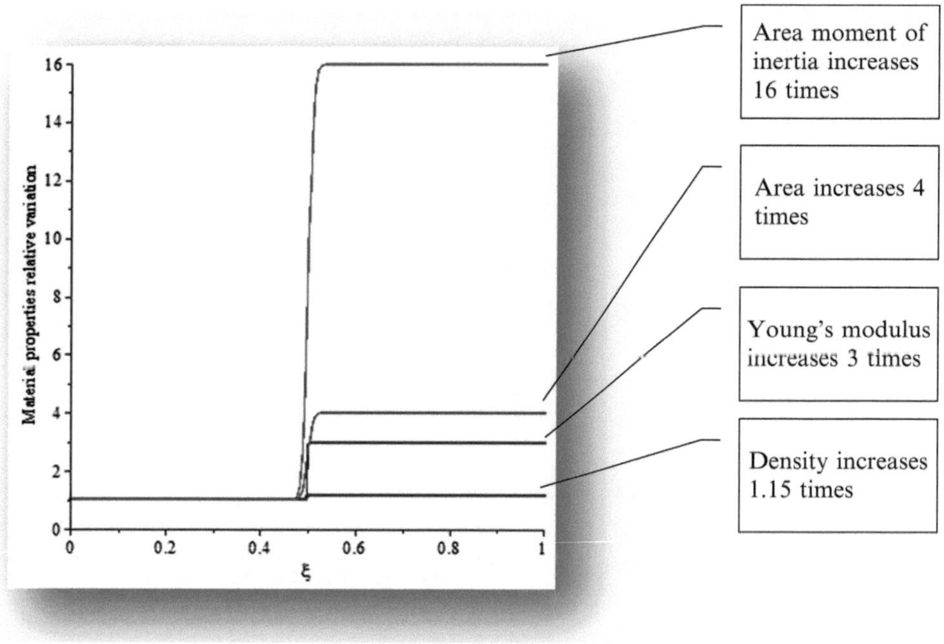

Fig. 20.6 Relative properties variation for non-uniform two-cell beam

In this case, for the assumed mode approach, the first polynomial of the orthogonal series is given by:

$$\phi_1(x) = c_1 x^2 \left(x^2 - 2x + 1\right) \tag{20.15}$$

which corresponds to a non-dimensional version of the deflection of a clamp-clamp beam under distributed load. Computation of the results gives: $\nu = 2.81$ (first eigenvalue).

As in the pin-pin case, excellent agreement between the methods is seen.

Consider next the pin-pin two-cell beam with same material variation but with a cross-section discontinuity at the interface of the cells. For this example a variation in area of four times is considered (see Fig. 20.6).

The logistic functions for the material variations are the same as above and the ones for the cross-section variations are:

$$\begin{aligned} f_2(\xi) &= 1 + \tfrac{I_2 - I_1}{I_1}\left(\tfrac{1}{2} + \tfrac{1}{2}\tanh\left(100\left(\xi - \tfrac{1}{2}\right)\right)\right), \\ f_4(\xi) &= 1 + \tfrac{A_2 - A_1}{A_1}\left(\tfrac{1}{2} + \tfrac{1}{2}\tanh\left(100\left(\xi - \tfrac{1}{2}\right)\right)\right) \end{aligned} \tag{20.16}$$

Results using the forced motion and the assumed mode approaches, for the first bending frequency, are *0.96* and *1.15*, respectively. Using ten polynomials, the assumed mode estimates the first frequency to be approximately 19.8 % larger than the one predicted by the forced motion approach.

Taking the clamp-clamp set of boundary conditions for this case leads to the following results. First frequency given by the forced motion approach is $\nu = 4.14$ and for the assumed mode is $\nu = 4.24$. In this case the estimation of the assumed mode is 2.4 % larger than the forced motion.

Next, consider a pin-pin two-cell beam with a circular cross-section and a sinusoidal profile.

For this example the radius in the center of the shaft is twice the one at the ends (radii at the ends = *0.0127 m* and L = *1 m*). The material variation functions are the same as above, and the cross-section functions are given by:

$$f_2(\xi) = 1 + \frac{I_2 - I_1}{I_1}\left(\frac{1}{2} + \frac{1}{2}\sin(\pi\xi)\right),\quad f_4(\xi) = 1 + \frac{A_2 - A_1}{A_1}\left(\frac{1}{2} + \frac{1}{2}\sin(\pi\xi)\right) \tag{20.17}$$

The variations can be seen in Fig. 20.7.

Here the forced motion approach estimates the first frequency to be at $\nu = 2.06$. The assumed mode approach leads to the following value: $\nu = 2.08$. As in the uniform case, very good agreement is seen.

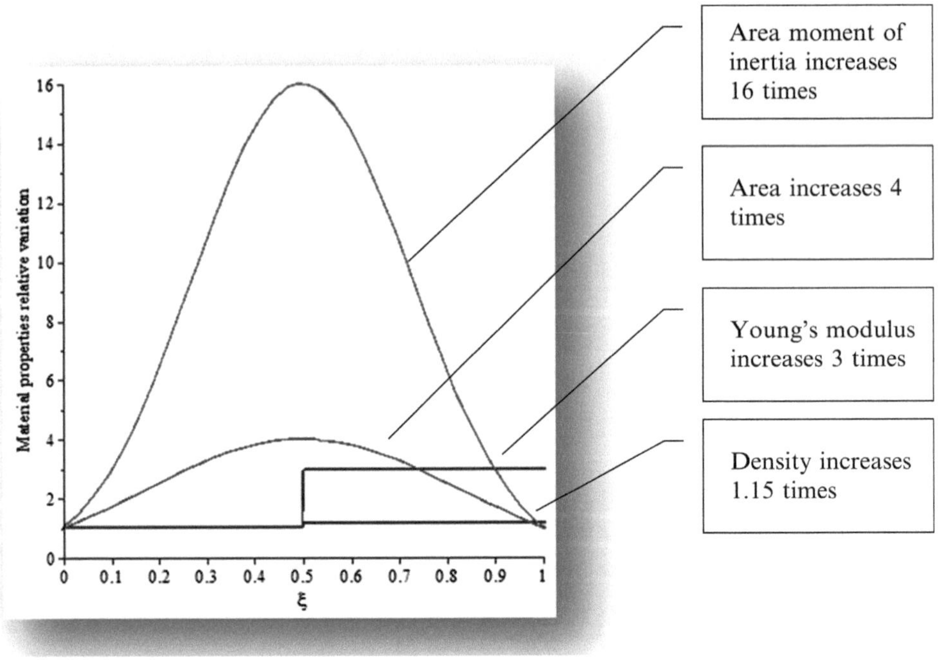

Fig. 20.7 Relative properties variation for sinusoidal two-cell beam

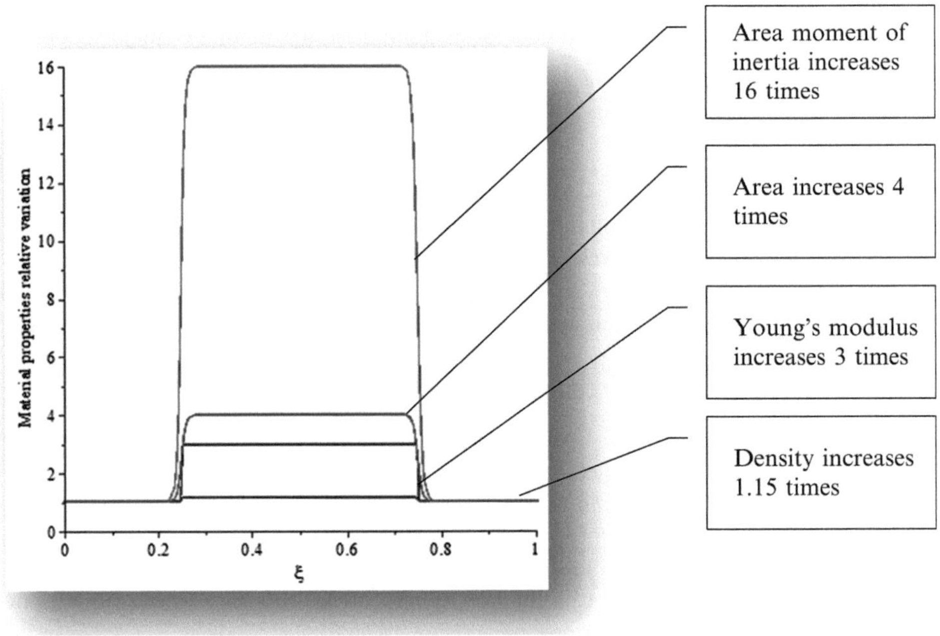

Fig. 20.8 Relative properties variation for three-cell beam

20.3.2 Three-Cell Beams

Consider the rectangular cross section non-tapered beam discussed above. For this example, three cells of materials are used in the following sequence: Aluminum, Silicon Carbide and Aluminum (properties given above). The material and geometric variations can be seen in Fig. 20.8.

Material and cross-section functions are given by:

$$\begin{aligned}
f_1(\xi) &= 1 + \tfrac{E_2-E_1}{E_1}\left(\tfrac{1}{2}+\tfrac{1}{2}\tanh\left(1000\left(\xi-\tfrac{1}{4}\right)\right)\right) - \tfrac{E_2-E_1}{E_1}\left(\tfrac{1}{2}+\tfrac{1}{2}\tanh\left(1000\left(\xi-\tfrac{3}{4}\right)\right)\right),\\
f_2(\xi) &= 1 + \tfrac{I_2-I_1}{I_1}\left(\tfrac{1}{2}+\tfrac{1}{2}\tanh\left(100\left(\xi-\tfrac{1}{4}\right)\right)\right) - \tfrac{I_2-I_1}{I_1}\left(\tfrac{1}{2}+\tfrac{1}{2}\tanh\left(100\left(\xi-\tfrac{3}{4}\right)\right)\right),\\
f_3(\xi) &= 1 + \tfrac{\rho_2-\rho_1}{\rho_1}\left(\tfrac{1}{2}+\tfrac{1}{2}\tanh\left(1000\left(\xi-\tfrac{1}{4}\right)\right)\right) - \tfrac{\rho_2-\rho_1}{\rho_1}\left(\tfrac{1}{2}+\tfrac{1}{2}\tanh\left(1000\left(\xi-\tfrac{3}{4}\right)\right)\right),\\
f_4(\xi) &= 1 + \tfrac{A_2-A_1}{A_1}\left(\tfrac{1}{2}+\tfrac{1}{2}\tanh\left(100\left(\xi-\tfrac{1}{4}\right)\right)\right) - \tfrac{A_2-A_1}{A_1}\left(\tfrac{1}{2}+\tfrac{1}{2}\tanh\left(100\left(\xi-\tfrac{3}{4}\right)\right)\right)
\end{aligned} \tag{20.18}$$

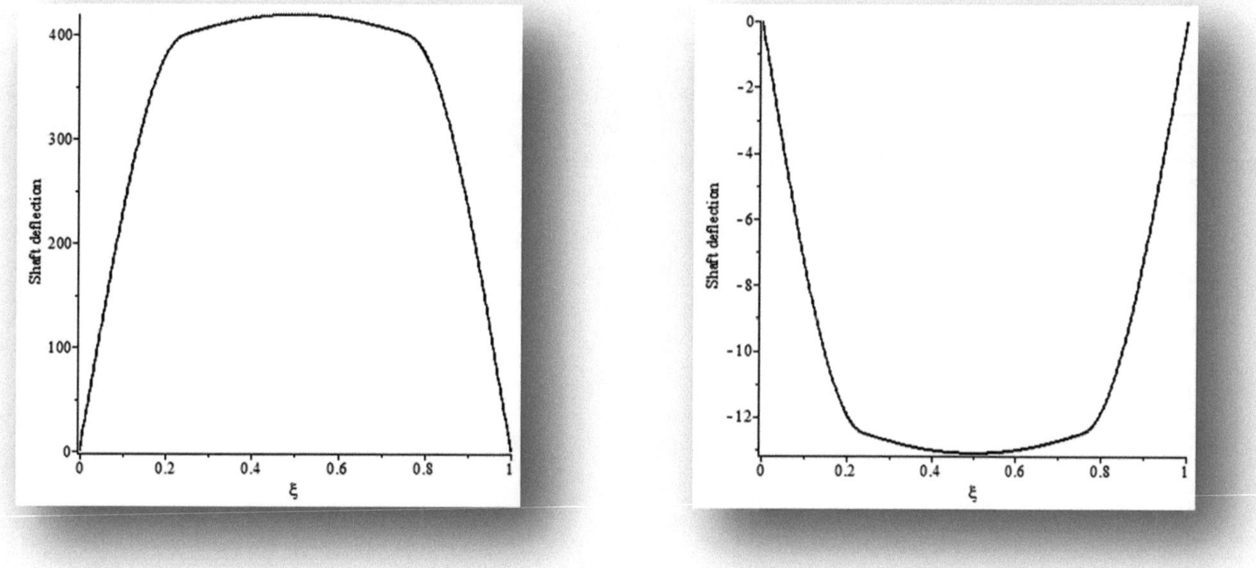

$\nu = 1.27$ $\nu = 1.28$

Fig. 20.9 Pin-pin beam deflections as a function of ν – three-cell beam

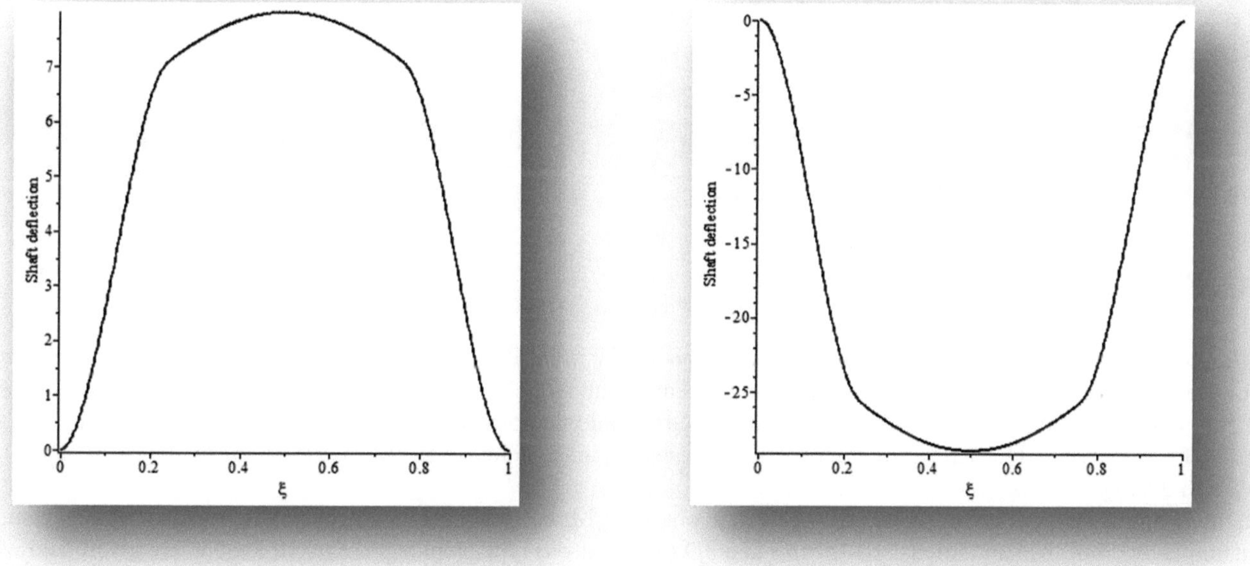

$\nu = 2.47$ $\nu = 2.48$

Fig. 20.10 Clamp-clamp beam deflections as a function of ν – three-cell beam

The forced motion approach gives the first frequency for this case as $\nu = 1.28$. This can be seen in Fig. 20.9.

Using the assumed mode approach, with the same initial polynomial given before and ten orthogonal polynomials, the method produces a first frequency $\nu = 1.63$, which is larger than the one given by the forced motion by 27.3 %. Differences start to increase as the assumed polynomial for the deflection and the actual deflection shapes become more distinct.

For clamp-clamp boundary conditions and using the forced motion, the deflections can be seen in Fig. 20.10. Resonance is observed at $\nu = 2.48$.

For this case the assumed mode gives: $\nu = 3.02$. The difference here is 21.8 %.

A summary of the results is given in Table 20.1.

Table 20.1 Results summary

First frequency ν	Uniform square section 2-cell $\Omega_0 = 146.10\ rad/s$		Non-uniform sinusoidal circular section 2-cell $\Omega_0 = 321.38\ rad/s$		Non-uniform stepped square section 2-cell $\Omega_0 = 146.10\ rad/s$		Non-uniform stepped square section 3-cell $\Omega_0 = 146.10\ rad/s$	
	Forced motion	Rayleigh-Ritz	Forced motion	Rayleigh-Ritz	Forced motion	Rayleigh-Ritz	Forced motion	Rayleigh-Ritz
Pin-Pin	*1.20(27.9 Hz)*	*1.21(28.1 Hz)*	*2.06(105.4 Hz)*	*2.08 (106.4 Hz)*	*0.96(22.3 Hz)*	*1.15(26.7 Hz)*	*1.28(29.6 Hz)*	*1.63 (37.9Hz)*
Clamp-Clamp	*2.80(65.1 Hz)*	*2.81(65.3 Hz)*	*3.20(163.7 Hz)*	*3.23 (165.2 Hz)*	*4.14(96.3 Hz)*	*4.24(98.6 Hz)*	*2.48(57.7 Hz)*	*3.02(70.2 Hz)*

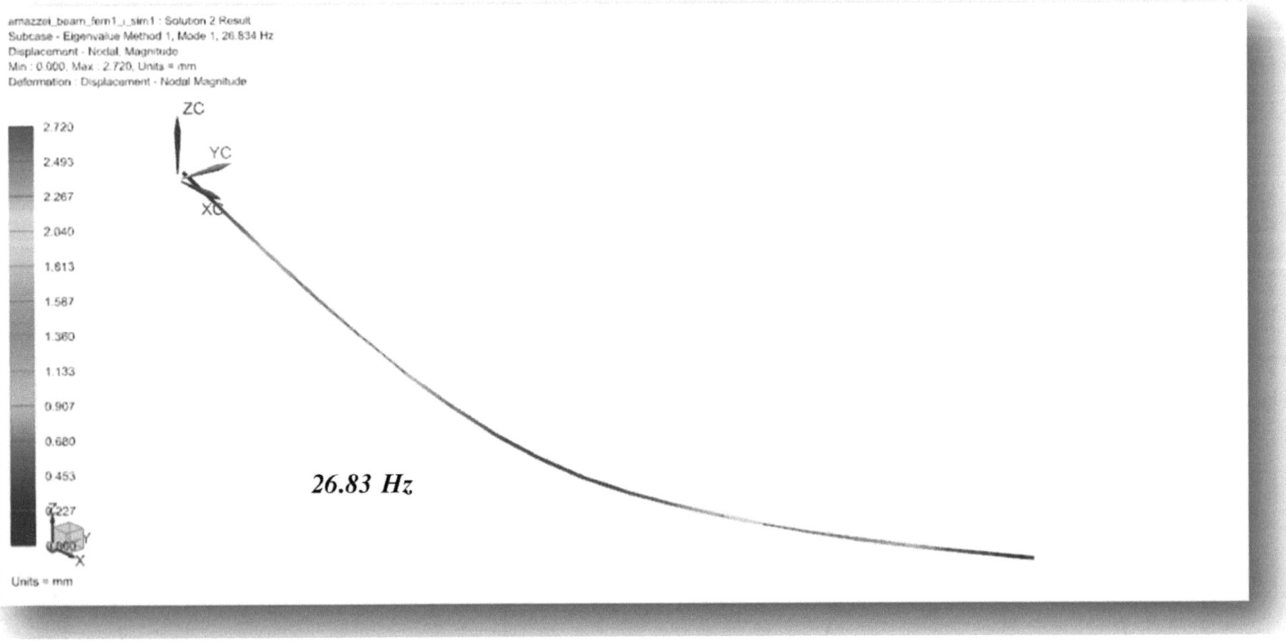

Fig. 20.11 First mode of square pin-pin two-cell beam via FEM

20.4 FEM

Next these problems are tackled via finite element analysis. Note that the forced motion approach provides a simpler set up for the problems, and MAPLE® worksheets can be generic, allowing for treatment of different cases by simply altering boundary conditions and different materials and profiles. It should be noted that FEM solutions can be obtained if one has access to FEM software. For reasons of completeness, these are pursed below(Figs. 20.11, 20.12, 20.13, 20.14, 20.15, 20.16, 20.17, and 20.18).

In this section, Siemens® NX[2] software is utilized to obtain FEM results for mode shapes and frequencies of the examples discussed above. For the simulations, 1D (beam) and 3D (tetrahedron) type elements are used. Results are given below and summarized in Table 20.2.

Note that excellent agreement is seen for the majority of cases.

20.5 Conclusions

Replacing discrete property variations with continuously varying ones, together with resonance monitoring of forced motion solutions of the resulting single ordinary differential equation, obtained by MAPLE®'s ODE solver, has been shown to lead to accurate solutions for the resonant frequencies of layered beams.

---

[2] http://www.plm.automation.siemens.com/en_us/products/nx/index.shtml

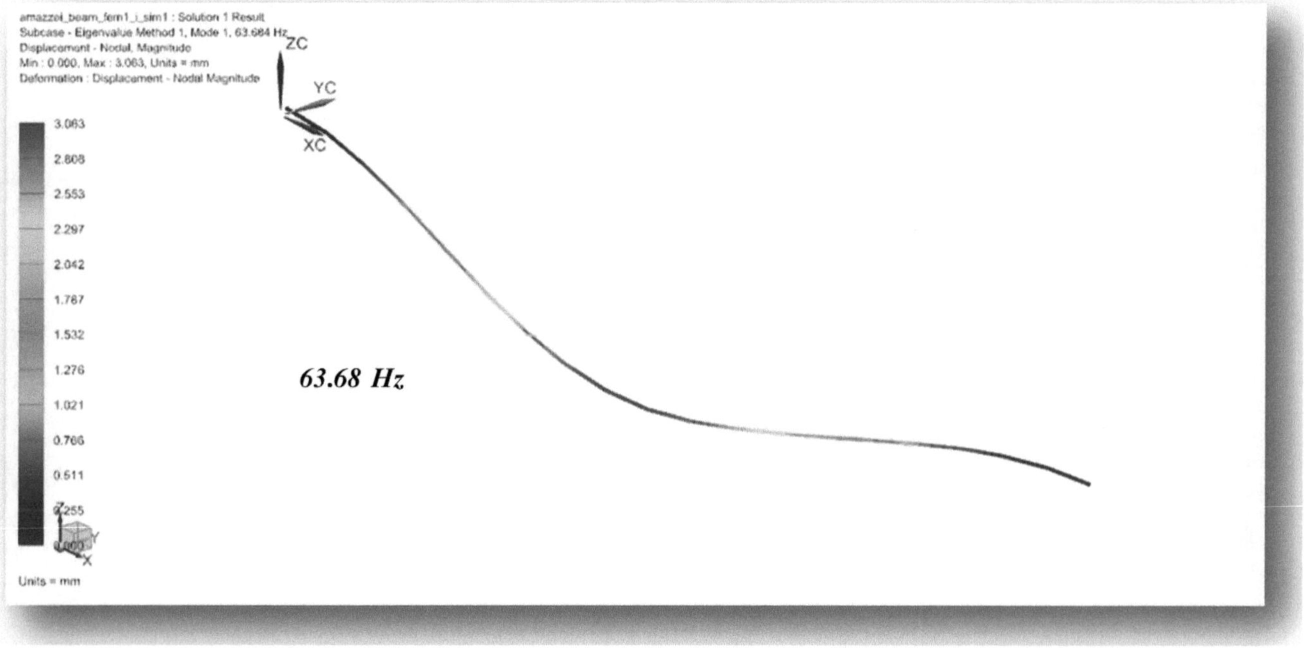

Fig. 20.12 First mode of clamp-clamp square two-cell beam via FEM

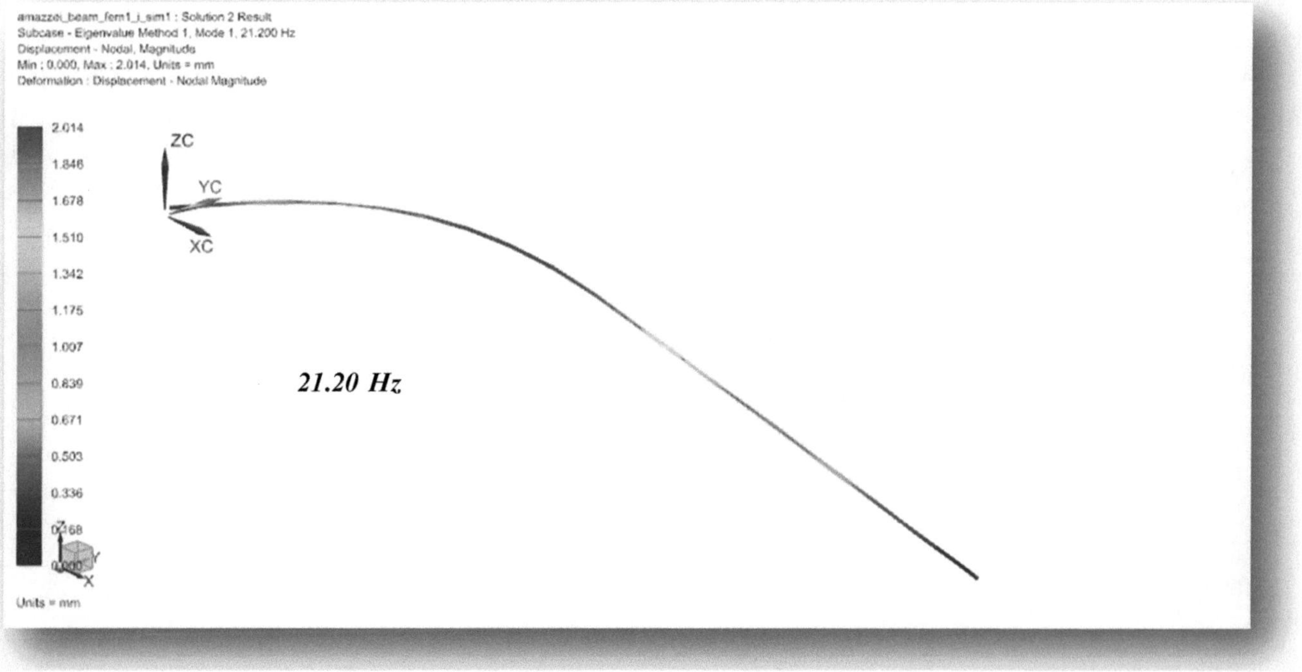

Fig. 20.13 First mode of pin-pin square non-uniform two-cell beam via FEM

The following cases were examined: two-cell beams (uniform and non-uniform), a three-cell beam (non-uniform) and a beam with a sinusoidal profile. Accuracy was assessed by comparison with results obtained using a Rayleigh-Ritz method and a commercial finite element code.

For uniform cases excellent agreement is seen. For non-uniform cases with smooth profile variation, agreement is also very good.

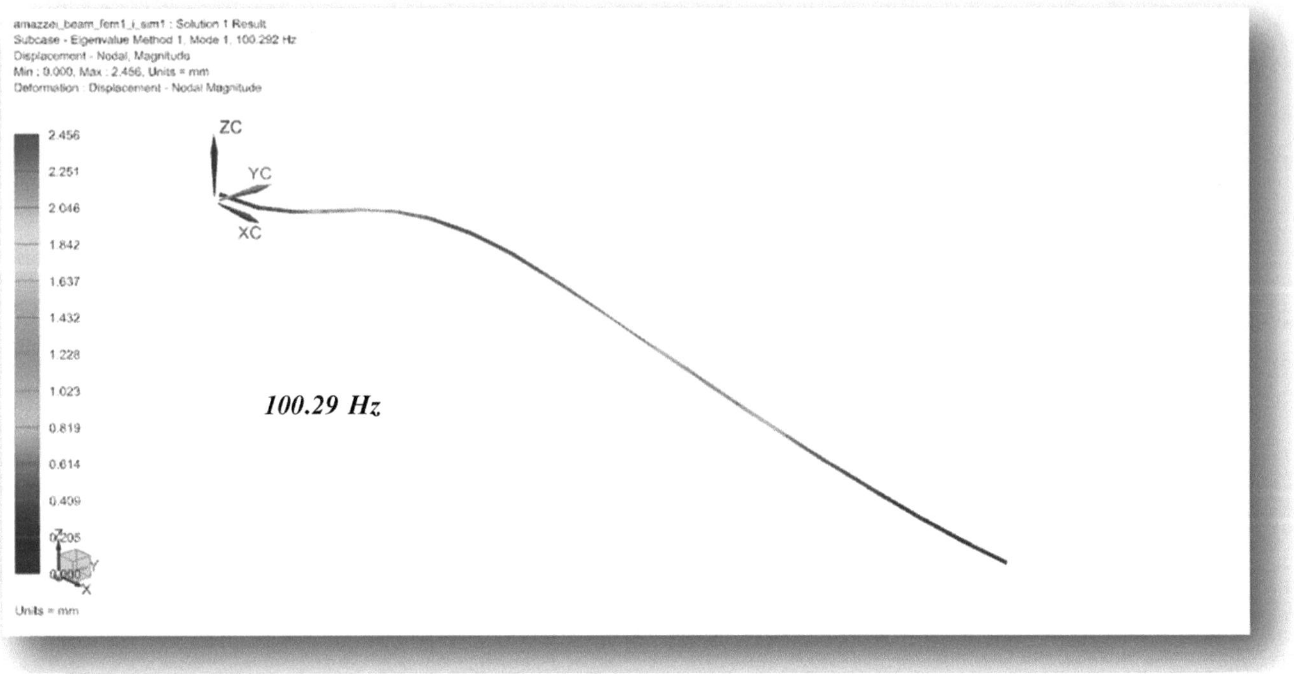

Fig. 20.14 First mode of clamp-clamp square non-uniform two-cell beam via FEM

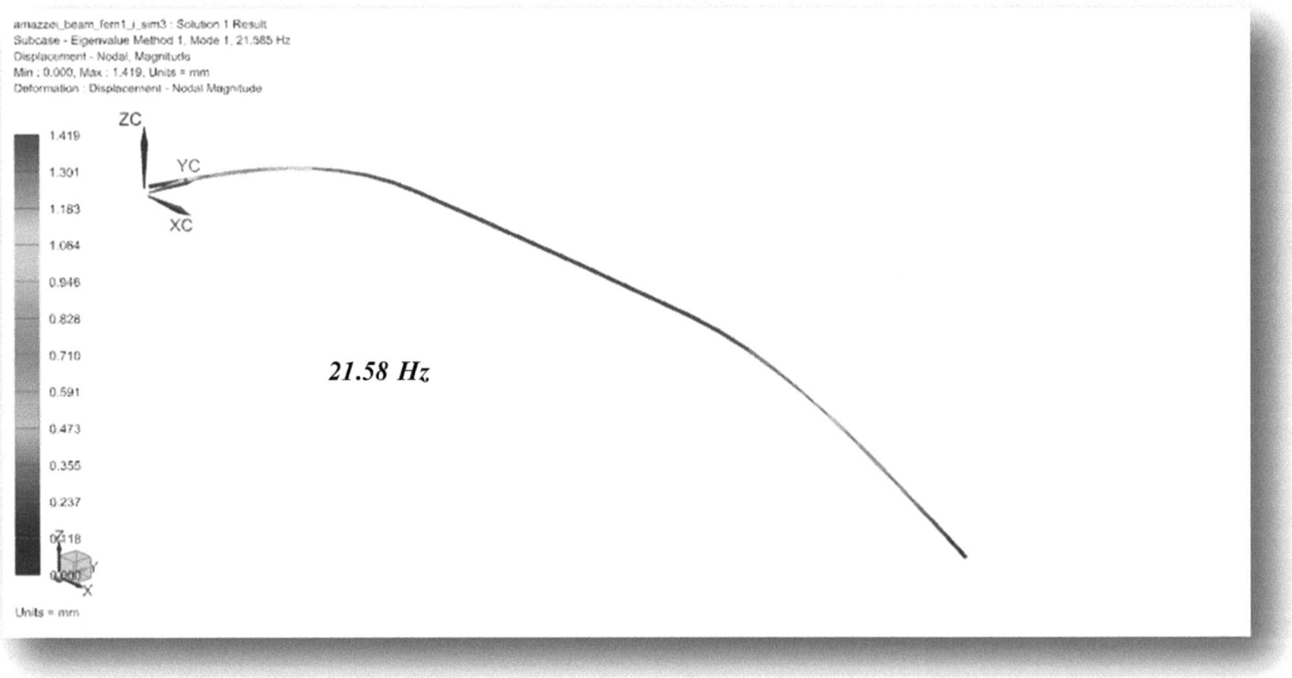

Fig. 20.15 First mode of pin-pin square non-uniform three-cell beam via FEM

In the case of non-uniform with more than two steps, results compare reasonably, with maximum differences around 30 % for the first frequency. For future work, for better accuracy, it is recommended that more polynomials be used in the assumed mode approach.

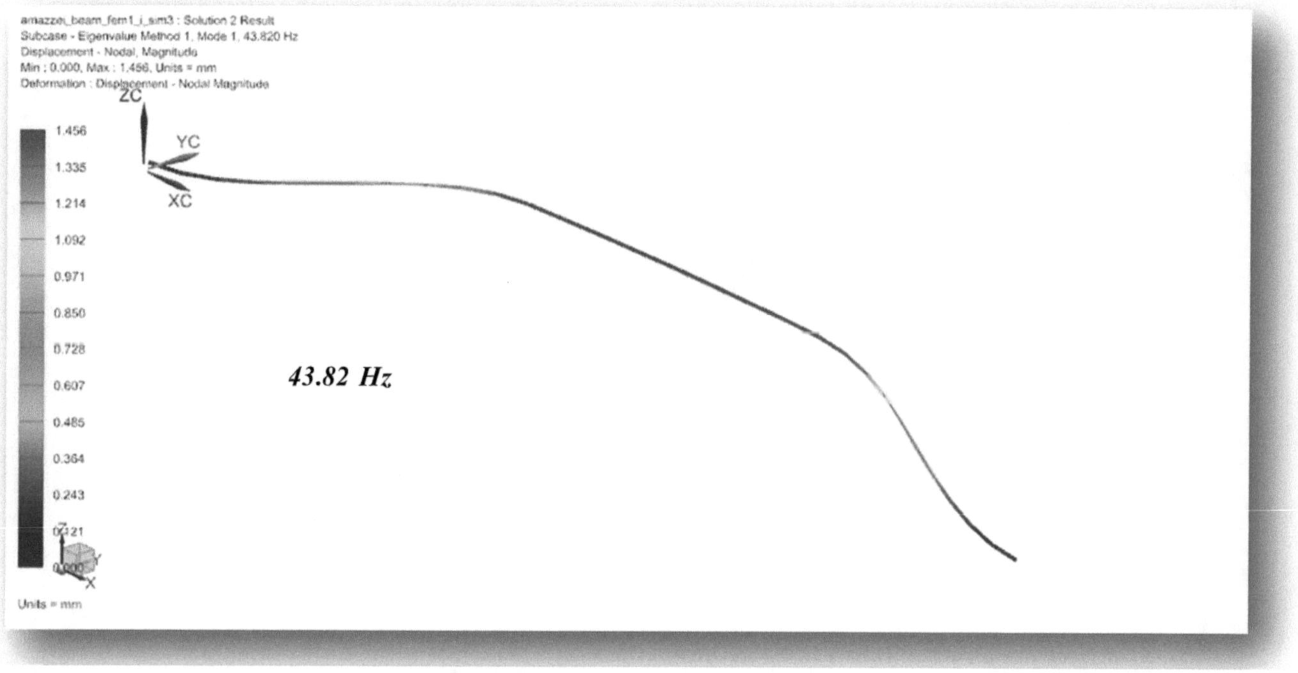

Fig. 20.16 First mode of clamp-clamp square non-uniform three-cell beam via FEM

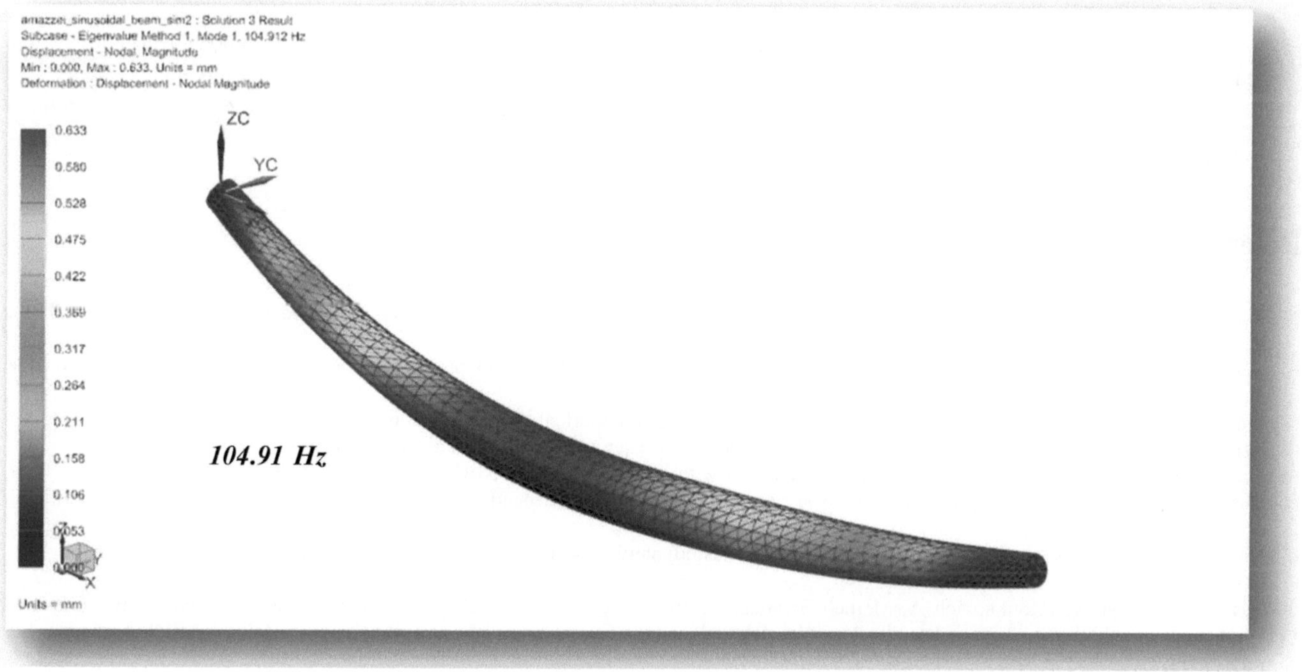

Fig. 20.17 First mode of pin-pin sinusoidal two-cell beam via FEM

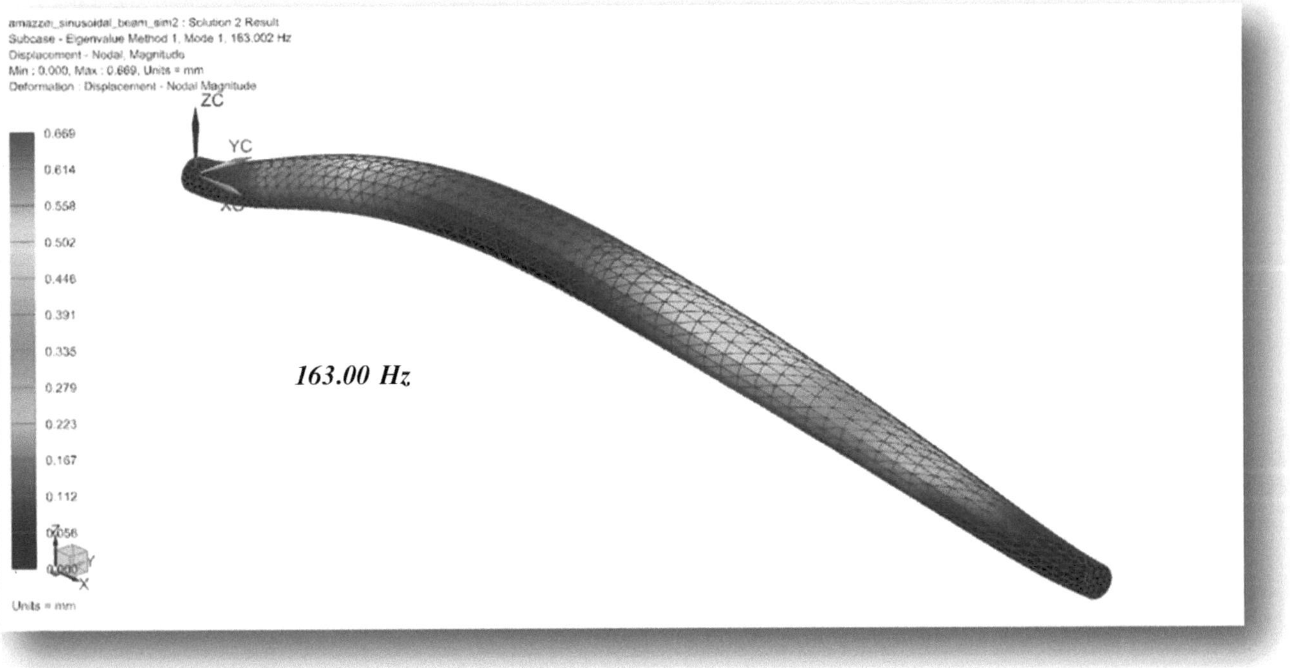

Fig. 20.18 First mode of clamp-clamp sinusoidal two-cell beam via FEM

Table 20.2 Results comparison: forced motion versus FEM

	Uniform square section 2-cell		Non-uniform sinusoidal circular section 2-cell		Non-uniform stepped square section 2-cell		Non-uniform stepped square section 3-cell	
First frequencyν	Forced motion	FEM	Forced motion	FEM	Forced motion	FEM	Forced motion	FEM
Pin-Pin	*27.9 Hz*	*26.8 Hz*	*105.4 Hz*	*104.9 Hz*	*22.3 Hz*	*21.2 Hz*	*29.6 Hz*	*21.6 Hz*
Clamp-Clamp	*65.1 Hz*	*63.7 Hz*	*163.7 Hz*	*163.0 Hz*	*96.3 Hz*	*100.3 Hz*	*57.7 Hz*	*43.8 Hz*

References

1. Mazzei AJ Jr, Scott RA (2014) Vibrations of discretely layered structures using a continuous variation model. In: Allemang R (ed) Topics in modal analysis II, vol 8. Springer International Publishing, pp. 385–396
2. Cekus D (2011) Use of Lagrange multiplier formalism to solve transverse vibrations problem of stepped beams according to Timoshenko theory. Sci Res Inst Math Comput Sci 2(10):49–56
3. Jang SK, Bert CW (1989) Free vibration of stepped beams: exact and numerical solutions. J Sound Vib 130(2):342–346
4. Bert CW, Newberry AL (1986) Improved finite element analysis of beam vibration. J Sound Vib 105:179–183
5. Lee EH, Yang WH (1973) On waves in composite materials with periodic structure. SIAM J Appl Math 25(3):492–499
6. Hussein MI, Hulbert GM, Scott RA (2006) Dispersive elastodynamics of 1d banded materials and structures: analysis. J Sound Vib 289 (4–5):779–806
7. Hussein MI, Hulbert GM, Scott RA (2007) Dispersive elastodynamics of 1d banded materials and structures: design. J Sound Vib 307 (3–5):865–893
8. Vasseur JO, Deymier P, Sukhovich A, Merheb B, Hladky-Hennion AC, Hussein MI (2013) Phononic band structures and transmission coefficients: methods and approaches. In: Deymier PA (ed) Acoustic metamaterials and phononic crystals, vol 173. Springer, Berlin, pp 329–372
9. Berezovski A, Engelbrecht J, Maugin GA (2008) Numerical simulation of waves and fronts in inhomogeneous solids. World Scientific Publishing Co. Pte. Ltd, Singapore
10. Lee BK, Carr AJ, Lee TE, Kim IJ (2006) Buckling loads of columns with constant volume. J Sound Vib 294(1–2):381–387
11. Mazzei AJ, Scott RA (2013) On the effects of non-homogeneous materials on the vibrations and static stability of tapered shafts. J Vib Control 19(5):771–786
12. Mazzei JAJ (2012) On the effect of functionally graded materials on resonances of rotating beams. Shock Vib 19(6):1315–1326
13. Chihara TS (1978) Introduction to orthogonal polynomials. Gordon and Breach, London
14. Bhat RB (1986) Transverse vibrations of a rotating uniform cantilever beam with tip mass as predicted by using beam characteristic orthogonal polynomials in the Rayleigh-Ritz method. J Sound Vib 105(2):199–210

Chapter 21
Analysis of H_∞ and H_2 Optimal Design Scheme for an Electromagnetic Damper with Shunt Resonant Circuit

Wai Kei Ao and Paul Reynolds

Abstract Electrodynamic actuators and electromagnetic dampers (EMD) are used extensively in mechanical systems. They utilise the electromagnetic induction concept to generate eddy current and Lorentz forces for providing vibration suppression forces. In this research, these principles will be developed for vibration serviceability control of a civil engineering structure.

An electrodynamic actuator is used in this research, together with a shunt resonant circuit, an RLC resonant circuit (a simple electronic circuit oscillator) which is needed to cascade with the EMD for closing the circuit and generating electrical damping forces. The EMD is set between the structure and the ground. The kinetic energy of the vibrating structure provides the input energy to activate the EMD.

H_∞ optimisation (minimisation of the maximum response) and H_2 optimisation (minimisation of kinetic energy) are used to obtain the vibration suppression performance, which is compared against a conventional tuned mass damper (TMD). The EMD with shunt resonant circuit (EMDS) can have similar dynamic performance, which achieves suppression of resonant vibration amplitude of the primary structure down to two lower amplitude peaks in the frequency domain. Hence, this work shows that the EMDS can in principle achieve satisfactory vibration suppression performance.

Keywords Tuned mass damper • Electromagnetic damper (EMD) • RLC resonant circuit • H_∞ optimisation • H_2 optimisation

21.1 Introduction

Electromagnetic dampers (EMD) are based on electromagnetic theory and are extensively used in the automobile industry and in other mechanical structures. These devices convert mechanical energy to electric energy through electromagnetic induction. From the theory of electromagnetism, damping forces are created by conductors moving within magnetic poles and cutting through magnetic lines. Mizuno [1] studied an electromagnetic induction vibration isolation system using active control and negative stiffness with a voice coil motor. Fleming et al. [2] proposed an electromagnetic transducer involving voice coil. The transducer can also create a damping force when used with a suitable controller for active vibration mitigation. Behrens et al. [3] studied an electromagnetic transducer and developed a passive control technique with shunt impedance.

From this point of view, the electromagnetic concept can satisfy some vibration control requirements. Some researchers have proposed and investigated an electromagnetic device with shunt circuit (or control circuit) to provide and improve system damping. It is called an electromagnetic shunt damper (EMDS) [4–7]

Over the last two decades, civil engineers have been developing an interest in such electromagnetic devices for structural vibration control. Inoue et al. [8] illustrated a system consisting of voice coil motor connected to an RC circuit that dealt with the vibration suppression problem using such EMDS as well as the theoretical development of expressions for optimal parameter design. McDaid and Mace [9] used an electromagnetic device (voice coil) with shunt impedance to attenuate structural vibration. Cheng and Oh [10] used a current-flowing method to apply an electromagnetic device with RLC shunt circuit to perform vibration control at relatively high frequencies.

A lot of numerical and experimental tests were performed using a small scale flexible structure by [2]. Their results showed reductions of the two peaks in structural frequency response functions (FRFs). This behaviour is similar to a classical

W.K. Ao (✉) • P. Reynolds
Vibration Engineering Section, College of Engineering, Mathematics and Physical Sciences, University of Exeter, North Park Road, Exeter EX4 4QF, UK
e-mail: wka203@exeter.ac.uk; p.reynolds@exeter.ac.uk

A. Wicks (ed.), *Shock & Vibration, Aircraft/Aerospace, and Energy Harvesting, Volume 9*,
Conference Proceedings of the Society for Experimental Mechanics Series, DOI 10.1007/978-3-319-15233-2_21

tuned mass damper (TMD). Some studies [4] quoted classical TMD H_∞ optimisation method [11] and reproved the optimal design scheme with EMDS. It is similar to conventional TMD and in this work the H_∞ optimisation method will be proved for connecting RLC resonant circuit (RLC circuit is a simple electronic circuit oscillator).

The concept of EMDS in this paper uses shunt circuit component-capacitance to provide virtual mass or equivalent mass and tunes the damping property by selection of circuit components. The optimal design parameters of H_∞ will be defined corresponding to equivalent mass ratio, frequency ratio and damping ratio. Former studies in this area implemented the H_∞ optimisation method rather than H_2 optimisation method. Hence, in this work the H_2 optimisation method with RLC resonant circuit will also be examined. The optimal parameters of H_2 design scheme minimising the system energy will be presented. The theoretical dynamic response will be presented in this paper to demonstrate the similarity or difference compared with a classical TMD.

This paper is structured as follows: in the first part is introduced the novel EMDS concept in the context of mitigation of civil engineering vibrations, which is the motivation of this study. The second section reviews the TMD H_∞ and H_2 optimisation methods for obtaining optimal tuning parameters. In the third section is presented the H_∞ and H_2 optimisation methods for optimal design of parameters for EMDS utilising an RLC resonant circuit. Finally, the last section shows potential application to a slab strip structure and derivation of EMD properties for vibration mitigation. Also show are some simulations and illustrations of EMDS using RLC resonant circuit.

21.2 Tuned Mass Damper (TMD)

A tuned mass damper (TMD) is sometimes called a dynamic or vibration absorber and is constructed using a mass, spring and viscous damper. A typical TMD and single degree of freedom system is shown in Fig. 21.1. Adding the TMD to the primary structure forms a two degree of freedom system and hence the equation of motion can be written as:

$$\begin{bmatrix} m_1 & 0 \\ 0 & m_2 \end{bmatrix} \begin{Bmatrix} \ddot{x}_1 \\ \ddot{x}_2 \end{Bmatrix} + \begin{bmatrix} c_1 + c_2 & -c_2 \\ -c_2 & c_2 \end{bmatrix} \begin{Bmatrix} \dot{x}_1 \\ \dot{x}_2 \end{Bmatrix} + \begin{bmatrix} k_1 + k_2 & -k_2 \\ -k_2 & k_2 \end{bmatrix} \begin{Bmatrix} x_1 \\ x_2 \end{Bmatrix} = \begin{Bmatrix} f \\ 0 \end{Bmatrix} \tag{21.1}$$

In Fig. 21.1 the single degree of freedom primary system has properties m_1, k_1 and c_1 representing its mass, stiffness (elastic coefficient) and damping coefficient. $f(t)$ is an external force acting on the primary structure. m_2, k_2 and c_2 represent the mass, stiffness and damping coefficient of the TMD. $x_1(t)$ and $x_2(t)$ represent the displacement vector of m_1 and m_2. Taking the Laplace transform of Eq. (21.1), the relationship between the displacement output $X_1(s)$ and input excitation $F(s)$ is given by the following transfer function:

$$\frac{X_1(s)}{F(s)} = \frac{s^2 + 2\xi_2\omega_2 s + {\omega_2}^2}{m_1 s^4 + (2\xi_2\omega_2 + m_2 + c_1 + c_2)\, s^2 + ({\omega_2}^2 m_1 + 2\xi_2\omega_2 c_1 + k_1 + k_2)\, s^2 + ({\omega_2}^2 c_1 + 2\xi_2\omega_2 k_1)\, s + {\omega_2}^2 k_1} \tag{21.2}$$

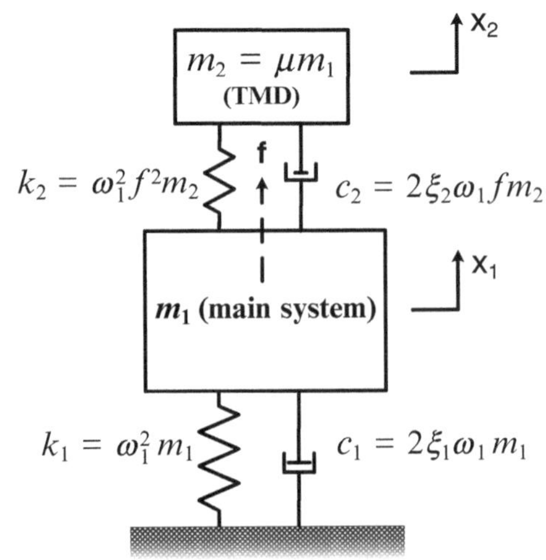

Fig. 21.1 TMD with single degree of freedom system under one external excitation

where $\omega_2 = \sqrt{\frac{k_2}{m_2}}$ is the natural frequency and $\xi_2 = \frac{c_2}{2\sqrt{k_2 m_2}}$ is the damping ratio of the TMD. The damping coefficient c_1 is often assumed to be zero. In his theoretical development, the den Hartog method [11] assumed zero damping for the primary system (i.e. $c_1 = 0$) and developed the classical expressions for optimal tuning of frequency and damping ratio.

For observing the displacement FRF of the system equation (21.2) can be simplified and rewritten as follows:

$$\begin{aligned} H(\lambda) &= \frac{X_1(\lambda)}{F(\lambda)/k_1} \\ &= \frac{\gamma^2 - \lambda^2 + j2\xi_2\gamma\lambda}{\lambda^4 + \gamma^2 - \gamma^2\lambda^2 - \lambda^2 - \mu\gamma^2\lambda^2 + j\left(-2\xi_2\gamma\lambda^3 - 2\xi_2\mu\gamma\lambda^3 + 2\xi_2\gamma\lambda\right)} \end{aligned} \tag{21.3}$$

where $\mu = \frac{m_2}{m_1}$ is the ratio between the mass of the TMD and the primary system mass. $\gamma = \frac{\omega_2}{\omega_1}$ is the frequency ratio, which is the ratio between TMD frequency to the structural natural frequency. $\lambda = \frac{\omega}{\omega_1}$ is the ratio of the excitation frequency to structural natural frequency. It is noted that the above FRF equation is dimensionless.

21.2.1 TMD $H\infty$ Optimisation

Den Hartog [11] used an optimisation procedure which disregarded the primary structure damping ($c_1 = 0$), to determine the optimal tuning of TMD parameters such as the optimal frequency ratio γ_{opt} and optimal damping ratio $\xi_{2,opt}$ as follows:

$$\gamma_{opt} = \frac{1}{1+\mu} \tag{21.4}$$

$$\xi_{2,opt} = \sqrt{\frac{3\mu}{8(1+\mu)}} \tag{21.5}$$

The optimal frequency ratio and damping ratio are functions of the mass ratio. It means that controlling of dynamic response is manipulated primarily by the mass ratio, which has a tuning implication.

21.2.2 TMD H_2 Optimisation

Another optimisation method is H_2 optimisation, which aims to minimise the total vibration energy or the mean square motion of a SDOF system under random force excitations. The FRF function of a system including a TMD was already defined in Eq. (21.3). For the following discussion, the norm of the FRF function can be written as follows:

$$\begin{aligned} |G(\lambda)| = |H(\lambda)| &= \left|\frac{X_1(\lambda)}{F(\lambda)/k_1}\right| \\ &= \left|\frac{\gamma^2 - \lambda^2 + j2\xi_2\gamma\lambda}{(1-\lambda^2)(\gamma^2-\lambda^2) - \mu\gamma^2\lambda^2 + j2\xi_2\gamma\lambda\left(1-\lambda^2-\mu\lambda^2\right)}\right| \end{aligned} \tag{21.6}$$

The definitions of all parameters are the same as in the previous section. The objective function of H_2 optimisation of vibration absorber can be achieved from minimisation of the integral of the square of the norm of the FRF function, which can be expressed as:

$$E\left[|G(\lambda)|^2\right] = \int_{-\infty}^{\infty} |G(\lambda)|^2 S_0 d\omega \tag{21.7}$$

where S_0 is the uniform power spectral density function. The exact solution of the H_2 optimal design parameter can be derived as:

$$\gamma_{opt} = \sqrt{\frac{\mu + 2}{2(\mu + 1)^2}} \tag{21.8}$$

$$\xi_{opt} = \frac{1}{2}\sqrt{\frac{\mu\,(3\mu + 4)}{2\,(\mu + 2)\,(\mu + 1)}} \tag{21.9}$$

It can also be observed that the optimal frequency and damping ratios of the TMD are functions of mass ratio as well as dimensionless leading to the global minimum mean square motion of the primary mass.

21.3 Electromagnetic Shunt Damper (EMDS)

A typical concept of an electromagnetic damping device is shown in Fig. 21.2a. A permanent magnet bar and conductive coil or material are contained within a cover shell. The conductive material is normally bonded with the cover shell. A permanent magnet bar moves inside within the conductive material. In this case the direction of motion is vertical and the purpose is to suppress vertical vibration. The electromagnetic damper can be expressed as in the circuit diagram shown in Fig. 21.2b. It is shown that one resistor, one inductor and a voltage source are required for this device.

21.3.1 H_∞ Optimisation Design

Figure 21.3a shows a single degree of freedom system model connected with an electromagnetic device in open circuit. For driving the device, the circuit should be closed. From an energy point of view of the damper, it converts mechanical energy from the primary structure into electrical energy. In this concept there is no need to apply any external energy source to drive the device. The structure vibration (mechanical energy) generates an electromotive force (emf) in a passive way. The emf is proportional to velocity of the motion, with the relationship between the two give by the machine constant K_{emV}. After multiplication of K_{emV} and velocity, mechanical energy is transferred to the device which produces the input voltage source (electrical energy) to the shunt circuit. After that the induced current (eddy current) will flow through the shunt circuit.

An EMD can be connected to an RLC resonant circuit consisting of resistors, inductors and capacitors, which is a basic oscillating circuit as seen in Fig. 21.3b. For different control objectives, it can be connected to more than one RLC circuit with frequencies close to the different vibration frequencies or structural resonant frequencies of the primary structure. For this study, only a single mode is investigated for vibration suppression. Therefore, one RLC circuit should be used.

When the vibration frequency is equal to natural frequency of the RLC circuit, the reactance of the shunt circuit will be equal to zero. Hence, it is implied that the circuit current tends to be maximum. It is known that the damper force is proportional to the current; therefore, when current is maximum, the damping force also evaluates to have maximum value. If the RLC oscillating frequency is closed to the fundamental frequency of primary structure, the electromagnetic damper

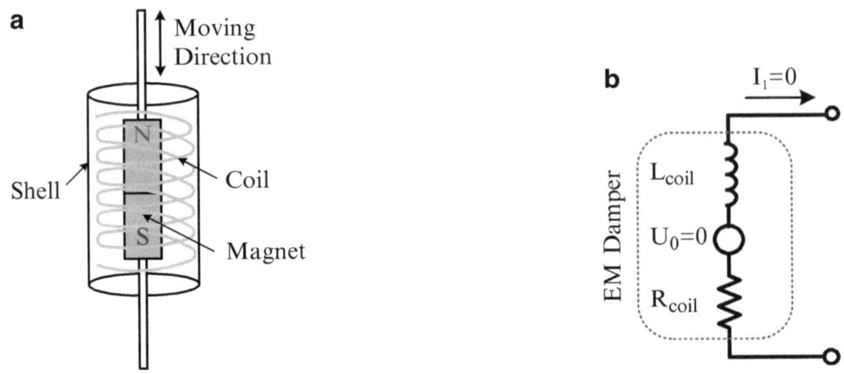

Fig. 21.2 Electromagnetic damper device. (**a**) Conceptual electromagnetic damper. (**b**) Circuit expression of electromagnetic damper

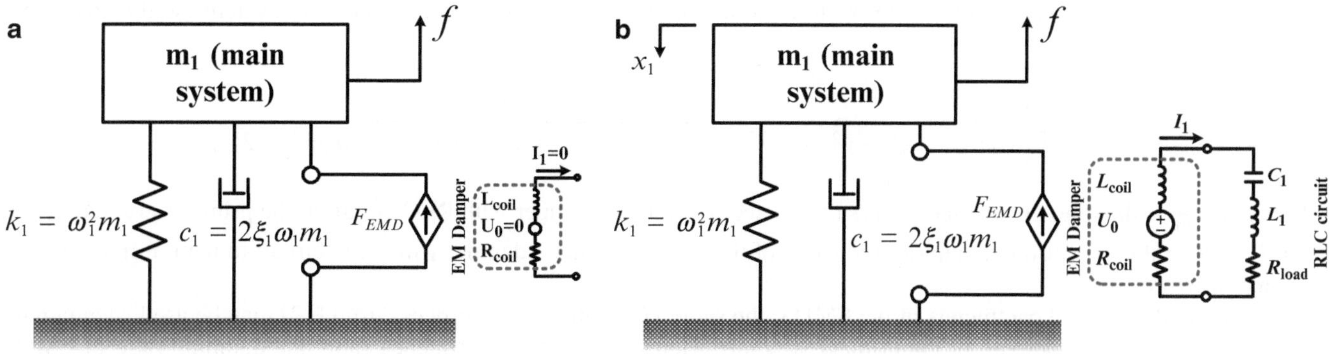

Fig. 21.3 Electromagnetic damper with shunt circuit. (**a**) SDOF structure with open circuit electromagnetic damper. (**b**) SDOF structure with electromagnetic device and RLC shunt circuit

will provide a resonant-type damping effect, which is able to suppress the structure single vibrational mode effectively. The dynamic equation of motion to this system with RLC shunt circuit as follows:

$$\begin{bmatrix} m_1 & 0 \\ 0 & L_{coil}+L_1 \end{bmatrix}\begin{Bmatrix} \ddot{x}_1 \\ \ddot{q} \end{Bmatrix} + \begin{bmatrix} c_1 & K_{emN} \\ -K_{emV} & R_{coil}+R_1 \end{bmatrix}\begin{Bmatrix} \dot{x}_1 \\ \dot{q} \end{Bmatrix} + \begin{bmatrix} k_1 & 0 \\ 0 & \frac{1}{C_1} \end{bmatrix}\begin{Bmatrix} x_1 \\ q \end{Bmatrix} = \begin{Bmatrix} f \\ 0 \end{Bmatrix} \tag{21.10}$$

The above equation can be rewritten in a more concise form as follows:

$$\begin{bmatrix} m_1 & 0 \\ 0 & L \end{bmatrix}\begin{Bmatrix} \ddot{x}_1 \\ \ddot{q} \end{Bmatrix} + \begin{bmatrix} c_1 & K_{emN} \\ -K_{emV} & R \end{bmatrix}\begin{Bmatrix} \dot{x}_1 \\ \dot{q} \end{Bmatrix} + \begin{bmatrix} k_1 & 0 \\ 0 & \frac{1}{C} \end{bmatrix}\begin{Bmatrix} x_1 \\ q \end{Bmatrix} = \begin{Bmatrix} f \\ 0 \end{Bmatrix} \tag{21.11}$$

where R, L and C are the total resistance, inductance and capacitance of EMDS. Taking the Laplace transform of Eq. (21.11) with zero initial conditions:

$$\frac{X_1(s)}{F(s)} = \frac{s^2L + sR + \frac{1}{C}}{s^4 m_1 L + (m_1R + c_1L)s^3 + \left(\frac{m_1}{C} + c_1R + k_1L + K_{emN}K_{emV}\right)s^2 + \left(\frac{c_1}{C} + Rk_1\right)s + \frac{k_1}{C}} \tag{21.12}$$

The RLC circuit is a simple oscillating circuit. Hence, the RLC circuit has its own natural frequency and damping ratio or damping factor. Using this concept it is possible to rewrite Eq. (21.12) as follows:

$$\frac{X_1(s)}{F(s)} = \frac{s^2 + 2\bar{\xi}_2\bar{\omega}_2 s + \bar{\omega}_2^2}{m_1s^4 + \left(m_1\frac{R}{L} + c_1\right)s^3 + \left(m_1\bar{\omega}_2^2 + c_1\frac{R}{L} + k_1 + \frac{K_{emN}K_{emV}}{L}\right)s^2 + \left(c_1\bar{\omega}_2^2 + k_1\frac{R}{L}\right)s + k_1\bar{\omega}_2^2} \tag{21.13}$$

where $\bar{\omega}_2$ is the natural frequency of the resonant circuit (RLC) and $\bar{\xi}_2$ is the damping ratio of the resonant circuit (RLC), which can be defined as:

$$\bar{\omega}_2 = \sqrt{\frac{1}{LC}} \quad \bar{\xi}_2 = \frac{R}{2}\sqrt{\frac{C}{L}} \tag{21.14}$$

Comparing Eqs. (21.2) and (21.13), it is not difficult to observe that these two equations have a similar form. Examining in particular the denominator, an equivalent damper stiffness may be defined as:

$$\bar{k}_2 = \frac{K_{emN}K_{emV}}{L} \tag{21.15}$$

Also, equivalent mass and damping coefficients of the electromagnetic damper can be expressed as:

$$\bar{m}_2 = K_{emN}K_{emV}C \tag{21.16}$$

$$\bar{c}_2 = \frac{K_{emN}K_{emV}RC}{L} \tag{21.17}$$

Similar to a conventional TMD system, with the assumption of zero damping of the primary system the dimensionless FRF form can be shown as follows:

$$\bar{H}(\lambda) = \frac{X_1(\lambda)}{F(\lambda)/k_1} = \frac{\bar{\gamma}^2 - \lambda^2 + j2\bar{\xi}_2\bar{\gamma}\lambda}{\lambda^4 + \bar{\gamma}^2 - \bar{\gamma}^2\lambda^2 - \lambda^2 - \bar{\mu}\bar{\gamma}^2\lambda^2 + j\left(-2\bar{\xi}_2\bar{\gamma}\lambda^3 - 2\bar{\xi}_2\bar{\mu}\bar{\gamma}\lambda^3 + 2\bar{\xi}_2\bar{\gamma}\lambda\right)} \tag{21.18}$$

where $\bar{\mu} = \frac{\bar{m}_2}{m_1}$ is the ratio of equivalent mass of the electromagnetic damper with RLC circuit to the primary structure mass. $\bar{\gamma} = \frac{\bar{\omega}_2}{\omega_1}$ is the ratio of equivalent natural frequency of the electromagnetic damper with RLC circuit to natural frequency of primary structure.

For finding the optimal parameters of the EMDS, the same H_∞ optimisation process from [12] may be used, again under the assumption of zero damping of the primary structure. Hence, the optimal frequency ratio and damping ratio of EMDS (RLC circuit) can be expressed as:

$$\bar{\gamma}_{opt} = \sqrt{\frac{2}{2+\bar{\mu}}} \quad \bar{\xi}_{2,opt} = \sqrt{\frac{3\bar{\mu}}{8}} \tag{21.19}$$

The equivalent frequency ratio and damping ratio are also functions of the equivalent mass ratio.

21.3.2 H_2 Optimisation Design

Cheung and Wong [13] proposed a new H_2 optimal design model of a dynamic vibration absorber which was derived for minimising the mean square motion of a SDOF system under random excitation. The results showed that this newly proposed concept of absorber can provide a larger suppression of the primary structure. This concept can also be applied to optimal design of an EMDS.

The FRF function of an EMDS system was shown in Eq. (21.18). Hence, the amplitude of EMDS FRF function is expressed as:

$$\begin{aligned}|G(\lambda)| = |H(\lambda)| &= \left|\frac{X_1(\lambda)}{F(\lambda)/_{k_1}}\right| \\ &= \left|\frac{\bar{\gamma}^2 - \lambda^2 + j2\bar{\xi}_2\bar{\gamma}\lambda}{(1-\lambda^2)(\bar{\gamma}^2-\lambda^2) - \bar{\mu}\bar{\gamma}^2\lambda^2 + j2\bar{\xi}_2\bar{\gamma}\lambda(1-\lambda^2)}\right|\end{aligned} \tag{21.20}$$

The objective function of H_2 optimisation is a minimisation of the integral of the square of the amplitude of the FRF function of the primary structure. It can be illustrated as:

$$\begin{aligned}E\left[|G(\lambda)|^2\right] &= \omega_n \int_{-\infty}^{\infty} |G(\lambda)|^2 S_0 d\lambda \\ &= \omega_n \int_{-\infty}^{\infty} \left|\frac{\bar{\gamma}^2 - \lambda^2 + j2\bar{\xi}_2\bar{\gamma}\lambda}{(1-\lambda^2)(\bar{\gamma}^2-\lambda^2) - \bar{\mu}\bar{\gamma}^2\lambda^2 + j2\bar{\xi}_2\bar{\gamma}\lambda(1-\lambda^2)}\right|^2 S_0 d\lambda\end{aligned} \tag{21.21}$$

The integration result as follows from Grahshteyn and Ryzhik [14] is used to find the exact solutions for the optimal damping ratio and frequency ratio:

$$\bar{\xi}_{2,opt} = \frac{1}{2}\sqrt{\bar{\gamma}^2 + \bar{\mu}\bar{\gamma}^2 - 2 + \frac{1}{\bar{\gamma}^2}} \tag{21.22}$$

$$\bar{\gamma}_{opt} = 1 \tag{21.23}$$

The optimal parameters of EMDS under H_2 optimisation are functions of the equivalent mass ratio.

21.4 Theoretical Comparison of the TMD and EMDS

In order to carry out an analysis of the potential performance of comparative TMD and EMDS devices, the properties of a laboratory slab strip structure were used as summarised in Table 21.1. The structure is an in-situ cast post-tensioned slab strip that is simply supported between knife-edge supports and is shown in Fig. 21.4. The slab strip has span 10.8 m, width 2 m and depth 0.275 m. The three measured mode shapes corresponding with the dynamic properties in Table 21.1 are shown in Fig. 21.5.

Sections 21.2 and 21.3 introduced the mathematical models for the conventional TMD and the electromagnetic damper with shunt RLC circuit (EMDS). From classical control theory, Fig. 21.6a shows the control law involving the TMD.

Figure 21.7 shows an electromagnetic actuator from Parker (Electric Tubular Motor—ETT) that can be configured to function as an electromagnetic damper. The technical characteristics of this motor are shown in Fig. 21.7. The two most important parameters are the force constant K_{emN} and the emf constant K_{emV}, which represent the relationships between current and force and between velocity and voltage. Section 21.2 mentioned that the electromagnetic damper can be used as a circuit component; hence the relative inherent resistance and inductance of the EMD are 44 Ω and 38 mH, respectively.

The passive EMD is operated by closing the shunt circuit. From the controlling system perspective as seen in Fig. 21.8a, the electromagnetic damper and control circuit are in series and put it in the feedback loop together. The control circuit can stand for different circuits. In this study, an RLC circuit is used. From the mathematical model presented in Sect. 21.3 it can be seen that the EMD with RLC circuit can be expressed as an equivalent TMD (virtual TMD). To verify that the predicted relative dynamic behaviour is similar, FRF curves may be calculated to show the separation of the primary structural mode into two lower magnitude peaks. For this study, the mass ratio of the TMD and the equivalent mass of the EMDS are selected to be the same (1 %).

In Sect. 21.3 it was shown that the electromagnetic damper has two key machine constants that determine its performance, K_{emN} and K_{emV}. Using these two parameters and the H_∞ optimisation design equations, a derived acceleration frequency response function (FRF) is shown in Figs. 21.8a (magnitude) and 21.8b (phase). From these figures can be seen that red

Table 21.1 Measured dynamic properties of the slap strip structure

Mode number	Natural frequency (Hz)	Damping ratio (%)	Modal mass (kg)
1	4.6	2.4	6610
2	16.9	0.51	7830
3	37.6	1.09	5100

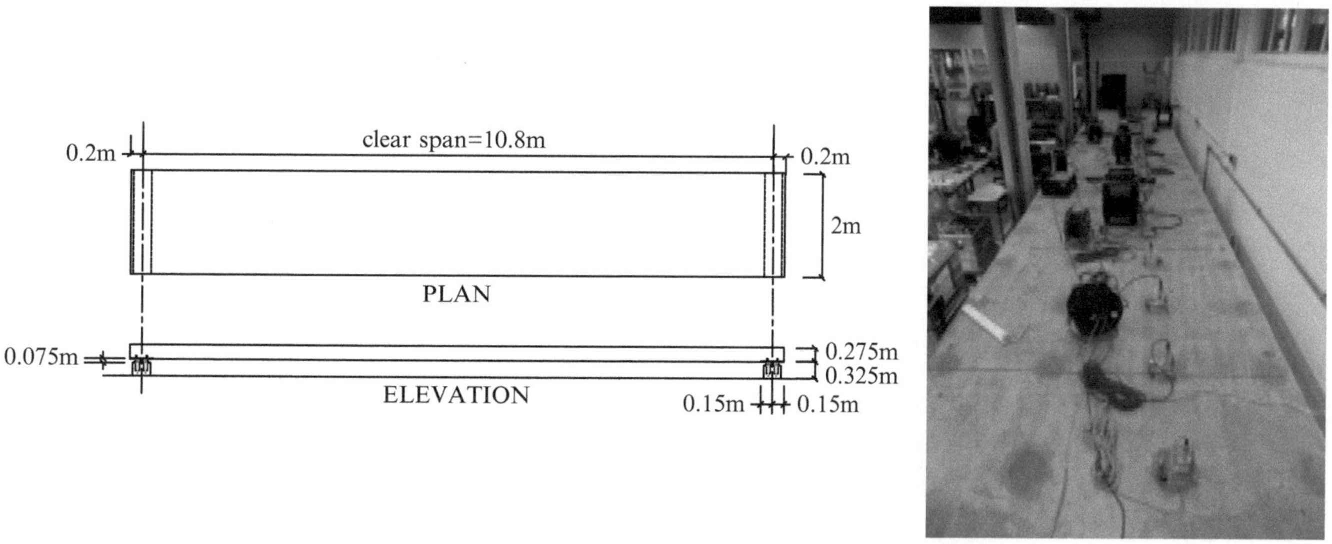

Fig. 21.4 Layout of the slap strip structure

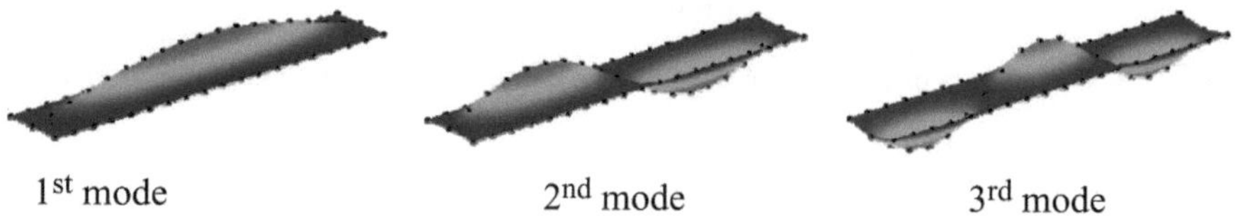

Fig. 21.5 Measured mode shapes of the slab strip structure

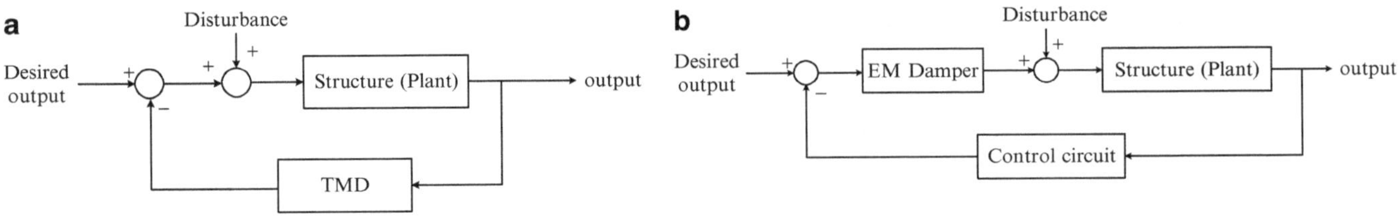

Fig. 21.6 Different control laws with TMD and EMDS. (**a**) Control law with passive TMD. (**b**) Control law with electromagnetic damper and control circuit

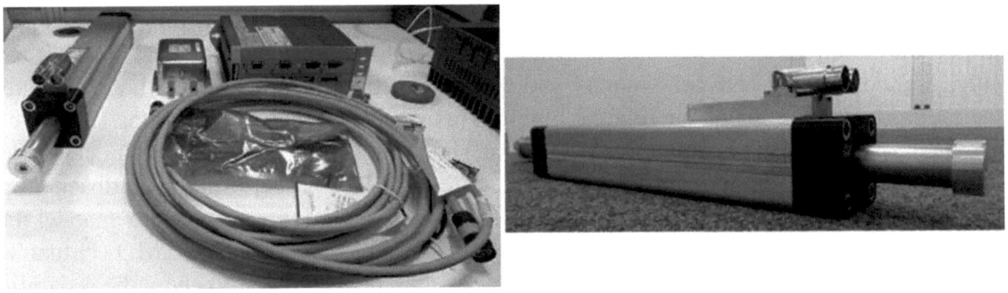

Fig. 21.7 Electromagnetic actuator

Table 21.2 Technical characteristics of electromagnetic damper (adopted from catalogue of Parker ETT—Electric Tubular Motor)

ETT060	Unit	ETT060S3
Effective stroke	mm	30 . . . 540
Rated force	N	128
Peak force for 10 s	N	512
Maximum speed	m/s	4
Peak acceleration	m/s^2	200
Actuator length	mm	386
Slider length w/o stop	mm	254 . . . 944
Slider weight	kg	0.56 . . . 2.12
Slider diameter	mm	25
Pole diameter	mm	60
Force constant	N/A	206.45
Back EMF	V/(m/s)	25.2
Phase resistance	ohm	44
Phase inductance	mH	38
Position repeatability	mm	± 0.05

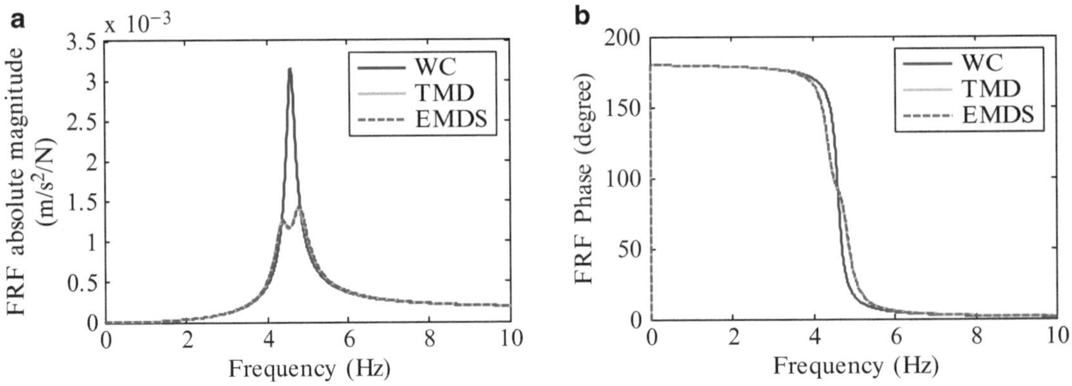

Fig. 21.8 Comparison of acceleration frequency response under H_∞ with machine constant $K_{emN} \neq K_{emV}$. (**a**) Acceleration response. (**b**) Phase angle

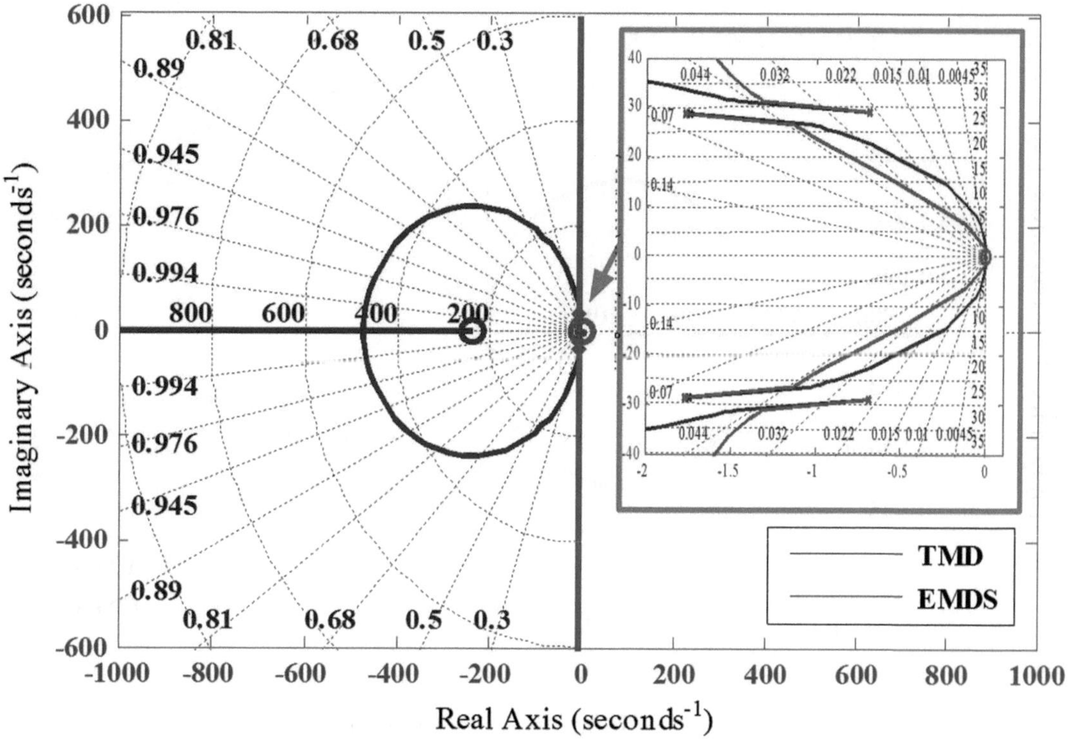

Fig. 21.9 Root locus comparison between TMD and EMDS (H_∞)

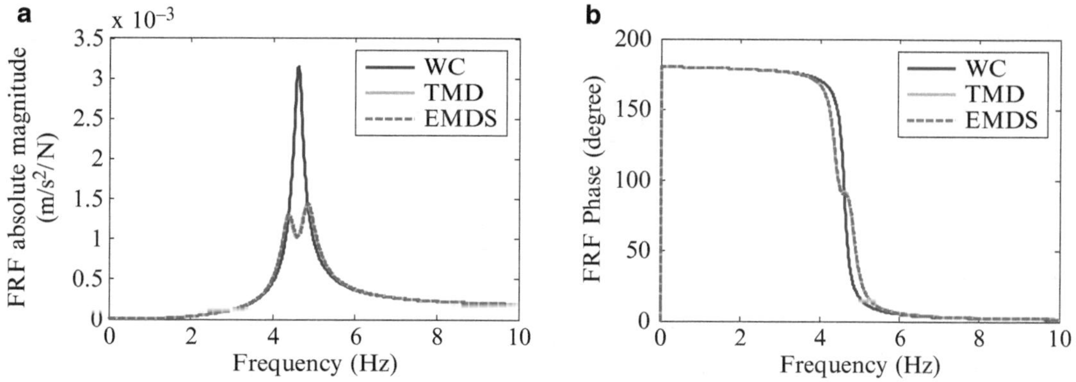

Fig. 21.10 Comparison of acceleration frequency response under H_2 with machine constant KemN $\neq$ KemV. (**a**) Acceleration response. (**b**) Phase angle

dashed line of the EMDS FRF curve is the same as the equivalent TMD under H_∞ optimisation. It clearly has the typical characteristic of separation of the primary structural mode into two lower magnitude peaks. From this point of view, the equivalent TMD (or EMDS) concept can be verified. In theory, the mass ratio could be increased to more than 1 %, but in this case the equivalent mass ratio was limited by the electromagnetic motor peak force.

Figure 21.9 shows the root locus comparison between the TMD and EMDS. Both have four poles which can be determined from Eq. (21.2) in which two complex conjugate poles are close to imaginary axis and another two are far from the imaginary axis. The different shapes of these root locus plots are caused by the equivalent stiffness term ($\frac{K_{emN} K_{emV}}{L}$) that is affected by force constant(K_{emN}) and emf constant (K_{emV}).

By contrast, the dynamic responses corresponding with H_2 optimisation using TMD and EMDS are shown in Fig. 21.10. It is noted that the mass ratio of this design was again chosen as 1 %, which is same as for the H_∞ design. Again it can be seen that the equivalent TMD (EMDS) response is almost identical to that of the TMD for these H_2 optimisation results.

Figure 21.11 shows the root locus plots corresponding with the TMD and EMDS. Again it can be seen that the root locus shapes are different causing by the equivalent stiffness term of transfer function.

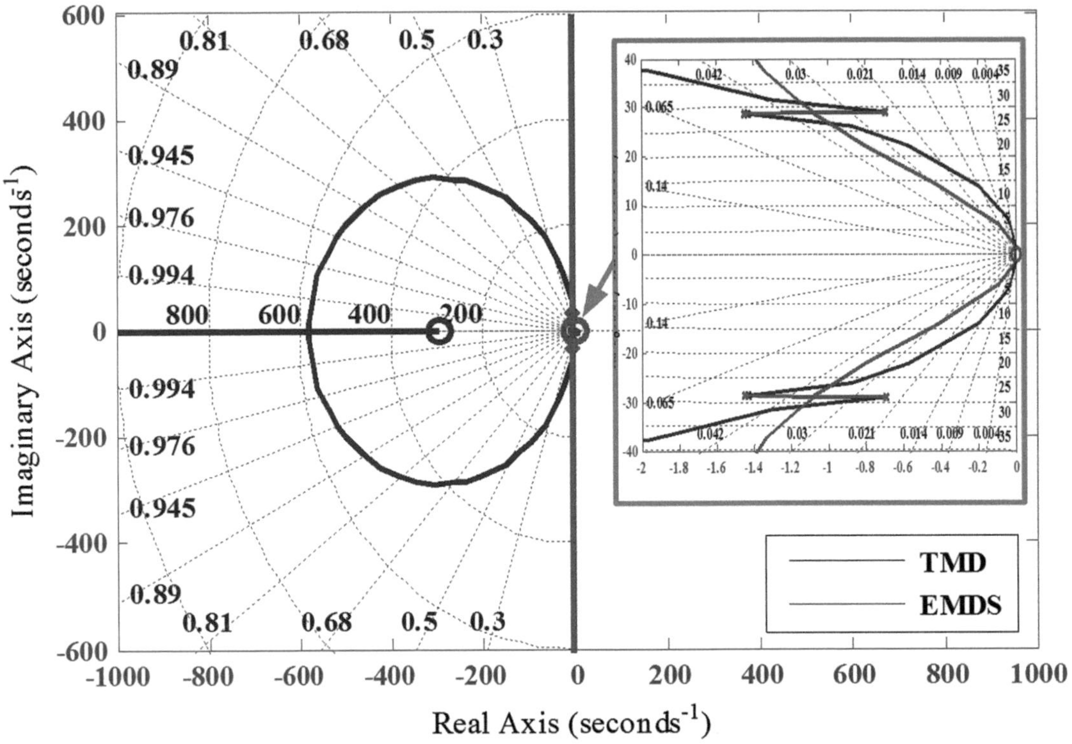

Fig. 21.11 Comparison root locus between TMD and EMDS (H_2)

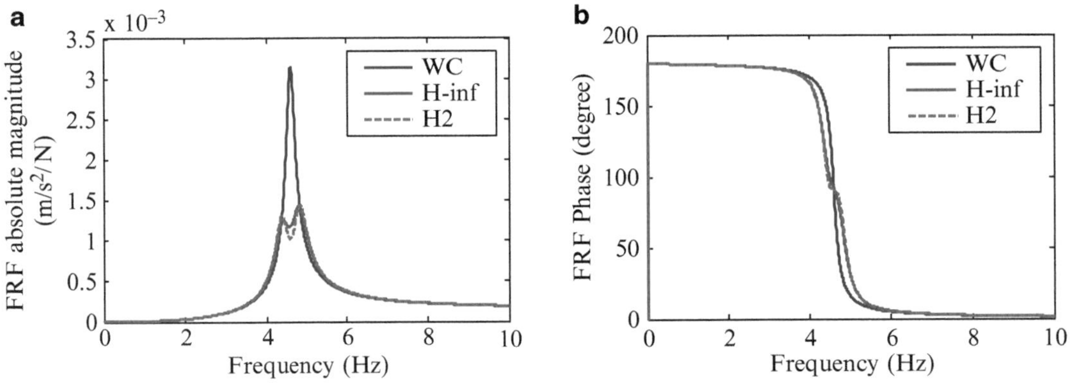

Fig. 21.12 Comparison of acceleration frequency response with H_∞ and H_2. (**a**) Acceleration response. (**b**) Phase angle

To further the comparison, Figs. 21.12 and 21.13 shows the FRF and root locus plots from both the H_∞ and H_2 methods. It can be seen that H_2 FRF has lightly lower peak than H_∞.

21.5 Discussion

The objective of this study has been to examine the use of an electromagnetic actuator or motor for vibration control of a civil engineering structure, by utilising it as an electromagnetic damper with a shunt circuit (EMDS). The H_∞ and H_2 optimisation methods using RLC resonant circuit (or EMDS) have been re-proved and amended and utilised in an analytical implementation study for comparison against an equivalent TMD. Since the inherent resistance of the actuator is higher than desired, a negative resistance circuit was used to scale down the total resistance.

From the frequency response, the peak magnitude of FRF curve of the bare structure can be alleviated by adding RLC resonant circuit. The cascaded RLC resonant circuit can have an equivalent TMD dynamic behaviour. The FRFs showed that

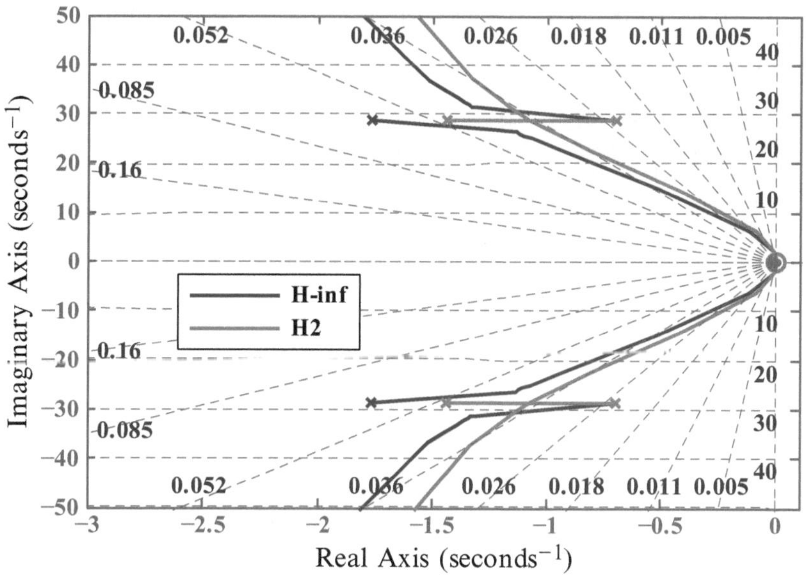

Fig. 21.13 Comparison root locus between H_∞ and H_2 for EMDS

the RLC shunt circuit produce the characteristic two lower peaks of the FRF of the whole system using either of the standard optimisation methods (H_∞ and H_2). It can also be seen that apart from the circuit components, the controlling effectiveness can also be affected by the two machine constants K_{emN} and K_{emV}.

The equivalent TMD does not involve any moving mass and theoretically can achieve a substantial virtual mass from circuit components (capacitance). It is known that larger TMD mass ratios results in better controlling performance and, in a comparative sense, increasing the equivalent mass ratio of an EMDS can also have a better controlling result. In a real application case, the equivalent mass ratio is dependent on the capacity of the force actuator.

Optimal design laws were presented in this paper. Optimal design can help a designer to find out the required parameters of optimal circuit components. From the analytical study it was shown that the optimal resistance and inductance are functions of the capacitance.

In this study a SDOF system was selected and vibration control of the single mode was carried out, which represented the first bending mode of the actual system. The approach pursued here can be extended to control of multiple modes by using multiple RLC resonant circuits with the system.

It is planned by the authors to carry out a practical implementation of the system, which will be presented in later publications.

Acknowledgements The authors would like to acknowledge the financial assistance provided by the UK Engineering and Physical Sciences Research Council (EPSRC) through a Leadership Fellowship Grant (Ref. EP/J004081/2) entitled "Advanced Technologies for Mitigation of Human-Induced Vibration".

References

1. Mizuno T, Toumiya T, Takasaki M (2002) Vibration isolation system using negative stiffness. JSME Int J 46:807–812
2. Fleming AJ, Moheimani SR, Member S (2006) Inertial vibration control using a shunted electromagnetic transducer. IEEE/ASME Trans Mechatron 11:84–92
3. Behrens S, Fleming A, Reza Moheimani S (2003) Electromagnetic shunt damping. In: Proceedings 2003 IEEE/ASME International Conference on Advanced Intelligent Mechatronics (AIM 2003), vol. 2, pp. 1145–1150
4. Marneffe BD (2007) Active and passive vibration isolation and damping via shunted transducers. Universite Libre de Bruxelles, Brussels
5. Fleming AJ, Hons BEE (2007) Synthesis and implementation of sensor-less shunt controllers for piezoelectric and electromagnetic vibration control. The University of Newcastle, Callaghan
6. Niu H, Zhang X, Xie S, Wang P (2009) A new electromagnetic shunt damping treatment and vibration control of beam structures. Smart Mater Struct 18:045009
7. Behrens S, Fleming AJ, Moheimani SR (2005) Passive vibration control via electromagnetic shunt damping. IEEE/ASME Trans Mechatron. 10:118–122
8. Inoue T, Ishida Y, Sumi M (2008) Vibration suppression using electromagnetic resonant shunt damper. J Vib Acoust 130:041003
9. McDaid AJ, Mace BR (2013) A self-tuning electromagnetic vibration absorber with adaptive shunt electronics. Smart Mater Struct 22:105013

10. Cheng TH, Oh IK (2009) Vibration suppression of flexible beam. IEEE Trans Magn 45:2758–2761
11. den Hartog J (1985) Mechanical vibrations. Dover, Mineola
12. Wong W, Cheung Y (2008) Optimal design of a damped dynamic vibration absorber for vibration control of structure excited by ground motion. Eng Struct 30:282–286
13. Cheung Y, Wong W (2011) H2 optimization of a non-traditional dynamic vibration absorber for vibration control of structures under random force excitation. J Sound Vib 330:1039–1044
14. Gradshteyn IS, Ryzhik IM (2007) Table of integrals, series, and products, 7th edn. The United States of America: Elsevier Academic Press, Boston

Chapter 22
Orbit Stability Determination of Satellites Using Harmonic Force Excitation Analysis

Joshua Johnson, William H. Semke, Matthew Zimmer, and Ronald Fevig

Abstract The focus of this paper is the determination of orbit stability of a small satellite around an asteroid with a complex gravitational field using harmonic force excitation analysis. The determination of stable orbits is critical in space mission planning, especially in deep space missions investigating asteroids having undetermined mass distribution. Many of these asteroids have complex gravitational fields that make close orbits, which are necessary for inspection, very difficult to predict and safely maintain. Simulations of the asteroid Itokawa, which was visited and analyzed by the Hayabusa Space Mission, have shown such complex gravitational fields. Orbit simulations using Systems Tool Kit (STK) software have demonstrated many interesting phenomena resulting from the nonlinear satellite/asteroid interaction. One behavior of special note was the influence of the rate of spin of the asteroid and the stability of the orbit. This behavior was found to be linked to the frequency and magnitude of the gravitational excitation force along with altitude of the orbiting satellite. A harmonic force excitation analysis of the system was shown to be an accurate predictor of orbital stability. The resulting frequency ratio determined is shown to predict regions where complex gravity effects are significant.

Keywords Harmonic analysis • Satellite control • Orbital stability • Itokawa • Space mission design

22.1 Introduction

The research presented in this paper is the result of a collaborative research project at the University of North Dakota between the Department of Mechanical Engineering and the Department of Space Studies. The main body of the project seeks to create an autonomous orbital control system for a spacecraft in close-proximity to a small Near-Earth Object (NEO). This is achieved by utilizing a MATLAB-STK interface that was developed in order to allow for complex orbital control. The findings presented here are focused on interesting dynamic behavior resulting from harmonic force excitation, discovered while studying the orbital mechanics of a satellite-asteroid system.

Prior work done by Church and Fevig investigated the feasibility of creating a highly detailed gravity map of small body asteroids (500 m and smaller) which allows for determination of the internal structure of the asteroid [1]. The proposed process involved having two satellites in orbit together around the body. Then observations are made, from one satellite to another, of their changes in trajectory. By observing these perturbations, a detailed gravity map can be derived. This method has proven to be successful during the GRACE, and GRAIL missions. It was found that two satellites put into orbit next to each other around a simulated model of asteroid 25143 Itokawa quickly diverge in their trajectories [1]. The satellites have the same initial conditions except that they are separated by 5° in true anomaly, the angle between their initial radius vectors. Figure 22.1 shows that quickly the satellite with the yellow trajectory crashes and the satellite with the red trajectory is ejected from the system, despite the fact that their initial conditions are quite similar (the white trajectory is a circular orbit, displayed for reference).

Shown in Fig. 22.1 are prograde orbits, meaning that the spacecraft's trajectory is pointed in the same direction as the rotation of the asteroid. It was observed that retrograde orbits, orbits where the satellite's trajectory is opposite the rotation of the asteroid, are much more stable (Fig. 22.2). This paper investigates this interesting phenomenon using a harmonic force excitation analysis to determine the orbital stability of a satellite around a NEO.

J. Johnson • W.H. Semke (✉)
Department of Mechanical Engineering, College of Engineering and Mines, University of North Dakota, Grand Forks, ND 58202, USA
e-mail: william.semke@engr.und.edu

M. Zimmer • R. Fevig
Department of Space Studies, John D. Odegard School of Aerospace Science, University of North Dakota, Grand Forks, ND 58202, USA

A. Wicks (ed.), *Shock & Vibration, Aircraft/Aerospace, and Energy Harvesting, Volume 9*,
Conference Proceedings of the Society for Experimental Mechanics Series, DOI 10.1007/978-3-319-15233-2_22

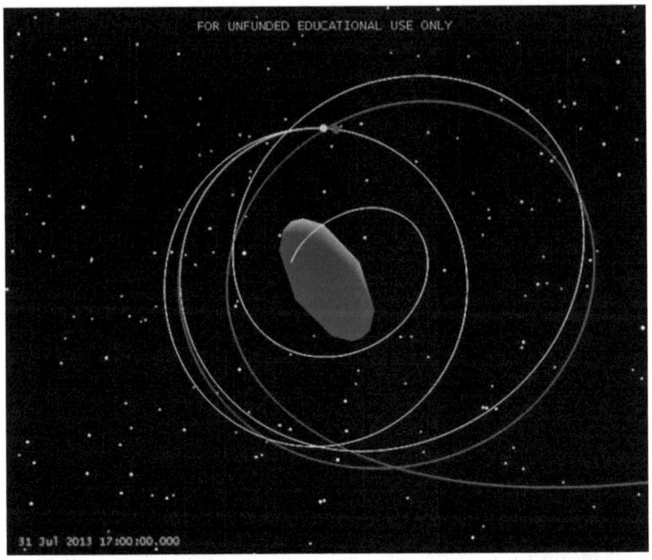

Fig. 22.1 Prograde orbits: unstable trajectories with similar initial conditions that quickly diverge from each other. One escapes the system (*red*), while the other crashes (*yellow*)

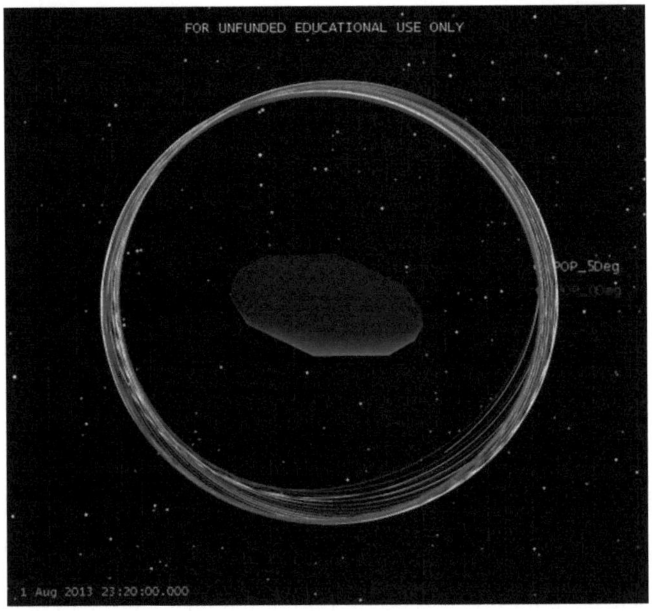

Fig. 22.2 Retrograde orbits: stable past 1 month while staying close to one another

22.2 Background

Near-Earth Objects are asteroids or comets in our Solar System that approach Earth's orbit. More specifically, their perihelion distance is less than 1.3 Astronomical Units (AU) [2]. The perihelion distance is the body's distance to the Sun at its closest point to the Sun it its orbit (Fig. 22.3). One AU is the average distance between the Sun and the Earth, and is used as a standard of measurement ($\approx 1.4960 \times 10^{11}$ m). Therefore, an asteroid or comet is considered an NEO if the closest point in its orbit around the Sun is less than 1.3 times the average distance between the Sun and the Earth. Figure 22.4 shows NEO types recognized by NASA which has a team of scientists dedicated to discovering and tracking NEOs in the Solar System.

There is a great interest in NEOs since (1) they are studied by the planetary science community, and provide clues for topics in this discipline such as the formation of our Solar System, (2) they posses raw materials that can be mined and used to collect in-situ resources for manned and unmanned spaceflight, and (3) they have the potential to collide with Earth, with an immense amount of kinetic energy, causing catastrophic destruction. At the time of this writing, there are over 1,500 Potentially Hazardous Asteroids (PHAs) that are being tracked and monitored by NASA [3]. The term potentially hazardous does not mean that an impact with Earth is imminent, only that the orbit and size of the object are cause for concern. As new observations of these PHAs become available, scientists can better predict their orbit and asses the likelihood of a future close approach to Earth.

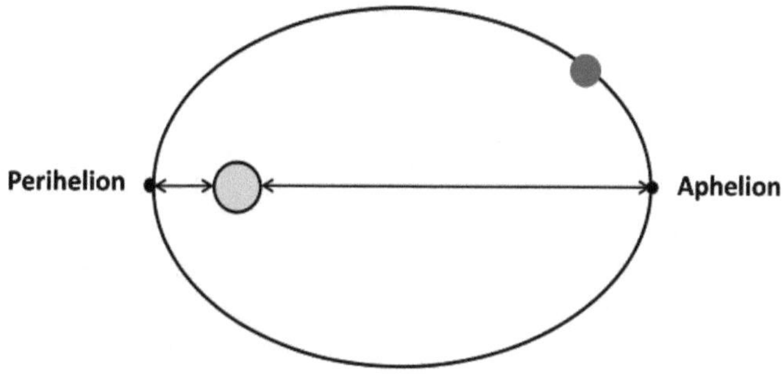

Fig. 22.3 Perihelion and aphelion points for a body orbiting the sun

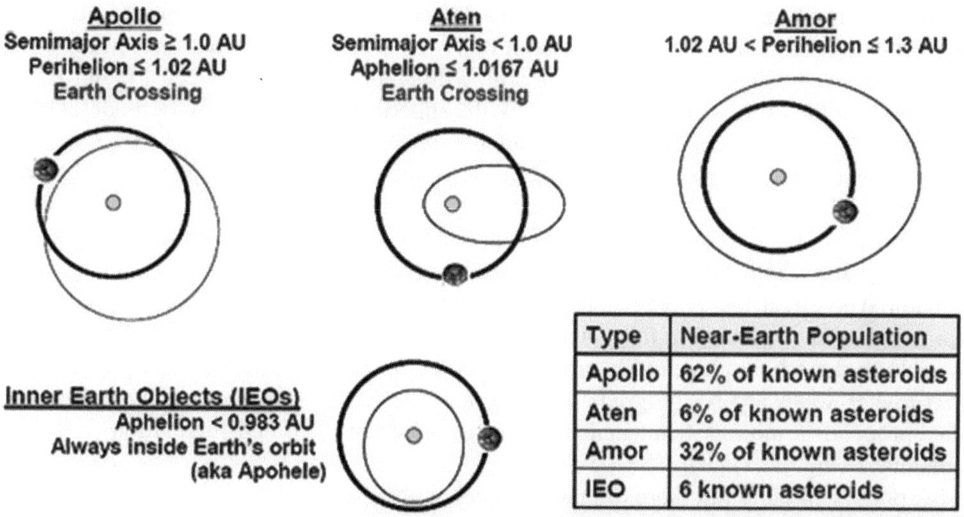

Fig. 22.4 NEO types as categorized by NASA [2]

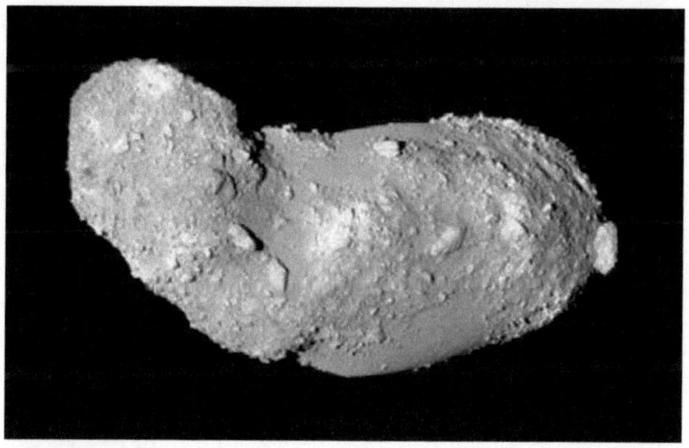

Fig. 22.5 Asteroid 25143 Itokawa as seen from Hayabusa [6]

The control problem in an NEO environment is being studied by researchers all around the world. NASA and the Japanese Aerospace Exploration Agency (JAXA) have sent unmanned spacecraft to operate in close-proximity to these objects [4, 5]. What makes the NEO environment so unique is the gravity field surrounding an NEO is tenuous and it is quite complex.

Gravitational force is a function of the spacecraft and NEO mass elements that decays proportionally to the inverse square of the distance between these element. Therefore, when dealing with small NEOs, the gravitational force exerted on an orbiting spacecraft is much weaker than it is for much larger bodies, like Earth or Mars. Furthermore, the gravitational field is correlated the shape of the body but can have variations based on varying mass distributions throughout. This study focuses on asteroid 25143 Itokawa, which the Hayabusa satellite visited in 2005 (Fig. 22.5). The shape of Itokawa resembles that of a potato, so a satellite in a circular orbit experiences higher gravitational force as it passes by either of the two protruding ends. This increased force can be enough to upset the orbital trajectory such that the satellite escapes the weak gravity field, or that the satellite crashes into the body.

In order to capture the irregularities of the gravity model in three dimensions, a spherical harmonic model of the gravitational potential is used, as shown in Eq. 22.1.

$$U = \frac{\mu}{r}\sum_{n=0}^{\infty} \sum_{m=0}^{n} \left\{ \left\{ \left(\frac{R_{CB}}{r} \right)^n P_{nm} sin\theta \cos(m\lambda) \right\} C_{nm} + \left\{ \left(\frac{R_{CB}}{r} \right)^n P_{nm} sin\theta \cos(m\lambda) \right\} S_{nm} \right\} \tag{22.1}$$

where U is gravitational potential. The coordinates r, θ, and λ are the radial distance, latitude, and longitude of the spacecraft in a coordinate system fixed to the object's center of mass. R_{CB} is the mean radius for the body and μ is the object's gravitational parameter. The functions P_{nm} are the normalized Legendre polynomials, and C_{nm} and S_{nm} are the gravity coefficients of degree n and order m [7].

Itokawa has been studied extensively and a degree and order four model was built based on the data captured by Hayabusa [8]. Studying the natural environment about Itokawa, with no control measures used, it was found that a stable prograde orbit can be achieved using a high-inclination orbit with an orbit radius between 1.0 and 1.5 km [9]. Any closer to the surface of Itokawa and the orbit becomes disrupted by the spherical harmonic gravity model. Outside of this range the orbit becomes disrupted by other effects, such as solar radiation pressure (SRP).

For Itokawa's gravity model, the C_{20}, C_{22}, C_{42}, and C_{44} coefficients are the most significant [9]. These values correspond with the effects of the asteroid's oblateness and its ellipcity (Fig. 22.6).

A body's oblateness and its effect on orbits has been characterized and studied extensively. For example, this perturbation is observed in Earth orbits. Due to the Earth's angular velocity as it spins about its polar axis, there is a bulge around equator. The effect this bulge has on orbits (known as the J_2 effect) causes a precession in the orbital plane, not unlike a spinning top that is about to fall; the orbital plane wobbles as it spins. An inclined orbit about an oblate body has the same wobble as the orbital plane twists around the body. Specifically, the orbit's right ascension of the ascending node (RAAN) (Ω) rotates westward for prograde orbits around the Earth, and the argument of periapsis (ω) rotates in the direction of the spacecraft's motion. Semi-major axis (a), eccentricity (e), and inclination (i) suffer no long-term perturbations from oblateness (Fig. 22.7) [10].

A body's ellipcity has more dramatic effects on the orbit and can cause the spacecraft to transition from a safe orbit into an impacting or ejecting orbit within a few periods. The ellipcity of the body causes changes in the orbit semi-major axis, eccentricity, and inclination while effecting both the orbit's energy $\varepsilon = -\mu/(2a)$ and angular momentum $h = [\mu a(1\text{-}e)]^{1/2}$ [11]. In previous studies, it has been observed that prograde orbits experience much larger changes in energy and angular momentum for each orbit, where retrograde orbits experience little, if any, changes per orbit [12].

This paper studies the difference in stability between prograde and retrograde orbits around a small asteroid with various rotation rates by looking at the excitation frequencies seen in the dynamic system.

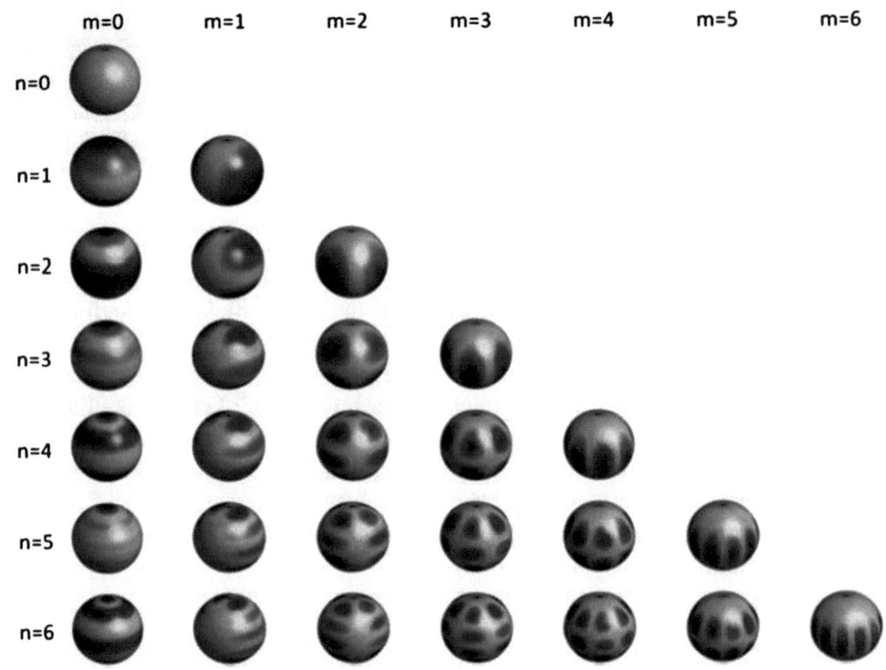

Fig. 22.6 Effects of degree (n) and order (m) on the spherical harmonic gravity model [8]

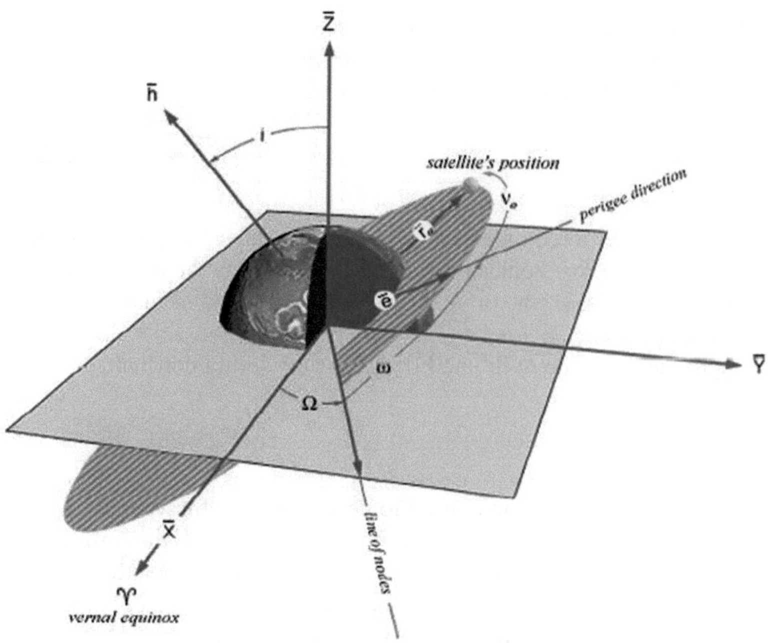

Fig. 22.7 Keplerian orbital elements [13]

Table 22.1 Initial conditions for satellite orbit used in the study

Keplerian element	Value
Semi-major axis	0.60 km
Eccentricity	0
Inclination	0° (prograde) or 180° (retrograde)
Argument of perigee	0°
Right-ascension of the ascending node	0°
True anomaly	0°

22.3 System Model

This project focuses on asteroid 25143 Itokawa as the asteroid-satellite system's central body. Itokawa was visited in 2005 and a significant amount of data was collected about its environment. This allows for computerized simulation of a "real-world" environment. The shape model used in our simulation is a visual approximation as a triaxial ellipsoid with axial dimensions 0.2741 km × 0.1561 km × 0.1376 km. These were calculated from values of the overall dimensions of Itokawa, given as 0.5481 km × 0.3122 km × 0.2751 km [9]. The rotational period of Itokawa is 12.132 h according the data provided by JPL Horizons [14]. The gravity model used is the degree and order four spherical harmonic model [9].

To conduct the study with a high degree of user manipulation and data capture, AGI's Systems Tool Kit (STK) software is used with data analysis done in MATLAB. An integration of STK and MATLAB has also been developed that allows for complex mathematical computation to be done in MATLAB and transferred into the STK simulation for control algorithm development, and other computationally intensive operations. The satellite used in the simulation scenario has a mass of 55 kg. The initial Keplerian elements of the satellite's orbit around Itokawa used in this study are given in Table 22.1.

22.4 Analysis

This project looks exclusively at the effects of the spherical harmonic gravity model. Other effects, such as SRP and third-body gravitation, are not considered. Gravity can be considered to be a "non-linear spring." The equation of motion for a linear spring-mass-damper system is defined as:

$$F = ma_x + cv_x + kx \tag{22.2}$$

where m is the mass (kg), a_x is the acceleration in the x-direction (m/s^2), c is the damping coefficient (N*s/m), v_x is the velocity in the x-direction (m/s), k is the spring constant (N/m), and x is the displacement in the x-direction (m).

And Newton's Law of Gravity is given as:

$$F_g = G\frac{m_1 m_2}{r^2} \tag{22.3}$$

where F_g is the force of gravity (N), G is the universal gravitational constant ($\approx 6.674 \times 10^{-11}$ N (m/kg)^2), m_1 and m_2 are the mass of the two bodies interacting (kg), and r is the distance between the two bodies' centers of mass (m). The terms G and m_1 (the mass of the orbited body) are combined to make the value μ (km^3/s^2) which is called the body's gravitational parameter.

This can be made to fit into Eq. 22.2 as the spring constant:

$$F = ma_x + cv_x + \frac{m\mu}{x^3}x \tag{22.4}$$

The damping of a system is used to decay the systems velocity with time. Without this, an ideal spring-mass system, once disturbed, would oscillate about its equilibrium point for an infinite amount of time. The damper slows the oscillations until the system eventually comes to rest at its equilibrium position. In the asteroid-satellite system, drag causes the velocity to decay with time. Typically drag is caused by air resistance, but in deep space missions, there is no atmosphere to resist the motion of the spacecraft so there is negligible drag. With these models applied, the equation of motion for the asteroid-satellite system becomes what is known in orbital mechanics as the simplified two-body problem:

$$F = ma_x + \frac{m\mu}{x^3}x \tag{22.5}$$

Now, some system parameters need to be defined in order to continue. From mechanical vibrations, the undamped natural frequency of a linear spring-mass system is defined as:

$$\omega_n = \sqrt{\frac{k}{m}} \tag{22.6}$$

This is the frequency at which an undamped system will naturally oscillate. This is an important parameter in vibrations because if a system is driven to oscillate at this resonant frequency, small excitations grow into large amplitudes of oscillation.

The amplification ratio at which the oscillations grow when an undamped system is excited at it natural frequency is given as:

$$\frac{X}{\delta_{st}} = \frac{1}{1 - \left(\frac{\omega}{\omega_n}\right)^2} \tag{22.7}$$

where X/δ_{st} is the ratio of the dynamic to the static amplitude of motion, and ω/ω_n is the ratio of the excitation frequency to the natural frequency of the system [15]. Figure 22.8 is a plot of ω/ω_n versus X/δ_{st} for an undamped linear system. From this we expect to see instabilities in the spacecraft's orbit as the excitation frequency experienced by the spacecraft from the rotating asteroid approaches the natural frequency of the system.

Replacing the spring constant (k) in Eq. 22.6 with the "non-linear spring" constant, it becomes:

$$\omega_n = \sqrt{\frac{\frac{m\mu}{x^3}}{m}} \tag{22.8}$$

which simplifies to:

$$\omega_n = \sqrt{\frac{\mu}{x^3}} \tag{22.9}$$

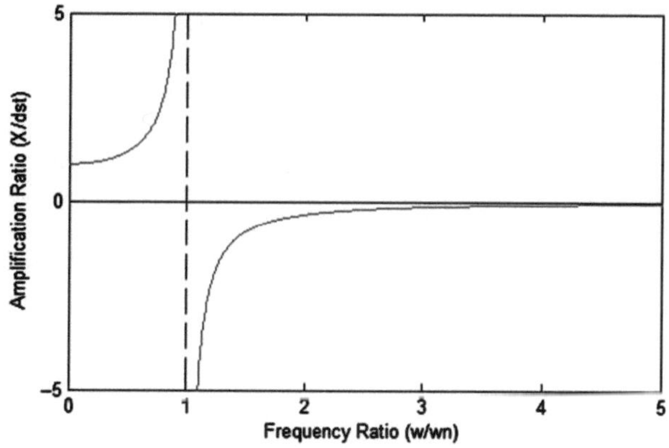

Fig. 22.8 Amplification ratio of an undamped linear system

where μ is the gravitational parameter (km^3/s^2) of Itokawa and x is the radial distance from the satellite to Itokawa's center (km). This leaves us with the natural frequency of the system in radians per second (rad/s). Scheeres provides that μ for Itokawa is 2.36×10^{-9} km^3/s^2 [9]. This gives a natural frequency of 1.05×10^{-4} rad/s when evaluated at our chosen initial orbital radius of 0.60 km.

In order to study the effects of the excitation frequency, we utilize the STK simulation. Through this, we can change the rotational rate of the asteroid to speed up or slow down the frequency of excitation experienced by the spacecraft from the spherical harmonic gravity model of asteroid Itokawa.

With the spherical harmonic model, the relative frequency between the satellite and the asteroid can be expressed as:

$$\omega_{sat/asteroid} = (\omega_{sat} \pm \omega_{asteroid}) \quad (22.10)$$

Where the frequencies add in the retrograde case, and subtract in the prograde case. With the C_{44} coefficient in the spherical harmonic model, we see that the gravitational force reaches a maximum value four times throughout a single rotation of Itokawa (Fig. 22.6). This results in the excitation frequency:

$$\omega = m\omega_{sat/asteroid} \quad (22.11)$$

where m is the highest order of the gravity model significantly impacting the excitation.

Multiple scenarios were simulated in STK varying the rate of rotation of the asteroid and the direction of the orbit. This allows for the study of the interaction between the excitation frequency (ω), which changes according to Eq. 22.11, and the natural frequency (ω_n). Observations are given in Table 22.2.

It is observed that when the ratio of the excitation frequency to the natural frequency (ω/ω_n) is $-1.3 < \omega/\omega_n < 4.3$, the orbit is unstable. Examples of unstable and stable orbits are shown in Fig. 22.9. Figure 22.9a shows the asteroid at its natural rate of rotation with a satellite travelling with the direction of rotation (prograde). The satellite diverges from its intended circular orbit path and crashes into Itokawa. Figure 22.9b shows a satellite travelling in the opposite direction (retrograde). The satellite maintains an orbit that is mostly circular, but the orbit rotates around the body as the Right Ascension of the Ascending Node undergoes precession caused by the changing gravitational magnitude. The orbit is perturbed, but the frequency ratio of the excitation frequency to the natural frequency of the system is outside of the amplified region that causes the orbit to become unstable.

The time history of the prograde system, seen in Fig. 22.10a, shows that the force of gravity pulling the satellite inwards increases (negatively) as the satellite crashes into the surface of the asteroid. The retrograde time-history (Fig. 22.10b) is interesting since it shows that there is more than one excitation frequency as a higher frequency oscillation is superimposed over the lower frequency motion.

Figure 22.11a, b show the Fast Fourier Transform (FFT) analysis of the radial acceleration. The FFT is useful because any peaks seen in the graph illustrate a dominant frequency at which the system was excited. The prograde system is unstable from the very beginning, so we don't expect to see any dominant frequencies. The retrograde FFT again shows two dominant excitation frequencies. These potentially correspond with the C_{22} and C_{44} gravity model coefficients, though further analysis is required.

Table 22.2 Orbit scenario observations with varying asteroid spin rates and changing orbital direction (prograde and retrograde)

	Asteroid spin rate multiplier	Prograde frequency ratios (ω/ωn)	Orbit quality	Event	Retrograde frequency ratios (ω/ωn)	Orbit quality	Event
Natural rate	1/32	−3.8	Stable		4.2	Unstable	Impact
	1/16	−3.7	Stable		4.3	Semi-stable	High precession
	1/8	−3.3	Stable		4.7	Stable	
	1/4	−2.6	Stable		5.4	Stable	
	1/3	−2.2	Stable		5.8	Stable	
	1/2	−1.2	Unstable	Impact	6.8	Stable	
	3/4	0.1	Unstable	Impact	8.1	Stable	
	7/8	0.8	Unstable	Impact	8.8	Stable	
	1	1.5	Unstable	Impact	9.5	Stable	
	1 1/2	4.3	Unstable	Impact	12.3	Stable	
	1 3/4	5.6	Semi-stable	Eject	13.6	Stable	
	2	7.0	Semi-stable	Eject	15.0	Stable	
	3	12.5	Stable		20.5	Stable	
	4	18.0	Stable		26.0	Stable	
	5	23.5	Stable		31.5	Stable	
	6	29.0	Stable		37.0	Stable	
	7	34.5	Stable		42.5	Stable	

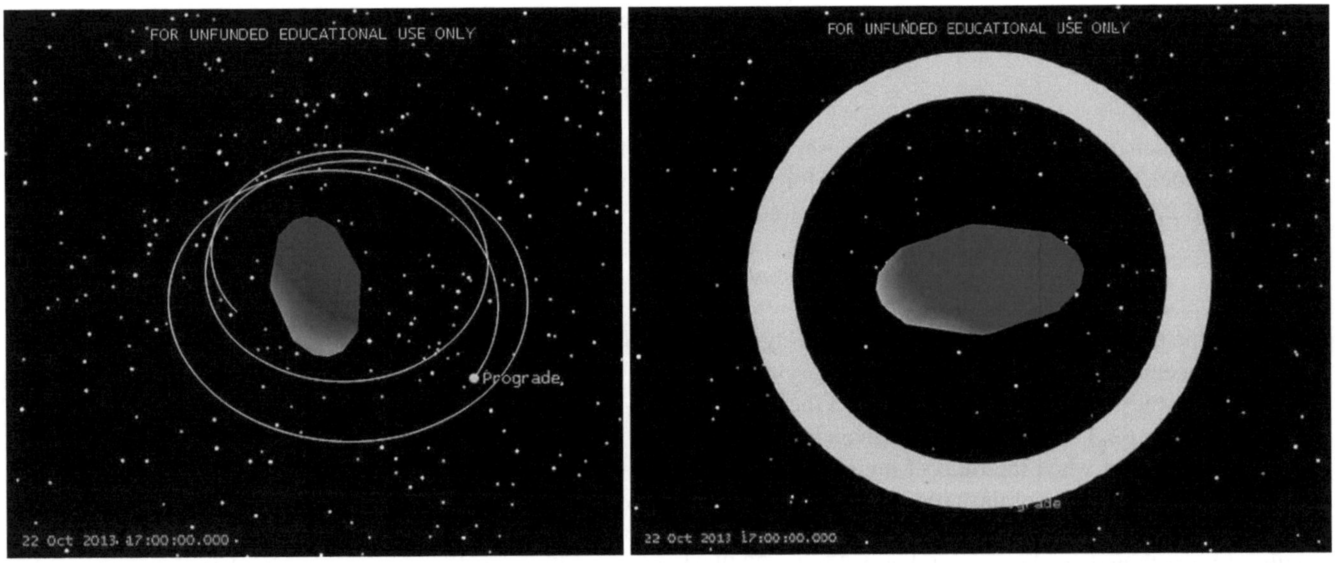

Fig. 22.9 (**a**) & (**b**). Example of an unstable prograde orbit that is perturbed enough to cause the orbit to diverge from its intended circular orbit and crash into the asteroid. Example of a stable retrograde orbit that maintains a mostly circular orbit, but experiences precession around the central body for upwards of 6 months

It is interesting to note that within $-1.3 < \omega/\omega_n < 4.3$ the orbit becomes unstable immediately and crashes into the asteroid within a few periods. However, at the values of 5.6 and 7.0, as seen in Table 22.2, the orbit is stable for some time and then at one point in the orbit is ejected from the system.

Figures 22.12a, b show the satellite at the point just before it is ejected from the system when ω/ω_n is 5.6 and 7.0, respectively. Just before the ejection, the satellite enters a highly eccentric orbit and has a close approach to the protruding end of the body near the periapsis. As the satellite passes by the asteroid, a small amount of energy is transferred to the satellite as it is then accelerated into a hyperbolic orbit and is ejected from the system [16].

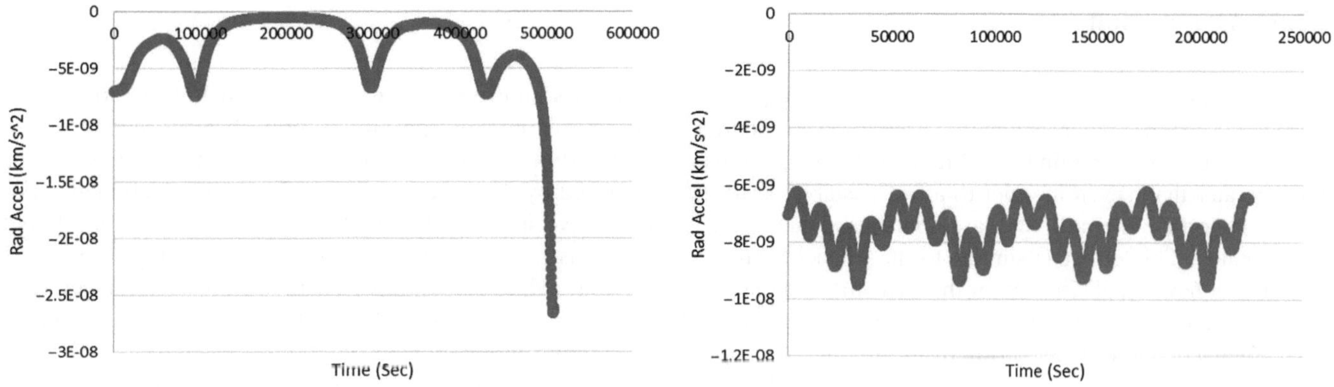

Fig. 22.10 (**a**) & (**b**). Time-history of the radial acceleration data for an unstable prograde orbit (*left*) and a stable retrograde orbit (*right*) at the asteroid's natural rotation rate

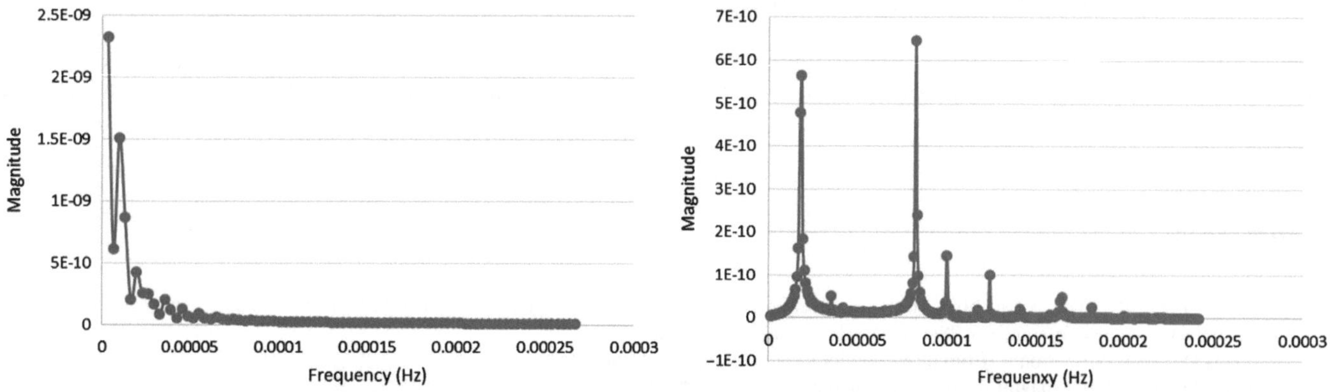

Fig. 22.11 (**a**) & (**b**). FFT of the radial acceleration data for an unstable prograde orbit (*left*) and a stable retrograde orbit (*right*) at the asteroid's natural rotation rate

Fig. 22.12 (**a**) & (**b**). Rotation of Itokawa sped up by 1 3/4 times the natural rate (*left*). Rotation rate of Itokawa sped up by 2 times the natural rate (*right*)

22.5 Conclusion

It was found that there is a range of excitation frequencies that causes an orbiting spacecraft around Itokawa to become unstable. There are other factors that need to be considered before attempting close-proximity orbits, but it is recommended that a spacecraft attempting to orbit small body asteroids (around 500 m in diameter) enter into an orbit that has a ω/ω_n ratio greater than seven in order to avoid unstable perturbations caused by the spherical harmonic gravity coefficients. For retrograde orbits, the ratio of ω/ω_n is increased due to the higher relative speed of the satellite to the surface of the asteroid as it rotates. This effect is similar to the impact harmonic excitations have on linear systems: relatively high frequency excitations have small impact on the dynamic response of the system, while excitations near the resonant frequency (ω_n) have a large impact on the dynamic response of the system. The same basic response is observed in satellite orbits around an asteroid where a non-linear spring restoring force is coupled with complex gravitational excitation.

The use of a harmonic excitation analysis on the linearized system can predict the regions where orbital stability is expected. By computing the excitation frequency as a function of the rotational spin rate and order of the gravity model, a comparison can be made to the natural frequency of the satellite-asteroid system. This provides important information regarding the likelihood of stable, long-duration orbits.

Acknowledgements This research was supported in part by the North Dakota Space Grant Consortium, the North Dakota NASA EPSCoR grant, the UND Seed/Planning Grant for Collaborative Research, the National Science Foundation (NSF Grant #EPS-081442), and AGI with STK Educational Licenses.

References

1. Church C, Fevig R (2013) A feasibility study on the characterization of the internal structure of small NEOs with small spacecraft. In: Lunar and planetary science conference (LPSC XLIV), Houston, March 2013
2. Yeomans D (2014) NEO groups, from http://neo.jpl.nasa.gov/neo/groups.html
3. Yeomans D (2014) Orbit diagrams, from http://neo.jpl.nasa.gov/orbits/
4. Netting R (2014) NEAR-shoemaker, from http://science.nasa.gov/missions/near/
5. Kawaguchi J, Fujiwara A, Uesugi T (2005) Hayabusa (MUSES-C) – rendezvous and proximity operation. In: IAC-05-A3.5.A.01, international astronautical congress, International Astronautical Federation, Fouduoka, 2005
6. Kuninaka H, Nishiyama K, Funaki I, Shimizu Y, Yamada T, Kawaguchi J (2006) Assessment of plasma Interactions and flight status of the Hayabusa asteroid explorer propelled by microwave discharge ion engines. IEEE Trans Plasma Sci 35(5):2125–2132
7. Montenbruck O, Gill E (2000) Satellite orbits: models, methods, and applications. Springer Science & Business Media, Heidelberg, pp 53–116
8. Church C (2014) A feasibility study on the implementation of satellite-to-satellite tracking around a small near-earth object. M.S. thesis, Department of Space Studies, University of North Dakota
9. Scheeres DJ, Gaskell R, Abe S, Barnouin-Jha O, Hashimoto T, Kawaguchi J, Kubota T, Saito J, Yoshikawa M, Hirata, Mukai T, Ishiguro M, Kominato T, Shirakawa K, Uo M (2006) The actual dynamical environment about Itokawa, AIAA 2006–6661. In: AIAA/AAS astrodynamics specialist conference and exhibit. American Institute of Aeronautics and Astronautics, Inc., Keystone, Colorado, AIAA is based in Reston, Virginia, USA 2006
10. Scheeres DJ, Ostro SJ, Hudson RS, Werner RA (1996) Orbits close to asteroid 4769 Castalia. Icarus 121:67–87, Elsevier Limited
11. Scheeres DJ, Marzari F, Tomasella L, Vanzani V (1998) Rosetta mission: satellite orbits around a cometary nucleus. Planet Space Sci 36(6/7):649–671, Elsevier Limited
12. Scheeres DJ, Williams BG, Miller JK (2000) Evaluation of the dynamic environment of an asteroid: applications to 433 eros. J Guid Control Dyn 23(3):466–475
13. Kauderer A (2012) Orbital elements, from http://spaceflight.nasa.gov/realdata/elements/graphs.html
14. JPL Horizon System, http://ssd.jpl.nasa.gov/horizons.cgi
15. Rao SS (2011) Mechanical vibrations, 5th edn. Prentice Hall, Upper Saddle River, pp 259–362
16. Sellers JJ, Astore WJ, Griffen RB, Larson WJ (2005) Understanding space: an introduction to astronautics, 3rd edn. McGraw-Hill Companies, Inc, New York, pp 153–288

Chapter 23
Energy Harvesting Perspectives from Parametric Resonant Systems

Maryam Ghandchi Tehrani, Elvio Bonisoli, and Matteo Scapolan

Abstract Parametric resonances can occur in internally stressed systems due to the periodic variation of the stiffness in time. Parametric resonance can lead to unstable dynamic behaviour; however, their response is limited by existing nonlinearities in the system, thus resulting in limit cycle oscillations (LCOs). This phenomenon can be exploited in the design of energy harvesters. The amplitude and frequency of the parametric excitation can be adjusted so that the vibration response of internally stressed systems is close to instability. In this paper, a cantilever beam is considered in vertical position and an axial excitation is applied to the base of the beam. The imposed kinematics of the base leads to internal stress along the beam, which produces a variation of the bending stiffness. If the frequency of the base excitation is twice the first natural frequency of the beam, the principal parametric resonances can occur. A quasi-linear FEM approach is adopted, together with a simplified single-degree-of-freedom model of the beam, in order to numerically simulate its dynamic behaviour, to identify unstable conditions and to obtain the Floquet diagram. An analytical approach is developed as well, using a multi-degree-of-freedom model of the beam, considering the system as autoparametric. Harmonic balance method is used to determine the Floquet diagram and to validate the numerical model.

Principal parametric resonance is observed experimentally. Harvesting energy from parametric resonance is therefore potentially very efficient, especially if the external source is not directly exploitable. Parametric resonance in this case acts as a power amplification.

Keywords Energy harvesting • Parametric resonance • Floquet diagram • Lupos

23.1 Introduction

Faraday [1] first observed parametric resonance phenomena in a vertically oscillating fluid container, developing horizontal surface waves. The first model that described this behaviour was presented by Mathieu [2].

Parametric resonance can occur in systems with periodic-time-varying parameter(s) in the differential equation such as,

$$m(t)\,\ddot{x}(t) + c(t)\,\dot{x}(t) + k(t)\,x(t) = f(t) \tag{23.1}$$

where $m(t)$, $c(t)$, $k(t)$ are the time varying mass, damping and stiffness coefficients respectively and $f(t)$ is the external forces acting on the system.

The excitation can be externally provided by a forcing function (*external excitation*) or can arise internally from the time variation of the parameters (*self-excitation*). In contrast to externally excited systems, self-excitation is linked with the homogeneous equation of motion ($f(t) = 0$). Thus, the presence of a periodic time varying parameter in the homogenous equation can act as an excitation, commonly referred to as *parametric excitation.*

Floquet theory is used to solve the homogeneous equation of this kind. For undamped systems, Equation 1 can be simplified to the well-known Mathieu equation,

$$\ddot{x}(t) + (\delta + \varepsilon \cos(\Omega t))\ x(t) = 0 \tag{23.2}$$

M.G. Tehrani
Institute of Sound and Vibration Research, University of Southampton, Highfield, Southampton SO171BJ, United Kingdom

E. Bonisoli (✉) • M. Scapolan
Department of Management and Production Engineering, Politecnico di Torino, Corso Duca degli Abruzzi, Torino 24-10129, Italy
e-mail: elvio.bonisoli@polito.it

A. Wicks (ed.), *Shock & Vibration, Aircraft/Aerospace, and Energy Harvesting, Volume 9*,
Conference Proceedings of the Society for Experimental Mechanics Series, DOI 10.1007/978-3-319-15233-2_23

where the variable stiffness, consists of a constant part δ, which relates to the natural frequency of the system, and a variable part, $\varepsilon \cos(\Omega t)$, which is due to the parametric excitation. The variation of the stiffness leads to an increase or decrease of the elastic potential energy; the same happen for a variation of the inertial term [3].

Under certain conditions, the energy in the system can grow exponentially in each cycle, leading to unstable behaviour. For undamped systems, when the excitation frequency is about twice the first natural frequency, small parametric amplitude can result in primary parametric resonance. The damping in the system may not be sufficient to stabilise the unstable system and it may only delay the onset of the instability [4].

Nonlinear structural vibrations can however limit the growth of the response [5, 6], as demonstrated by a model including quadratic and cubic geometric nonlinearities.

Oueini and Nayfeh [7] used this concept to suppress structure vibrations using nonlinear velocity feedback control, on a cantilever beam subjected to primary parametric resonance. Linear velocity and position feedbacks are also implemented experimentally, by Ghandchi Tehrani et al. [8], to avoid unstable behaviour.

The aim of this paper is to take advantage of parametric instability rather than to avoid it by providing parametric resonance conditions for energy harvesting. References [9–12] also have demonstrated that under parametric resonance the system can harvest ten times more power compared to direct resonance.

Numerical approach based on quasi-linear FEM is developed using Lupos, which is an open source FEM software. Harmonic balance method is applied to obtain the Floquet diagram and to validate the numerical results. Experiments are conducted to demonstrate the instability due to parametric resonance. Geometric nonlinearities are observed in the response of the beam and their effects are investigated on the stability curves. An analysis based on the energy flow is carried out to explain how the potential energy varies when the parametric resonance occurs.

23.2 A Parametrically Excited Beam

A beam in vertical position subject to an imposed kinematics of the base is considered as shown in Fig. 23.1. The imposed kinematics is the axial displacement, along the z-axis. The axial acceleration leads to inertial effects. The distribution of the axial load along the beam is also shown in Fig. 23.1.

The distribution of the normal force leads to the variation of the bending stiffness of the beam. The axial excitation is sinusoidal with the frequency of Ω. Hence, the bending stiffness is also periodic with the same period as the base acceleration Ω, representing a parametrically excited system.

The characteristics of the beam are provided in Table 23.1.

The first natural frequency of the beam without parametric excitation is found to be $f_1 = 3.9978$ Hz.

The beam can be modelled as a single-degree-of-freedom system as described in [8], which is an approximate model and it does not capture the interaction between the axial and the bending motion of the beam, which is the main concern of this paper.

It is possible to derivate analytically the normal force acting along the beam and its influence, as done in the following.

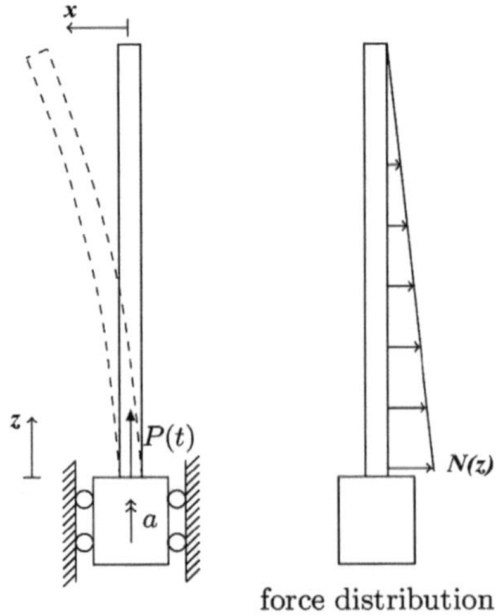

Fig. 23.1 Vertical beam subject to imposed kinematics along the beam and its normal force distribution due to inertial effects

Table 23.1 Beam characteristics

Characteristic	Property	Value
Beam geometry	l, length z [mm]	400
	b, width x [mm]	1
	h, height y [mm]	10
Beam material	ρ, density $\left[\frac{\text{kg}}{\text{m}^3}\right]$	7,800
	E, Young modulus [GPa]	207
	ν, Poisson ratio [−]	0.29

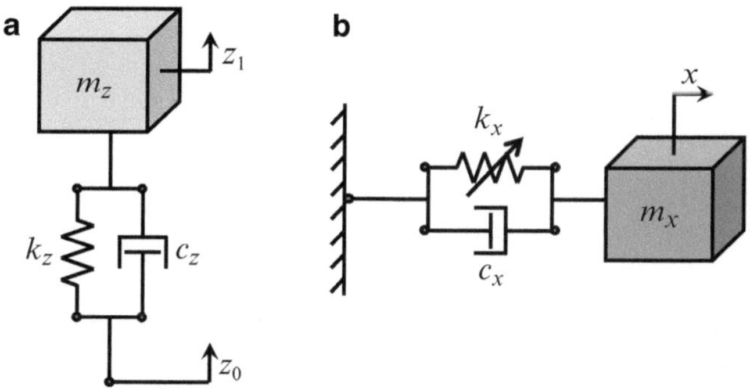

Fig. 23.2 Two degree-of-freedom system, vertical dof (**a**) and bending dof (**b**)

The beam is modelled as a two degree-of-freedom system; it can however been generalised to an n-degree-of- freedom (MDOF) system. The degrees of freedom correspond to the axial and the bending motion as shown in Fig. 23.2.

The system can be regarded as an autoparametric system [4, 13] and consists of two subsystems, a primary and a secondary subsystem. The primary system is externally forced in axial direction, while the secondary subsystem, which describes the bending motion of the beam, is coupled to the primary subsystem due to the internal stresses. The coupling between the two subsystems is represented by the geometric stiffness, which depends on the normal force. The bending stiffness k_x is therefore dependent on the imposed displacement along the axial direction.

The dynamic equation of motion in x direction can be written as:

$$m_x\ddot{x} + c_x\dot{x} + k_x(t)x = 0 \tag{23.3}$$

where m_x, c_x and $k_x(t)$ are the mass, damping and stiffness of the bending DOF.

Imposing the axial displacement, results in the geometric stiffness:

$$k_x = k_{x0} + N_0\Delta k_{x0}\cos\Omega t \tag{23.4}$$

Neglecting the damping, yields,

$$\ddot{x} + \left(\frac{k_{x0}}{m_x} + \frac{N_0\Delta k_{x0}}{m_x}\cos\Omega t\right)x = 0 \tag{23.5}$$

The bending stiffness in Eq. 23.4 has a constant part (k_{x0}) and a variable part $N_0\Delta k_{x0}\cos\Omega$, which depends on the amplitude of the normal force N_0,the excitation frequency Ω, and the variation of stiffness due to the normal unitary force Δk_{x0}.

Considering the axial direction as a SDOF system, the dynamic equation along the z-axis can be written as:

$$m_z\ddot{z}_1 + c_z(\dot{z}_1 - \dot{z}_0) + k(z_1 - z_0) = 0 \tag{23.6}$$

where m_z, c_z and k_z are the mass, the damping and the stiffness of the axial DOF of the beam.

The normal force can be obtained from Equation 7:

$$N_0 = \frac{m_z\Omega^2(k_z + ic_z\Omega)}{k_z - m_z\Omega^2 + ic_z\Omega}Z_0 \tag{23.7}$$

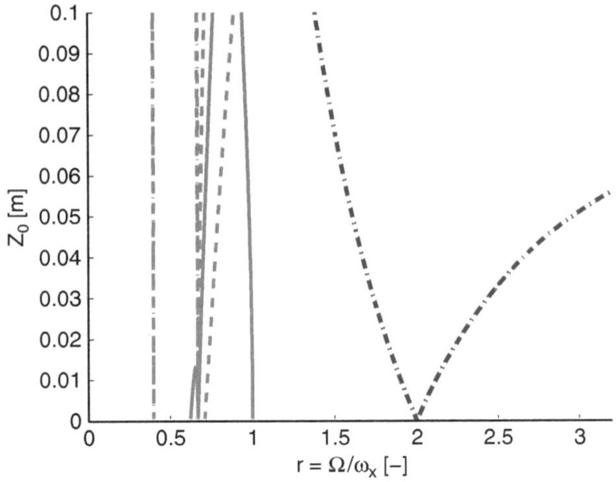

Fig. 23.3 Floquet diagram, analytically obtained

The transition curves in the Floquet diagram can therefore be obtained from solving Eq. 23.5 for ε and Ω, by using the harmonic balance method. They can also be obtained in terms of Z_0 and Ω, as [14] shown.

Figure 23.3 shows the analytical transition curves for the primary parametric resonance (blue dashed line), together with parametric resonances of higher orders (red solid line).

23.3 Analytical Analysis of Energy Flows

The instability phenomena can be analysed based on the energy flow in parametrically excited systems. In unstable conditions, the energy increases in each cycle. The total energy in the system is the sum of the kinetic energy and the potential energy. For the bending subsystem, the kinetic energy is proportional to the mass m_x as well as the square of the velocity . The potential energy is proportional to the stiffness k_x and the displacement x squared. The stiffness k_x consists of a constant term and a time-varying term. Therefore, the elastic potential energy results in a constant term U and a time-varying term ΔU.

$$\Delta U = \frac{1}{2}\Delta k_x x^2 \tag{23.8}$$

Since ΔU is dependent on the displacement and the variable part of the stiffness, combining the two periodic functions leads to a non-zero mean value for the energy:

$$\begin{cases} \Delta k_x = m\varepsilon \cos(\Omega t) \\ x = X\cos(\omega t) \end{cases} \Rightarrow \Delta U = \frac{1}{2} m\varepsilon X^2 \cos(\Omega t)\cos^2(\omega t) \tag{23.9}$$

Figure 23.4 shows the elastic energy ΔU for two different excitation frequencies, $\Omega = \omega_0$ (blue dashed line) and $\Omega = 2\omega_0$ (red solid line). When $\Omega = \omega_0$ the net energy in one cycle is zero, however, the elastic energy increases in each cycle when the system is at parametric resonance $\Omega = 2\omega_0$. This energy could be harvested and stored by an opportune device, instead of increasing the energy of the system, leading to unstable conditions.

The stiffness variation leads to an increase or a decrease of the elastic potential energy. If the frequency of the periodic stiffness is twice the natural frequency, the potential energy increases and the growth of the energy does not depend on the amplitude of the external excitation. The parametric amplitude, however, determines how fast the energy increases.

Since even the energy losses caused by friction are proportional to the energy already stored, they cannot stop the growth of the amplitude, if it begins. The effect of the losses is to establish a threshold, since the increment of energy provided by the external source must be larger than the amount of energy dissipated during the same time. This means that damping avoids unstable conditions, if the amplitude does not exceed the threshold.

The energy dissipated by damping could be harvested for other purposes, as [9–12] and [15–17] propose.

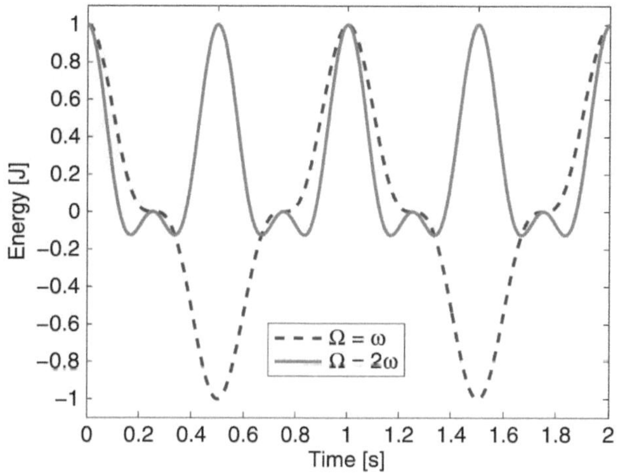

Fig. 23.4 Second term of elastic potential energy ΔU, with $\Omega = \omega$ (*blue dashed line*) and $\Omega = 2\omega$ (*red solid line*)

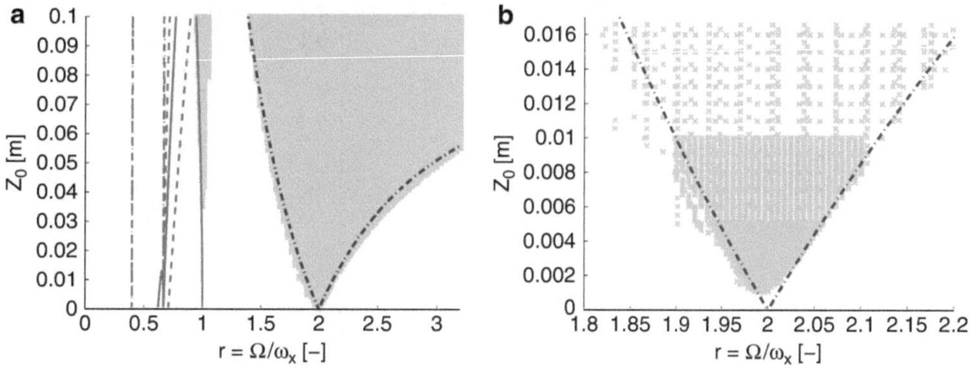

Fig. 23.5 Floquet diagram, numerical (*cyan dots*), analytical (*blue or red lines*). Detailed in (**b**)

23.4 Numerical Approach: Quasi-linear FEM

A numerical approach is considered to validate the analytical results. The numerical simulations are carried out on the complete model of a cantilever beam (see Table 23.1). The stiffness variation is as a consequence of the imposed kinematics and therefore a normal force distributes along the beam due to inertial effects as discussed previously.

A parametric software (Lupos), which is built in Matlab environment, is used to model the system.

Simulations are carried out for different values of Z_0 and Ω, by using Simulink, to provide a map of stable and unstable regions numerically as shown in Fig. 23.5. The cyano area are the points that are obtained from the numerical simulation, which is in agreement with the analytical approach in the previous section. Using the numerical method, the transition curves for the higher bending modes of the beam can also be obtained.

23.5 Numerical Analysis of Energy Flows

Considering the energy flows in these numerical simulations, by defining the instantaneous powers involved as:

$$\begin{cases} P_c = -c_x \dot{x}^2 \\ P_{\Delta k} = -\Delta k_x x \dot{x} \end{cases} \tag{23.10}$$

it is possible to appreciate that in stable conditions, as Fig. 23.6 shows, the damped instantaneous power tends to zero, so that the dissipated/harvested energy reaches an asymptote. The power introduced in the system is due to the stiffness variation is not enough, however, to increase the amplitude response and obtain an unstable system.

By using $\Omega = 2\omega_x$, with the same amplitude excitation, the system becomes unstable. The instantaneous damped power increases and the energy dissipated/harvested increases as well during the simulation, as shown in Fig. 23.7a. On the other

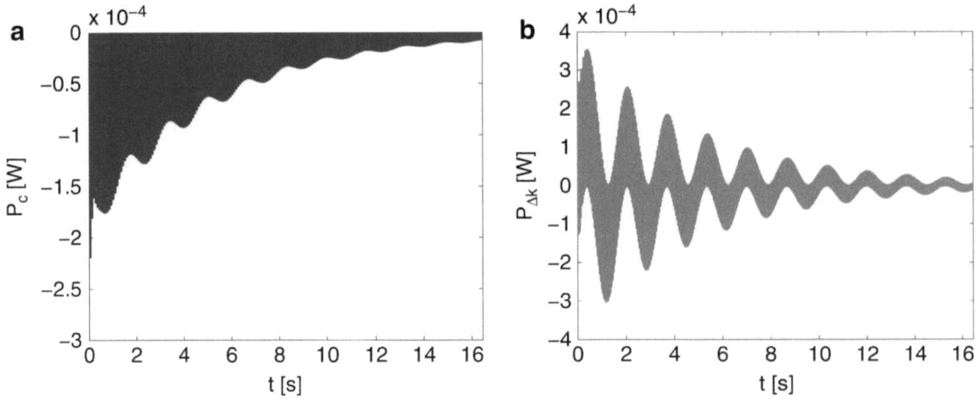

Fig. 23.6 Instantaneous power associated to damping (**a**) and stiffness variation (**b**) related to the damping, for $Z_0 = 1 \times 10^{-3}$ m and $\frac{\Omega}{\omega_x} = 1.9$

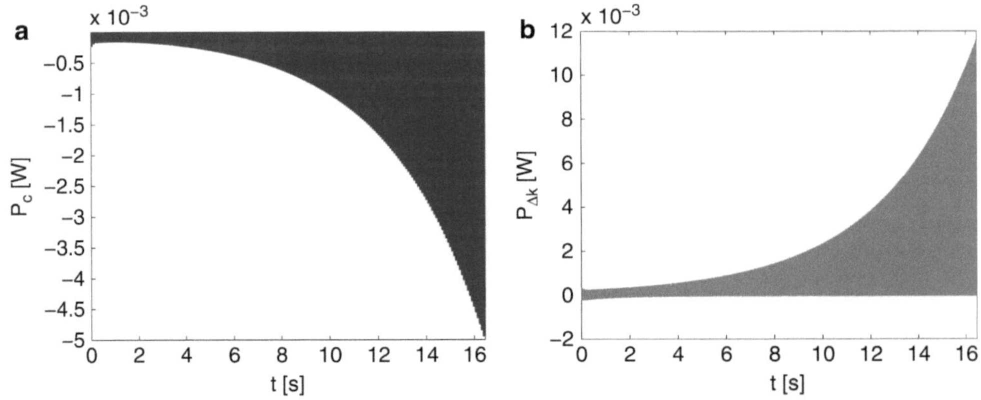

Fig. 23.7 Instantaneous power associated to damping (**a**) and stiffness variation (**b**) related to the damping, for $Z_0 = 1 \times 10^{-3}$ m and $\frac{\Omega}{\omega_x} = 2$

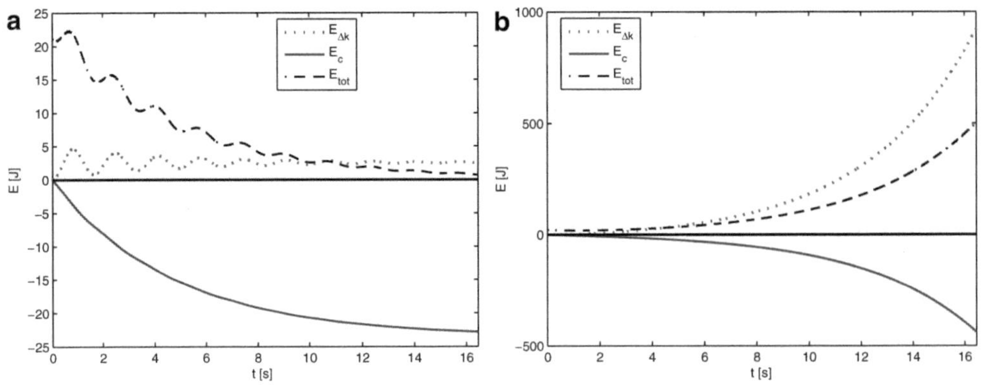

Fig. 23.8 Energy flows, with $Z_0 = 1 \times 10^{-3}$ m and $\frac{\Omega}{\omega_x} = 1.9$ (**a**) or $\frac{\Omega}{\omega_x} = 2$ (**b**)

side, according to Fig. 23.7b, the amount of energy introduced in the system is greater: part of this energy is dissipated and part of it increases the energy of the system, leading to unstable conditions.

Figure 23.8 provides the energy flow evolution in the two previous examples, highlighting that the energy of the system increases only in the second one, because the introduced power for each cycle is greater than the amount of energy dissipated/harvested.

By increasing the damping coefficient, the growth is slower, as Figs. 23.9 and 23.10 show.

Despite that, the energy in the system always increases and, in all the considered examples, part of the introduced energy is dissipated/harvested. A greater damping coefficient slows the growth by absorbing more energy.

It can be assume that by increasing the damping coefficient a larger part of the introduced energy is dissipated/harvested, but only if parametric resonance occurs, as Fig. 23.11 shows: the ratio $\frac{E_{\Delta k}+E_c}{E_{\Delta k}}$ becomes negative if the dissipated/harvested energy is greater than the introduced one. Obviously, in these conditions, the system is stable and the amplitude of motion decreases to zero; the amount of energy dissipated but not introduced is related to the initial energy of the system.

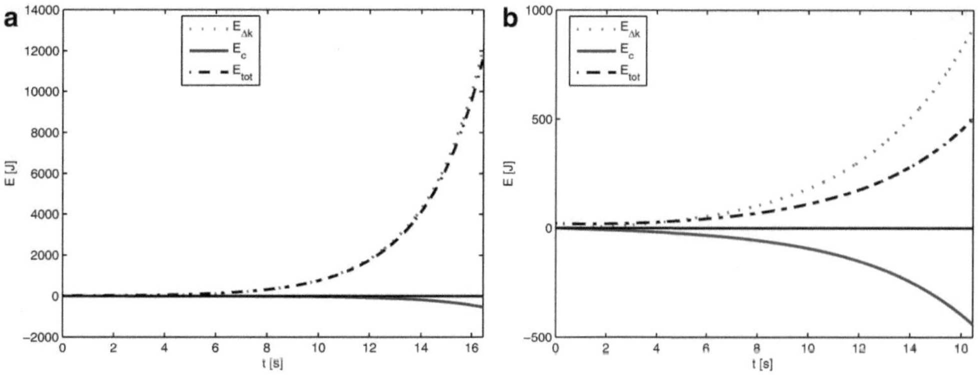

Fig. 23.9 Energy flows, with $Z_0 = 1 \times 10^{-3}$ m, $\frac{\Omega}{\omega_x} = 2$ and $c_x = 0.1457 \times 10^{-3} \frac{\text{Ns}}{\text{m}}$ (**a**) or $c_x = 1.5733 \times 10^{-3} \frac{\text{Ns}}{\text{m}}$ (**b**)

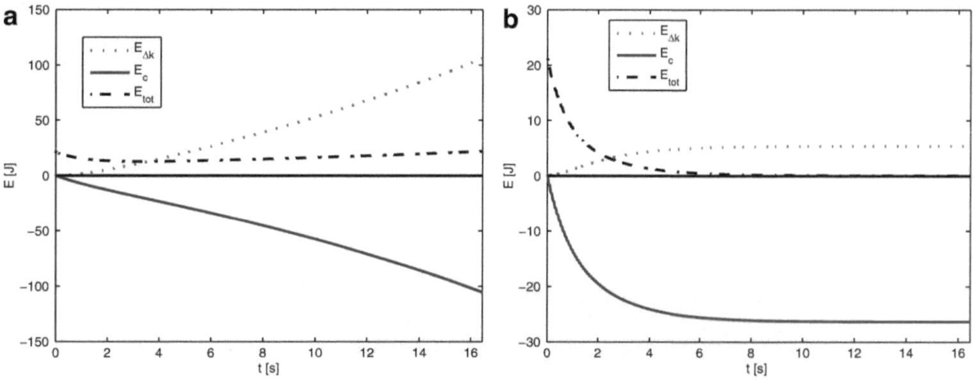

Fig. 23.10 Energy flows, with $Z_0 = 1 \times 10^{-3}$ m, $\frac{\Omega}{\omega_x} = 2$ and $c_x = 3.0010 \times 10^{-3} \frac{\text{Ns}}{\text{m}}$ (**a**) or $c_x = 7.2840 \times 10^{-3} \frac{\text{Ns}}{\text{m}}$ (**b**)

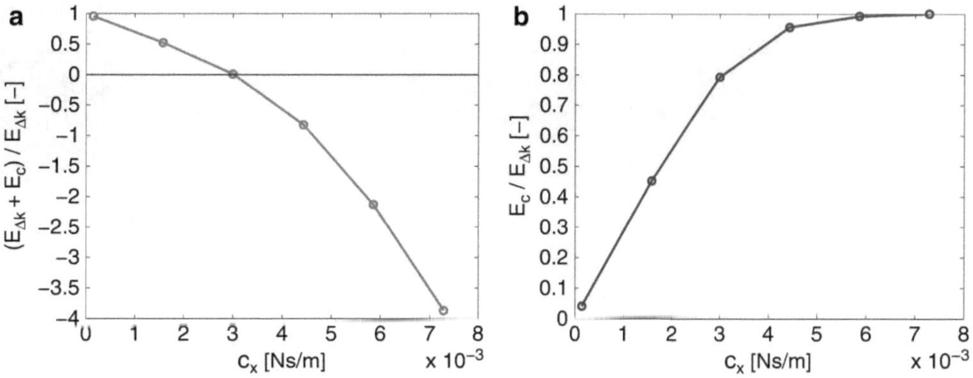

Fig. 23.11 Energy ratios: $\frac{E_{\Delta k}+E_c}{E_{\Delta k}}$ (**a**) and $\frac{|E_c+E_0|}{E_{\Delta k}}$ (**b**)

Figure 23.11a shows when parametric resonance occurs (positive sign) and when the system is simply a damped system (negative sign). Figure 23.11b, on the other hand, gives information about the ratio between the damped energy (without the initial energy contribution) and the introduced energy.

23.6 Experiment

A beam is attached vertically to an electromagnetic shaker, which provides the base displacement along the beam axis as shown in Fig. 23.1. The beam is attached at the centre of the shaker through a plate and two accelerometers are attached to the beam at each side to maintain symmetry and measure the bending acceleration, while one accelerometer is attached to the shaker to measure the axial acceleration (Fig. 23.12).

To generate parametric resonance, the excitation frequency should be equal to twice the first bending natural frequency of the beam. Modal analysis was carried out to obtain the natural frequencies and the mode shapes of the beam using impact hammer. The first natural frequency of the beam is $f_1 = 3.2031$ Hz and the second is; the first two mode shapes were identified using *Polymax* method [18].

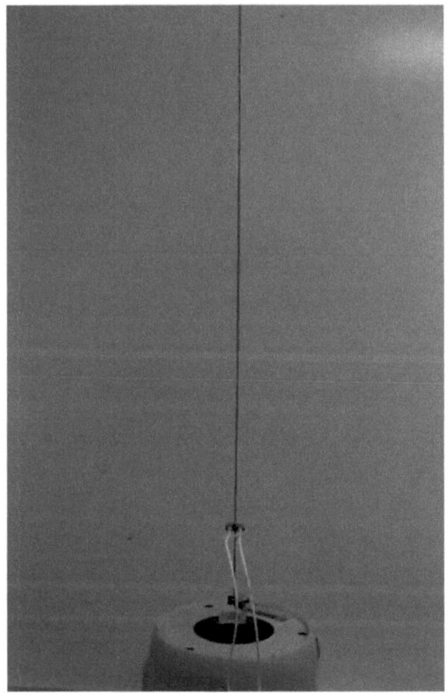

Fig. 23.12 Experimental beam layout

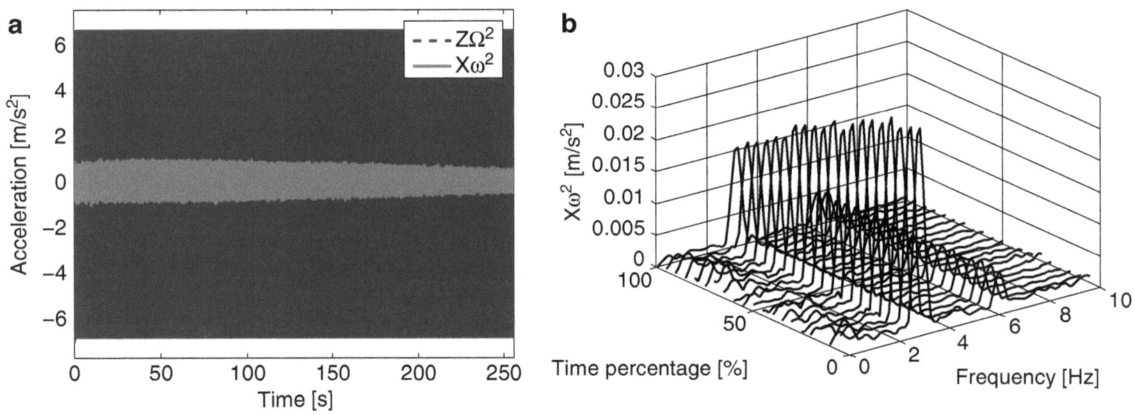

Fig. 23.13 Accelerations with $A = 1.0$ V and $f = 6.400$ Hz

To observe parametric instability experimentally, several experiments were carried out with different excitation frequencies (f [Hz]) and magnitudes (A [V]). An initial deflection was given to the beam and the accelerations were measured when the shaker was exciting the beam in the axial direction.

The experimental results show that the system is very sensitive to the frequency of the external excitation, as [14] shown. Fast Fourier transform of the measured acceleration signals are also plotted to monitor the frequency contents of the signal.

Figure 23.13 shows the evolution of the response with no parametric while Fig. 23.14 shows strong instability, with limit cycle due to non-linear effects.

23.7 Discussion and Further Applications

Jia et al [10] presented that significant amount of energy can be harvested when parametric resonance is used, about ten times more than the non-parametric case. In addition the frequency bandwidth in which the energy can be harvested is extended as well, especially when parametric resonance is applied together with the direct resonant principle. Reducing the threshold due to damping phenomena is a key spot to been able to harvest even small vibration amplitudes.

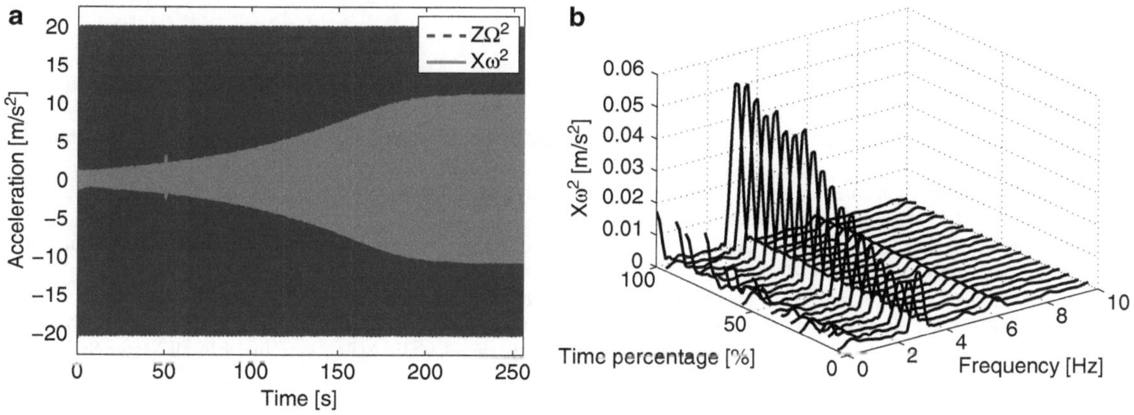

Fig. 23.14 Acceleration, with $A = 3.0$ V and $f = 6.380$ Hz

Using piezoelectric actuators, in order to change the bending stiffness of the system, could be helpful to tune the device according to excitation characteristics. This can possibly result in a wider harvestable frequency band and in a more efficient exploitation of the provided energy. Considering the nonlinearities and their effects, in fact, it could be possible to tune the device increasing the bending amplitude in steady state conditions, besides increasing the range of working frequencies.

Energy can be harvested using piezoelectric, electromagnetic or electrostatic approach. The last approach, according to [9–12], gives a magnitude lower peak power density.

Using piezoelectric approach the harvesting part should be implemented close to the base of the beam, where the strains are maxima. On the contrary, using electromagnetic approach, it is more profitable to use the free end of the beam, where the displacement is maximum and rare-earth materials allow high power density.

23.8 Conclusions

The paper proposed an analytical/numerical and experimental validation of a parametric resonant system with periodic variation of the stiffness in time. It has been shown that parametric resonance can lead to unstable dynamic behaviour and, if there are some nonlinearities, it results in limit cycle oscillations.

A quasi-linear FEM approach is adopted, together with a simplified single-degree-of-freedom model of the beam, in order to numerically simulate its dynamic behaviour, to identify unstable conditions and to obtain the Floquet diagram.

An analytical approach is also developed using a multi-degree-of-freedom model of the beam, which is based on partitioning the axial and the bending motions of the beam, according with the fact that the system can be considered as autoparametric. Harmonic balance method is used to determine the Floquet diagram analytically and to validate the numerical model. The effect of geometric nonlinearities on the stability curves in the Floquet diagram is also investigated.

Finally, principal parametric resonance is observed experimentally. Nonlinear behaviour in the form of softening stiffness is observed in the experimental results, leading to limit-cycle oscillations.

The unstable behaviour of the beam is explained using the energy flow. It is demonstrated that the external power provided to the beam flows from the axial direction to the bending direction, resulting in an exponential growth of the response. This phenomenon can be exploited in the design of energy harvesters.

Harvesting energy from parametric resonance is therefore potentially very efficient, especially if the external source is not directly exploitable. Parametric resonance in this case acts as a power amplification [9–12, 15–17].

Acknowledgements Dr Ghandchi Tehrani would like to acknowledge the support provided by EPSRC for her first grant (Ref: EP/K005456/1). The authors would like to acknowledge the support provided for Matteo Scapolan as a visiting student at ISVR, University of Southampton.

References

1. Faraday M (1831) On a peculiar class of acoustical figures and on certain forms assumed by a group of particles upon vibrating elastic surfaces. Philos Trans R Soc Lond 121:299–318
2. Mathieu E (1868) Mémorie sur le movement vibratoire d'une membrane de forme elliptique. Journal de Mathématiques Pures et Appliquées 13:137–203
3. Butikov E (2005) Parametric resonance in a linear oscillator at square-wave modulation. Eur J Phys 26:157–174
4. Nayfeh AH, Mook DT (1979) Nonlinear oscillations. John Wiley & Sons Inc, New York
5. Zavodney LD, Nayfeh AH, Sanchez NE (1989) The response of a single degree-of-freedom system with quadratic and cubic non-linearities to a principal parametric resonance. J Sound Vib 129:417–442
6. van Laarhoven BJH (2009) Stability analysis of parametric roll resonance, DCT 2009.062, 2009
7. Oueini SS, Nayfeh AH (1999) Single-mode control of a cantilever beam under principal parametric excitation. J Sound Vib 224(1):33–47
8. Ghandchi Tehrani M, Kalkowski MK, Elliott SJ (2014) Active control of parametrically exited systems. J Intell Mater Syst Struct, submit on July 2014
9. Jia Y, Seshia AA (2013) Directly and parametrically excited bi-stable vibration energy harvester for broadband operation. In: Solid-state sensors, actuators and microsystems, IEEE, Barcelona, pp 454–457
10. Jia Y, Yan J, Soga K, Seshia AA (2012) Parametrically excited MEMS vibration energy harvesters. Dig Tech Pap Power MEMS 12:215–218
11. Jia Y, Yan J, Soga K, Seshia AA (2013) Multi-frequency operation of a MEMS vibration energy harvester by accessing five orders of parametric resonance. J Phys Conf Ser, 476. doi:10.1088/1742-6596/476/1/012126
12. Jia Y, Yan J, Soga K, Seshia AA (2014) Parametric resonance for vibration energy harvesting with design techniques to passively reduce the initiation threshold amplitude. Smart Mater Struct 23:1–10
13. Nayfeh AH (1987) Parametric excitation of two internally resonant oscillators. J Sound Vib 119(1):95–109
14. Bonisoli E, Scapolan M, Ghandchi Tehrani M (2014) Internal stresses and unstable dynamics for harvesting design. Mech Syst Signal Process, submit on Oct 2014
15. Zaghari B, Ghandchi Tehrani M, Rustighi E (2014) Mechanical modeling of a vibration energy harvester with time-varying stiffness. In: Proceedings of the 9th international conference on structural dynamics, EURODYN 2014. Porto, 30 June–2 July, pp 2079–2085
16. Zaghari B, Rustighi E, Ghandchi Tehrani M (2014) Experimental study on harvesting energy from a parametrically excited system In: The 12th international conference on motion and vibration, MOVIC 2014, Sapporo, pp 1–13
17. Zaghari B, Ghandchi Tehrani M, Rustighi E (2014) Energy harvesting from a parametrically excited cantilever beam. J Sound Vib, submit on July 2014
18. Peeters B, Van der Auweraer H, Guillaume P, Leuridan J (2004) The PolyMAX frequency-domain method: a new standard for modal parameter estimation? J Shock Vib 11(3–4):395–409

MIX
Papier aus verantwortungsvollen Quellen
Paper from responsible sources
FSC® C105338

If you have any concerns about our products,
you can contact us on
ProductSafety@springernature.com

In case Publisher is established outside the EU,
the EU authorized representative is:
Springer Nature Customer Service Center GmbH
Europaplatz 3, 69115 Heidelberg, Germany

Printed by Libri Plureos GmbH
in Hamburg, Germany